Heat Transfer 1

Series Editor
Alain Dollet

Heat Transfer 1

Conduction and Convection

Yves Jannot
Christian Moyne
Alain Degiovanni

iSTE

WILEY

First published 2023 in Great Britain and the United States by ISTE Ltd and John Wiley & Sons, Inc.

ISTE Ltd
27-37 St George's Road
London SW19 4EU
UK

www.iste.co.uk

John Wiley & Sons, Inc.
111 River Street
Hoboken, NJ 07030
USA

www.wiley.com

Library of Congress Control Number: 2023932776

British Library Cataloguing-in-Publication Data
A CIP record for this book is available from the British Library
ISBN 978-1-78630-927-3

Contents

Preface

In addition to its obvious interest, which the current energy transition will certainly not contradict, thermal energy has a special place among the engineering sciences. Often taught at the beginning of engineering courses, it teaches how to analyze the phenomena involved in a real situation, identify the important aspects and develop a description that is likely to provide, before any more sophisticated approach, some solution elements, or even if possible, a first approximate solution that is easy to calculate.

Each of the three modes of heat transfer – conduction, convection and radiation – are associated with three different ways of thinking. Conduction, due to the apparent simplicity of Fourier's law and the heat equation, will allow different modeling and resolution methods depending on the simplifications used. The complexity of flowing fluids will generally make such an approach impractical in thermal convection without using numerical calculation: to remain relatively simple, we will resort to dimensional analysis approaches without, however, giving some elementary notions about essential approaches, especially the concept of boundary layer. Radiation opens onto the vast universe of physics, requiring simplifying assumptions that must be taken into account by the thermal engineer.

In all cases, the understanding of heat transfer processes requires the resolution and, above all, the meditation of a few exercises ranging from elementary cases (to begin with) to more complex situations, ultimately requiring a simultaneous approach to the different transfer modes that are often coupled.

This heat transfer course is intended for graduate students in universities and engineering schools, as well as practicing engineers. It presents the main modes of heat transfer: conduction, convection (Volume 1) and radiation (Volume 2). Volume 2 also includes two chapters on systems in which several modes of heat transfer occur simultaneously: heat exchangers and solar collectors. The main methods for solving the heat equation are presented and illustrated by using the Laplace transform, the separation of variables, the integral transformation, the quadrupole method and the numerical methods.

The chapters on steady-state conduction, convection, radiation between surfaces and on heat exchangers, as well as their application exercises can be approached by undergraduate students.

The proposed exercises are all corrected in detail; they present practical applications covering all the theoretical aspects of the course. Some long exercises are real case studies, and show that the assimilation of this course makes it possible to solve concrete problems in many fields of interest: for example, calculation of the evolution of the temperature and thermal losses of a building, calculation of losses in double glazing, measurement of the thermal properties of solids, sizing of a heat exchanger and sizing of a solar collector.

The appendices contain all the physical data and correlations needed to solve the exercises and problems presented and will be a valuable source of information for the engineer.

January 2023

Nomenclature

a	thermal diffusivity ($m^2\ s^{-1}$)
Bi	Biot number
c	specific heat ($J\ kg^{-1}\ K^{-1}$)
C_f	friction coefficient
D	diameter (m)
D_h	hydraulic diameter (m)
e	thickness (m)
E	thermal effusivity ($W\ m^{-2}\ K^{-1}\ s^{1/2}$)
F	conduction form coefficient (m)
Fo	Fourier number
g	acceleration of gravity ($m\ s^{-2}$)
Gr	Grashof number
h	convection heat transfer coefficient ($W\ m^{-2}\ K^{-1}$)
H	enthalpy (J)

ΔH	latent heat of phase change (J kg^{-1})
I	electrical energy intensity (A)
I_n	modified Bessel function of the first kind of order n
J_n	unmodified Bessel function of the first kind of order n
K_n	modified Bessel function of the second kind of order n
ℓ	width (m)
L	length (m)
$\dot{m}$	mass flow rate (kg s^{-1})
$\vec{n}$	unit vector normal to a surface
Nu	Nusselt number
p	Laplace variable (s^{-1})
p_e	perimeter (m)
Pr	Prandtl number
$\dot{q}$	energy volume density (W m^{-3})
Q	heat quantity (J)
q_c	heat flow (W K^{-1})
r, R	radius (m), resistance (Ω)
Ra	Rayleigh number
R_c	contact resistance (K W^{-1})
Re	Reynolds number
s	curvilinear abscissa (m)
S	area (m^2)

t	time (s)
T	temperature (K)
$\bar{T}$	average temperature (K)
u	velocity (m s^{-1})
V	volume (m^3)
x, y, z	space variables (m)
Y_n	unmodified Bessel function of the second kind of order n

Greek letters

α	absorption coefficient of a wall
β	cubic expansion coefficient (K^{-1})
δ	declination (°)
ε	emissivity
ϕ	heat flux (W m^{-2})
Φ	Laplace transform of the heat flow rate
φ	heat flow rate (W)
λ	thermal conductivity (W m^{-1} K^{-1}), wavelength (m)
μ	dynamic viscosity (Pa s^{-1})
ν	kinematic viscosity (m^2 s^{-1}), frequency (Hz)
η	yield or efficiency (%), emission coefficient of a medium (W m^{-2})
κ	absorption coefficient of a medium (m^{-1})
Ω	solid angle

τ	transmission coefficient of a wall
θ	Laplace transform of the temperature

1

Introduction to Heat Transfer

1.1. Introduction

Thermodynamics enables us to predict the total amount of energy that a system must exchange with the outside to pass from one state of equilibrium to another.

Thermics (or thermokinetics) aims to quantitatively describe (both in space and time) the evolution of the characteristic parameters of the system, in particular the temperature, between the initial and final states of equilibrium.

1.2. Definitions

1.2.1. *Temperature field*

Energy transfers are determined from the evolution in space and time of the temperature: $T = f(x, y, z, t)$. The instantaneous value of the temperature at any point of space is a scalar called the temperature field. Two cases can be distinguished:

– time-independent temperature field: the regime is said to be permanent or stationary;

– evolution of the temperature field over time: the regime is said to be variable or transient.

1.2.2. *Temperature gradient*

If we join all the points in space that have the same temperature, we obtain a surface called an isothermal surface. The temperature variation per unit length is maximum along the normal to the isothermal surface. This variation is characterized by the temperature gradient.

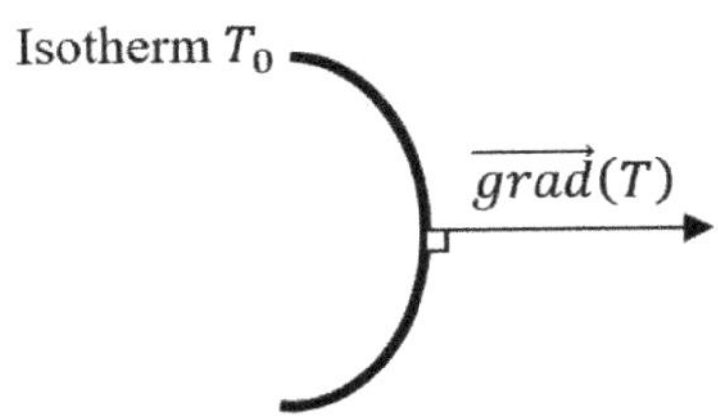

Figure 1.1. *Isotherm and thermal gradient*

$$\overrightarrow{grad}(T) = \vec{n}\frac{\partial T}{\partial n} \qquad [1.1]$$

with:

– $\vec{n}$, the normal unit vector;

– $\frac{\partial T}{\partial n}$, the derivative of the temperature along the normal.

1.2.3. *Heat flux*

Heat flows under the influence of a temperature gradient from high to low temperatures. The amount of heat transmitted per unit time and per unit area of the isothermal surface is called the heat flux ϕ (W m^{-2}):

$$\phi = \frac{1}{S}\frac{dQ}{dt} \qquad [1.2]$$

where S is the surface area (m^2).

We define the heat flow rate φ (W) as the amount of heat transmitted by the surface S per unit time:

$$\varphi = \frac{dQ}{dt} \qquad [1.3]$$

1.3. Formulation of a heat transfer problem

1.3.1. *Energy balance*

It is first necessary to define a system (S) by its limits in space and then to establish the inventory of the various heat flow rates which influence the state of the system (see Figure 1.2).

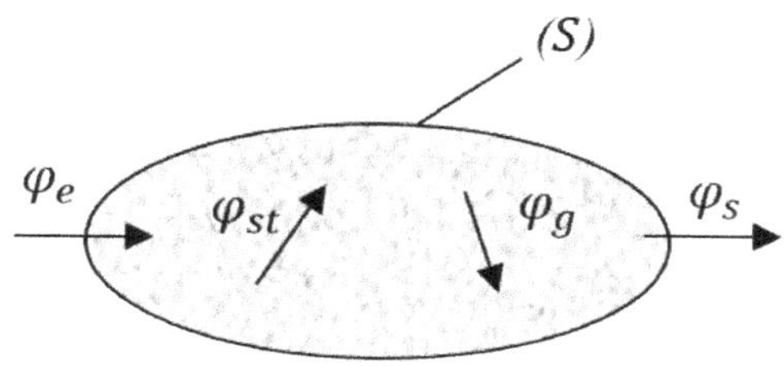

Figure 1.2. *System and energy balance*

These flows can be of several types:

– φ_{st}, stored heat flow rate;

– φ_g, generated heat flow rate;

– φ_e, input heat flow rate;

– φ_s, output heat flow rate.

We then apply the first principle of thermodynamics to establish the energy balance of the system (S):

$$\varphi_e + \varphi_g = \varphi_s + \varphi_{st} \qquad [1.4]$$

1.3.2. *Expression of energy flows*

It is necessary to establish the expressions of the various energy flows. By substituting these expressions in the energy balance, we obtain the differential equation whose resolution makes it possible to know the temperature evolution at each point of the system.

1.3.2.1. *Conduction*

Conduction is the transfer of heat within an opaque medium, without displacement of matter, under the influence of a temperature difference. The propagation of heat by conduction inside a body takes place by two distinct mechanisms: transmission by the vibrations of atoms or molecules and transmission by free electrons.

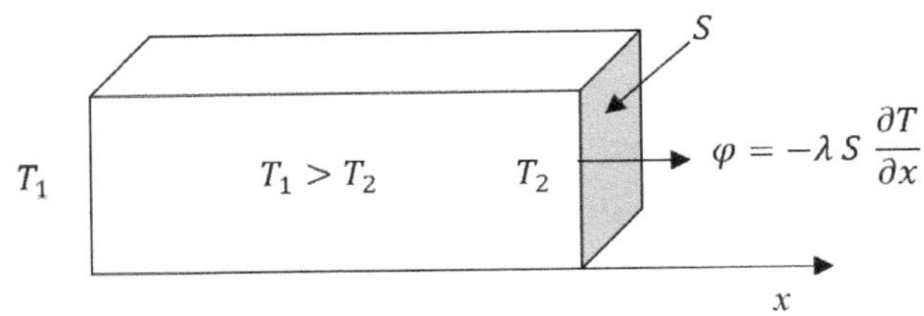

Figure 1.3. *Diagram of conductive heat transfer*

The theory of conduction is based on the Fourier hypothesis: the heat flux ϕ is proportional to the temperature gradient:

$$\vec{\phi} = -\lambda \overrightarrow{grad}(T) \qquad [1.5]$$

The heat flow rate transmitted by conduction in the direction Ox can therefore be written in algebraic form:

$$\varphi = -\lambda S \frac{\partial T}{\partial x} \qquad [1.6]$$

with:

- ϕ, the heat flux transmitted by conduction (W m^{-2});
- φ, the heat flow rate transmitted by conduction (W);
- λ, the thermal conductivity of a medium (W m^{-1} K^{-1});
- x, the space variable in the direction of the flow (m);
- S, the cross-sectional area (m^2).

Table 1.1 gives the values of the thermal conductivity λ of some of the most common materials. A more complete table is given in section A.1 of the Appendix.

Material	$\lambda\ (\mathrm{Wm^{-1}K^{-1}})$	Material	$\lambda\ (\mathrm{Wm^{-1}K^{-1}})$
Silver	419	Plaster	0.48
Copper	386	Asbestos	0.16
Aluminum	204	Wood (hardwood–softwood)	0.12–0.23
Mild steel	45	Cork	0.044–0.049
Stainless steel	15	Rockwool	0.038–0.041
Ice	1.9	Glass wool	0.035–0.051
Concrete	1.4	Expanded polystyrene	0.036–0.047
Clay brick	1.1	Polyurethane (foam)	0.030–0.045
Glass	1.0	Extruded polystyrene	0.028
Water	0.60	Air	0.026

Table 1.1. *Thermal conductivity of some materials*

1.3.2.2. *Convection*

Convection is the transfer of heat between a solid and a fluid, the energy being transmitted by the fluid displacement. This transfer mechanism is governed by Newton's law:

$$\varphi = hS\left(T_p - T_\infty\right) \quad [1.7]$$

with:

– φ, the heat flow rate transmitted by convection (W);

– h, the convective heat transfer coefficient (W m^{-2} K^{-1});

– T_p, the solid surface temperature (K);

– T_∞, the temperature of the fluid away from the surface of the solid (K);

– S, the surface area of the solid/fluid contact (m^2).

REMARK.– The value of the convective transfer coefficient h depends on the nature of the fluid, its temperature, its velocity and the geometric characteristics of the solid/fluid contact surface.

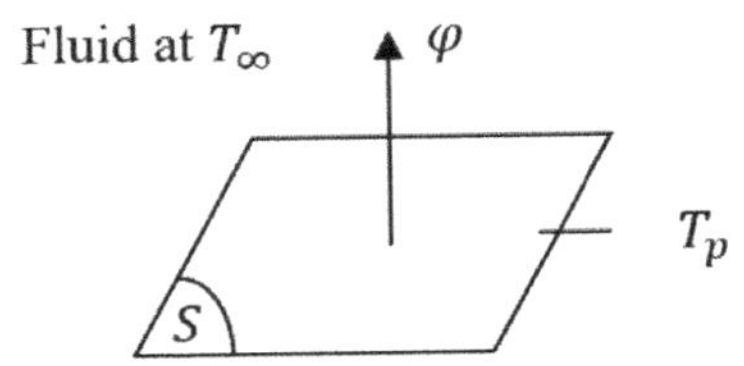

Figure 1.4. *Diagram of convective heat transfer*

1.3.2.3. *Radiation*

Radiation is the transfer of electromagnetic energy between two surfaces (even in vacuum). In conduction problems, we take into account the radiation between a solid surface and the surrounding medium. In this case, we have the relationship:

$$\varphi = \sigma\varepsilon S\left(T_p{}^4 - T_\infty{}^4\right) \qquad [1.8]$$

with:

– φ, the radiation heat flow rate (W);

– σ, the Stefan–Boltzmann constant (5.67×10^{-8} W m^{-2} K^{-4});

– ε, the surface emissivity;

– T_p, the surface temperature (K);

– T_∞, the temperature of the medium surrounding the surface (K);

– S, the surface area (m^2).

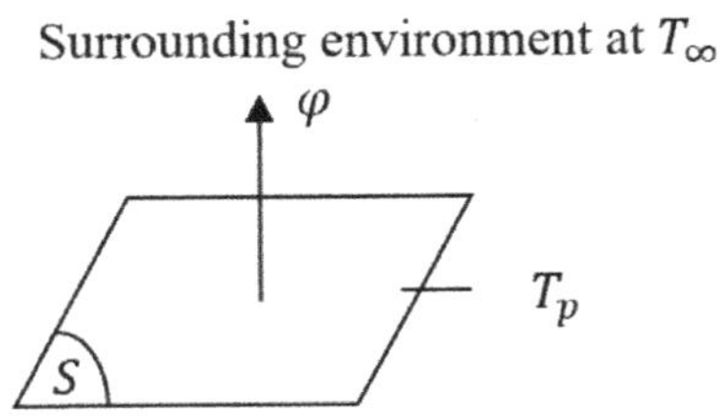

Figure 1.5. *Diagram of radiative heat transfer*

REMARK.– In equations [1.6] and [1.7], the temperatures can either be expressed in °C or K because they only appear in the form of differences. However, in equation [1.8], the temperature must imperatively be expressed in K.

1.3.2.4. *Heat flow linked to a mass flow*

When a mass flow $\dot{m}$ of matter enters the system at temperature T_1 and comes out at temperature T_2, we must consider in the balance [1.4] a corresponding input heat flow rate:

$$\varphi_e = \dot{m}c(T_1 - T_2) \quad [1.9]$$

with:

- φ_e, the heat flow rate entering the system (W);
- $\dot{m}$, the mass flow rate (kg s^{-1});
- c, the specific heat (J kg^{-1} K^{-1});
- T_1, T_2, the inlet and outlet temperatures (K).

1.3.2.5. *Energy storage*

Energy storage in sensible form in a body corresponds to an increase in its enthalpy over time, from which (at constant pressure and in the absence of a change of state):

$$\varphi_{st} = \rho V c \frac{\partial T}{\partial t} \quad [1.10]$$

with:

- φ_{st}, the stored heat flow rate (W);
- ρ, the density (kg m^{-3});
- V, the volume (m^3);
- c, the specific heat (J kg^{-1} K^{-1});
- T, the temperature (K);
- t, the time (s).

The product $\rho V c$(J K^{-1}) is called the thermal capacitance of the body.

1.3.2.6. *Thermal energy generation*

It occurs when another form of energy (chemical, electrical, mechanical or nuclear) is converted into thermal energy. It can be written in the form:

$$\varphi_g = \dot{q}V \quad [1.11]$$

with:

– φ_g, the thermal power generated (W);

– $\dot{q}$, the generated power density (W m^{-3});

– V, the volume (m^3).

2

Steady-State Conduction Heat Transfer

2.1. The heat equation

In one-dimensional form, the heat equation describes the unidirectional heat transfer through a plane wall, as shown in Figure 2.1.

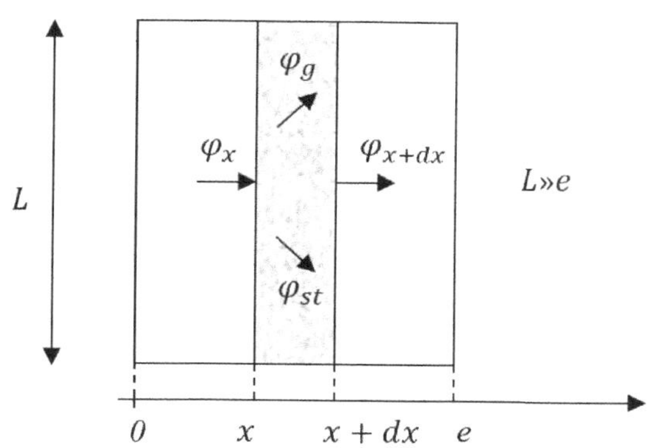

Figure 2.1. *Thermal balance on an elementary system*

Consider a system of thickness dx in the direction x and area section S normal to direction Ox. The energy balance on this system is written as:

$$\varphi_x + \varphi_g = \varphi_{x+dx} + \varphi_{st} \quad [2.1]$$

with:

$$-\varphi_x = -\left(\lambda S \frac{\partial T}{\partial x}\right)_x;$$

$$-\varphi_{x+dx} = -\left(\lambda S \frac{\partial T}{\partial x}\right)_{x+dx};$$

$$-\varphi_g = \dot{q} S dx;$$

$$-\varphi_{st} = \rho c S dx \frac{\partial T}{\partial t}.$$

where $\dot{q}$ is the energy source per unit volume and time.

By reporting in the energy balance and dividing by dx, we obtain:

$$\frac{\left(\lambda S \frac{\partial T}{\partial x}\right)_{x+dx} - \left(\lambda S \frac{\partial T}{\partial x}\right)_x}{dx} + \dot{q} S = \rho c S \frac{\partial T}{\partial t}$$

thus:

$$\frac{\partial}{\partial x}\left(\lambda S \frac{\partial T}{\partial x}\right) + \dot{q} S = \rho c S \frac{\partial T}{\partial t}$$

In the three-dimensional case, we get the heat equation in the most general form:

$$\frac{\partial}{\partial x}\left(\lambda_x \frac{\partial T}{\partial x}\right) + \frac{\partial}{\partial y}\left(\lambda_y \frac{\partial T}{\partial y}\right) + \frac{\partial}{\partial z}\left(\lambda_z \frac{\partial T}{\partial z}\right) + \dot{q} = \rho c \frac{\partial T}{\partial t} \quad [2.2]$$

This equation can be simplified in a number of cases:

a) if the medium is isotropic: $\lambda_x = \lambda_y = \lambda_z = \lambda$;

b) if there is no energy generation inside the system: $\dot{q} = 0$;

c) if the medium is homogeneous and λ only function T with the hypotheses a) + b) +c), the heat equation can be written as:

$$\lambda\left(\frac{\partial^2 T}{\partial x^2} + \frac{\partial^2 T}{\partial y^2} + \frac{\partial^2 T}{\partial z^2}\right) + \frac{d\lambda}{dT}\left[\left(\frac{\partial T}{\partial x}\right)^2 + \left(\frac{\partial T}{\partial y}\right)^2 + \left(\frac{\partial T}{\partial z}\right)^2\right] = \rho c \frac{\partial T}{\partial t}$$

d) if moreover, λ is constant (moderate difference in temperature), we obtain the following equation:

$$\frac{\partial^2 T}{\partial x^2} + \frac{\partial^2 T}{\partial y^2} + \frac{\partial^2 T}{\partial z^2} = \frac{1}{a}\frac{\partial T}{\partial t} \qquad [2.3]$$

The ratio $a = \frac{\lambda}{\rho c}$ is called the thermal diffusivity (m^2 s^{-1}); it characterizes the diffusion of the heat flow rate through a material. Some values for different types of materials can be found in section A.1 of the Appendix;

e) in the steady state, we obtain the Laplace equation:

$$\nabla^2 T = \frac{\partial^2 T}{\partial x^2} + \frac{\partial^2 T}{\partial y^2} + \frac{\partial^2 T}{\partial z^2} = 0 \qquad [2.4]$$

Moreover, hypotheses a), c) and d) allow us to write:

– the heat equation in cylindrical coordinates (r, θ, z):

$$\frac{\partial^2 T}{\partial r^2} + \frac{1}{r}\frac{\partial T}{\partial r} + \frac{1}{r^2}\frac{\partial^2 T}{\partial \theta^2} + \frac{\partial^2 T}{\partial z^2} + \frac{\dot{q}}{\lambda} = \frac{1}{a}\frac{\partial T}{\partial t} \qquad [2.5]$$

In the case of a problem with cylindrical symmetry where the temperature depends only on r and t, equation [2.5] can be written in a simplified form:

$$\frac{\partial^2 T}{\partial r^2} + \frac{1}{r}\frac{\partial T}{\partial r} + \frac{\dot{q}}{\lambda} = \frac{1}{a}\frac{\partial T}{\partial t} \qquad [2.6]$$

– the heat equation in spherical coordinates (r, θ, ψ):

$$\frac{1}{r}\frac{\partial^2}{\partial r^2}(rT) + \frac{1}{r^2 sin(\theta)}\frac{\partial}{\partial \theta}\left[sin(\theta)\frac{\partial T}{\partial \theta}\right] + \frac{1}{r^2 sin^2(\theta)}\frac{\partial^2 T}{\partial \psi^2} + \frac{\dot{q}}{\lambda} = \frac{1}{a}\frac{\partial T}{\partial t} \qquad [2.7]$$

2.2. Unidirectional transfer

2.2.1. *Simple wall*

We consider the case where the heat transfer is unidirectional and where there is neither generation nor storage of energy.

Consider a plane wall of thickness e, thermal conductivity λ and large cross-sectional dimensions whose extreme faces are at temperatures T_1 and T_2 (see Figure 2.2).

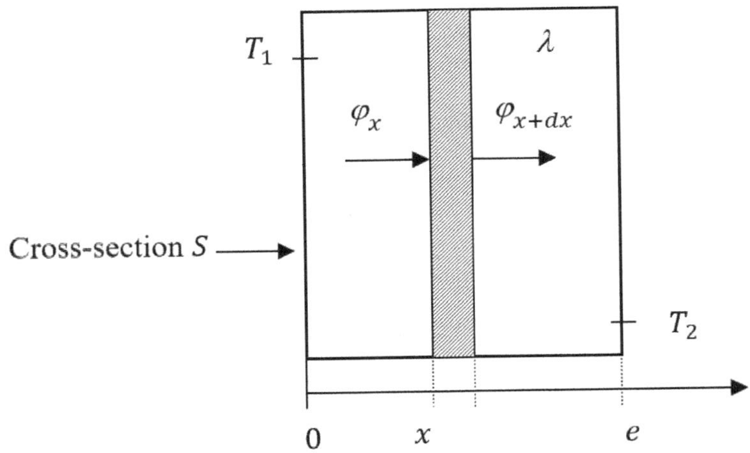

Figure 2.2. *Elementary heat balance on a simple wall*

By carrying out a thermal balance on the system (S) constituted by a slice of wall between the abscissas as x and $x + dx$, we get:

$$\varphi_x = \varphi_{x+dx} \Rightarrow \left(\lambda S \frac{dT}{dx}\right)_x = \left(\lambda S \frac{dT}{dx}\right)_{x+dx}$$

And therefore $\frac{dT}{dx} = A$ and: $T(x) = Ax + B$.

with the boundary conditions $T(x = 0) = T_1$ and $T(x = e) = T_2$:

$$T(x) = T_1 - \frac{x}{e}(T_1 - T_2) \qquad [2.8]$$

The temperature profile is linear. The heat flux rate crossing the wall is deduced by the equation $\phi = -\lambda \frac{\partial T}{\partial x}$:

$$\phi = \frac{\lambda}{e}(T_1 - T_2) \qquad [2.9]$$

Equation [2.9] can also be put in the form: $\varphi = \frac{T_1 - T_2}{\frac{e}{\lambda S}}$; this equation is analogous to Ohm's law in electricity which defines the intensity of the current as the ratio of the difference of electric potential divided by the electric resistance. The temperature T thus appears as the thermal potential and the term $\frac{e}{\lambda S}$ as the thermal resistance of a plane wall of thickness e, thermal conductivity λ and section S. This system can be represented by the equivalent diagram shown in Figure 2.3.

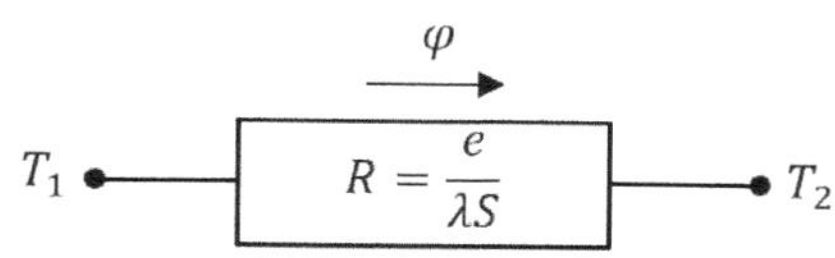

Figure 2.3. *Equivalent electrical diagram of a simple wall*

REMARK.– The heat flow rate is constant; this is a general result for any steady-state heat flow tube (conservative flux system).

2.2.2. *Multilayer wall*

Real walls (as schematized in Figure 2.4) are generally made up of several layers of different materials. More only the temperatures T_{f1} and T_{f2} of the fluids in contact with the two faces of the wall of section S are known.

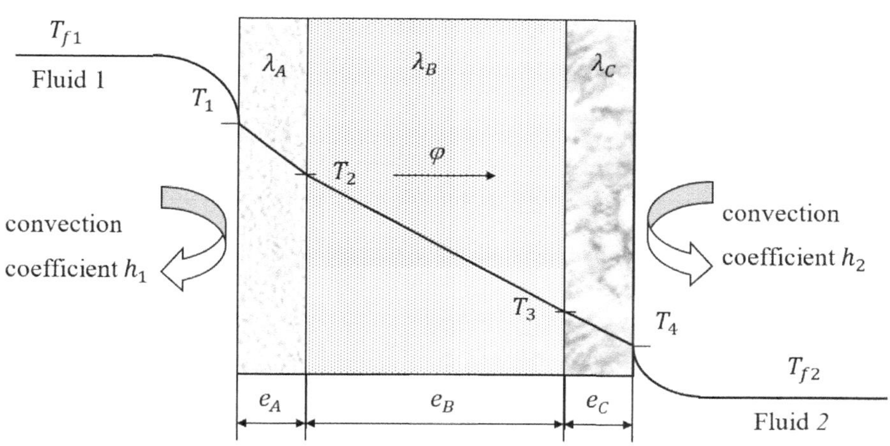

Figure 2.4. *Schematization of flows and temperatures in a multilayer wall*

In the steady state, the heat flow rate is conserved when passing through the wall and is written as:

$$\varphi = h_1 S(T_{f1} - T_1) = \frac{\lambda_A S(T_1 - T_2)}{e_A} = \frac{\lambda_B S(T_2 - T_3)}{e_B} = \frac{\lambda_C S(T_3 - T_4)}{e_C}$$
$$= h_2 S(T_4 - T_{f2})$$

hence:

$$\varphi = \frac{T_{f1} - T_{f2}}{\frac{1}{h_1 S} + \frac{e_A}{\lambda_A S} + \frac{e_B}{\lambda_B S} + \frac{e_C}{\lambda_C S} + \frac{1}{h_2 S}} \qquad [2.10]$$

It has been considered that the contact between the different layers is perfect and that there is no temperature discontinuity at the interfaces.

In fact, given the roughness of the surfaces, a microlayer of air exists between the facing surfaces which contributes to the creation of a thermal resistance (air is an insulator) called thermal contact resistance.

The previous formula is then written as:

$$\varphi = \frac{T_{f1} - T_{f2}}{\frac{1}{h_1 S} + \frac{e_A}{\lambda_A S} + R_{AB} + \frac{e_B}{\lambda_B S} + R_{BC} + \frac{e_C}{\lambda_C S} + \frac{1}{h_2 S}} \qquad [2.11]$$

The equivalent electrical diagram is shown in Figure 2.5.

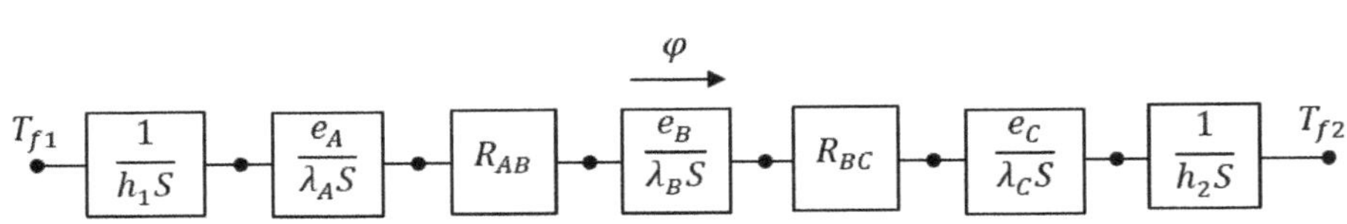

Figure 2.5. *Equivalent electrical diagram of a multilayer wall*

REMARKS.–

– A thermal resistance can only be set on a heat flow tube.

– The thermal contact resistance can be neglected if the wall has an insulating layer or if the walls are joined by welding.

2.2.3. *Composite wall*

This is the most common case when the walls are not homogeneous. Consider as an example a wall of thickness L made of hollow bricks (Figure 2.6).

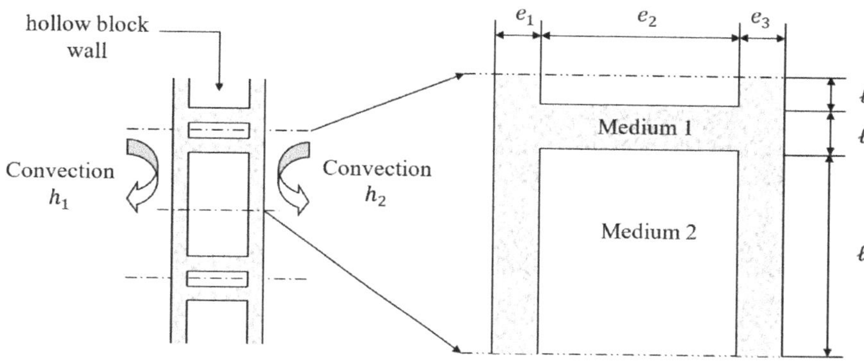

Figure 2.6. *Diagram of a composite wall*

Assuming a unidirectional transfer and taking into account the symmetry of the problem, the thermal resistance of the wall can be reduced to the calculation of the heat flow rate through the insulated element on the right of the figure and the equivalent thermal resistance R of a portion of the wall of width L and height $\ell = \ell_1 + \ell_2 + \ell_3$ using the laws of association of resistances in series and in parallel by the equation can be calculated:

$$R = R_1 + R_2 + \frac{1}{\frac{1}{R_3} + \frac{1}{R_4} + \frac{1}{R_5}} + R_6 + R_7 \qquad [2.12]$$

with:

$$R_1 = \frac{1}{h_1 \ell L}; R_2 = \frac{e_1}{\lambda_1 \ell L}; R_3 = \frac{e_2}{\lambda_2 \ell_1 L}; R_4 = \frac{e_2}{\lambda_1 \ell_2 L}; R_5 = \frac{e_2}{\lambda_1 \ell_3 L}; R_6 = \frac{e_3}{\lambda_1 \ell L};$$

$$R_7 = \frac{1}{h_2 \ell L}$$

This can be schematized by the equivalent electrical diagram shown in Figure 2.7.

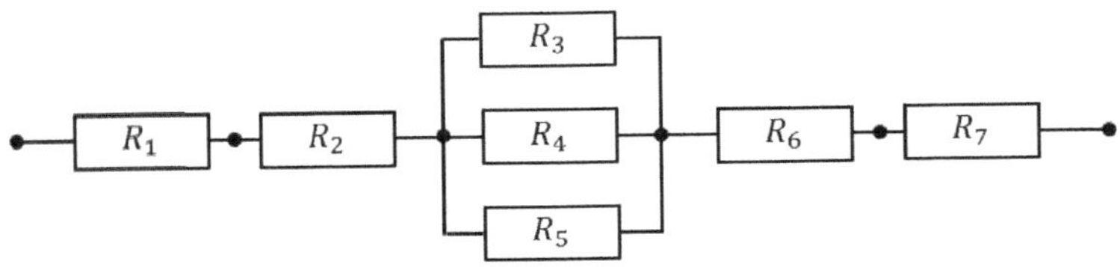

Figure 2.7. *Equivalent electrical diagram of the composite wall*

It should be noted that this solution is only approximate because the transfer is in fact 2D due to the differences between the thermal conductivities $\lambda_1$and λ_2.

2.2.4. *Long hollow cylinder (tube)*

We consider a hollow cylinder of thermal conductivity λ, interior radius r_1, exterior radius r_2, length L, the temperatures of the internal and external faces being respectively T_1 and T_2. It is assumed that the longitudinal temperature gradient is negligible compared to the radial gradient.

Let us calculate the heat balance of the system consisting of the part of the cylinder between the radii r and $r + dr$ (see Figure 2.8), $\varphi_r = \varphi_{r+dr}$, with:

$$\varphi_r = -2\pi r L\lambda\left(\frac{dT}{dr}\right)_r \text{ and } \varphi_{r+dr} = -2\pi(r+dr)L\lambda\left(\frac{dT}{dr}\right)_{r+dr}$$

or:

$$-2\pi r L\lambda\left(\frac{dT}{dr}\right)_r = -2\pi(r+dr)L\lambda\left(\frac{dT}{dr}\right)_{r+dr} \quad \text{hence: } r\frac{dT}{dr} = C$$

with the boundary conditions: $T(r_1) = T_1$ and $T(r_2) = T_2$.

$$\frac{T(r) - T_1}{T_2 - T_1} = \frac{ln\left(\frac{r}{r_1}\right)}{ln\left(\frac{r_2}{r_1}\right)} \quad [2.13]$$

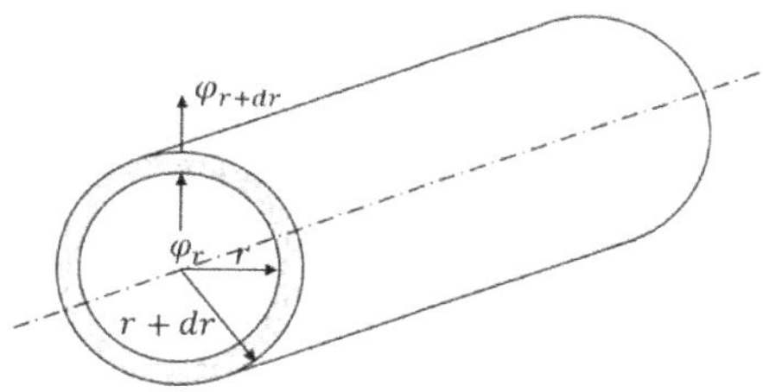

Figure 2.8. *Heat transfer in a hollow cylinder*

By applying the equation $\varphi = -2\pi r L\lambda\left(\frac{dT}{dr}\right)$, we get:

$$\varphi = \frac{2\pi\lambda L(T_1 - T_2)}{ln\left(\frac{r_2}{r_1}\right)} \quad [2.14]$$

This equation can also be put in the form: $\varphi = \frac{T_1 - T_2}{R_{12}}$ with $R_{12} = \frac{ln\left(\frac{r_2}{r_1}\right)}{2\pi\lambda L}$ and be represented by the equivalent electric diagram of Figure 2.9.

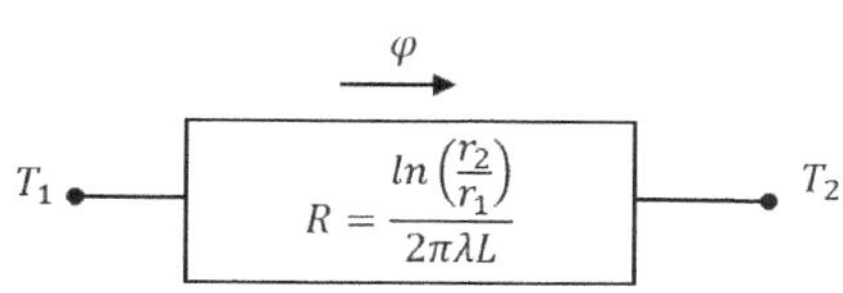

Figure 2.9. *Equivalent electric diagram of a hollow cylinder*

2.2.5. *Multilayer hollow cylinder*

This is the practical case of a tube covered with one or more layers of different materials and where only the temperatures T_{f1} and T_{f2} of the fluids in contact with the internal and external faces of the cylinder are known; h_1 and h_2 are the coefficients of heat transfer by convection between the fluids and the internal and external faces (see Figure 2.10).

In steady state, the heat flow rate φ is conserved while crossing the different layers and is written as:

$$\varphi = h_1 2\pi r_1 L(T_{f1} - T_1) = \frac{2\pi\lambda_A L(T_1 - T_2)}{ln\left(\frac{r_2}{r_1}\right)} = \frac{2\pi\lambda_B L(T_2 - T_3)}{ln\left(\frac{r_3}{r_2}\right)}$$
$$= h_2 2\pi r_3 L(T_3 - T_{f2})$$

hence:

$$\varphi = \frac{T_{f1} - T_{f2}}{\frac{1}{h_1 2\pi r_1 L} + \frac{ln\left(\frac{r_2}{r_1}\right)}{2\pi\lambda_A L} + \frac{ln\left(\frac{r_3}{r_2}\right)}{2\pi\lambda_B L} + \frac{1}{h_2 2\pi r_3 L}} \quad [2.15]$$

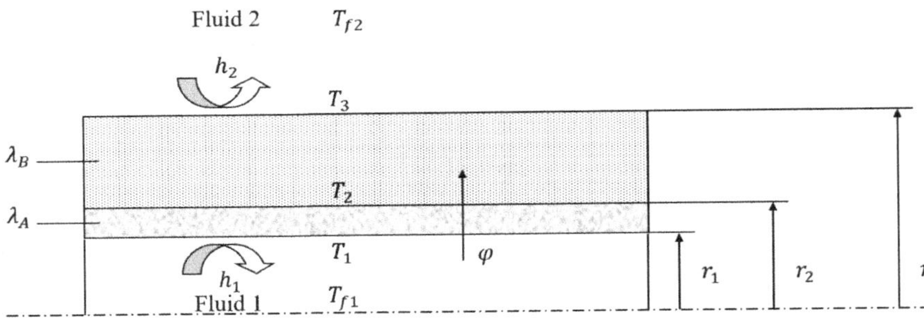

Figure 2.10. *Diagram of transfers in a multilayer hollow cylinder*

This can be represented by the equivalent electrical diagram of Figure 2.11.

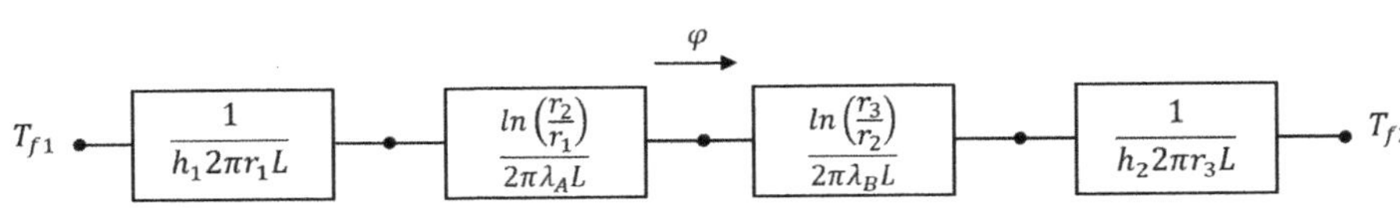

Figure 2.11. *Equivalent electrical diagram of a multilayer hollow cylinder*

2.2.6. *General case*

We have just given the value of the thermal resistance for some particular cases of practical interest. More generally, the thermal resistance of a heat flow tube (see Figure 2.12) is written as:

$$R = \int_{r_1}^{r_2} \frac{dr}{\lambda(r)S(r)} \quad [2.16]$$

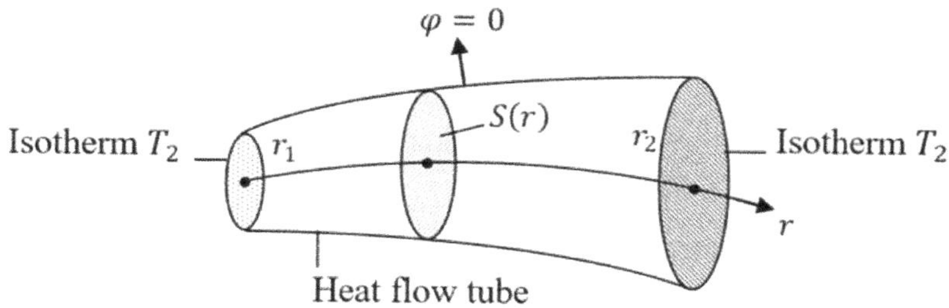

Figure 2.12. *Diagram of a heat flow tube*

2.2.7. Consideration of radiative transfer

In the preceding examples, the transfer of heat between a surface at temperature T and the surrounding medium is considered as purely convective. When the fluid in contact with the surface is a gas with only natural convection, the heat transfer by radiation with the walls (at the average temperature T_r) surrounding the surface can become of the same order of magnitude as the heat transfer by natural convection and can no longer be neglected.

The heat flow rate transferred by radiation is written according to equation [1.8]:

$$\varphi_r = \sigma\varepsilon S\left(T^4 - T_r{}^4\right)$$

which can be simplified in the linear form $\varphi_r = h_r S(T - T_r)$, h_r being called the radiation heat transfer coefficient $h_r = \sigma\varepsilon\left(T^2 + T_r{}^2\right)(T + +T_r)$.

The two transfers, convection and radiation, take place in parallel, and the corresponding electrical diagram is shown in Figure 2.13.

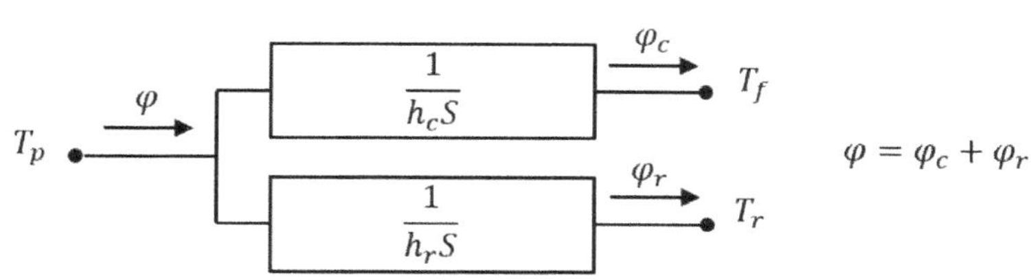

Figure 2.13. *Equivalent electrical diagram with simultaneous convection and radiation transfers*

The radiation heat transfer coefficient h_r varies slightly for limited variations of the temperatures T and T_r, for a first simplified calculation, can be considered as constant. For example, with $\varepsilon = 0.9$, $T = 60°\text{C}$ and $T_r = 20°\text{C}$, the exact value is $h_r = 6.28 \text{ W m}^{-2}\text{K}^{-1}$. If T becomes equal to 50°C (instead of 60°C), the value of h_r becomes equal to 5.98 W m^{-2}K^{-1}, i.e. a variation of only 5%.

2.3. Multi-directional transfer

When the heat transfer does not take place along a single direction, several resolution methods can be applied.

2.3.1. *Method of separation of variables*

A heat transfer problem on a system defined by its spatial boundaries is characterized by the heat equation and the boundary conditions on the system limits. These boundary conditions can be:

– homogeneous if they are of the type $\lambda \frac{\partial T}{\partial n} + h\,T = 0$;

– non-homogeneous if they are of the type $\lambda \frac{\partial T}{\partial n} + h\,T \neq 0$.

The method of resolution by separation of the variables applies if only one boundary condition is not homogeneous. This method is based on a decomposition of temperature into a product of functions and on the solution superposition theorem. It applies globally as follows:

– search for functions of the form $T_k(x,y,z) = X_k(x)Y_k(y)Z_k(z)$ satisfying the system formed by the heat equation and the homogeneous boundary conditions;

– search for the unique solution of the complete system in the form $T(x,y,z) = \sum_{k=0}^{\infty} A_k T_k(x,y,z)$, the coefficients A_k being determined using the inhomogeneous boundary condition.

Note that the solution $T(x,y,z)$ of the complete system is not a product of separate functions of x, y and z.

We will illustrate this method through the resolution of a 2D problem. Consider a plate with three sides maintained at one temperature T_1 and the fourth at T_0 (see Figure 2.14). By a change of variable, we can reduce to the case where $T_1 = 0$.

The heat equation is written as:

$$\frac{\partial^2 T}{\partial x^2}+\frac{\partial^2 T}{\partial y^2}=0 \quad \text{(a)}$$

with the boundary conditions:

$$T(0,y)=0 \quad \text{(b)}$$

$$T(e,y)=0 \quad \text{(c)}$$

$$T(x,0)=0 \quad \text{(d)}$$

$$T(x,e)=T_0 \quad \text{(e)}$$

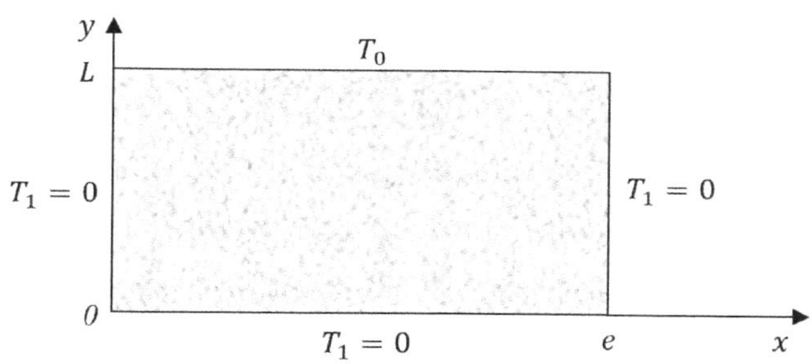

Figure 2.14. *Diagram of a plate with imposed temperatures on the sides*

We carry out a decomposition of the temperature into a product of functions in the form:

$$T(x,y)=X(x)Y(y)$$

The heat equation leads to: $Y\frac{d^2X}{dx^2}+X\frac{d^2Y}{dy^2}=0$ or again: $\frac{1}{X}\frac{d^2X}{dx^2}+\frac{1}{Y}\frac{d^2Y}{dy^2}=0$.

We deduce that:

$$\frac{1}{X}\frac{d^2X}{dx^2}=-\frac{1}{Y}\frac{d^2Y}{dy^2}=K$$

where K is a constant because the two functions X and Y depend one on x and the other on y.

Let us analyze the case where the constant K is positive and equal to ω^2.

We have:

$$X" - \omega^2 X = 0 \Rightarrow X = Aexp(\omega x) + Bexp(-\omega x)$$

and:

$$Y" + \omega^2 X = 0 \Rightarrow Y = Ccos(\omega y) + Dsin(\omega y)$$

The homogeneous boundary conditions impose:

(b): $T(0, y) = 0$ hence: $A + B = 0$

(c): $T(e, y) = 0$ hence: $Aexp(\omega e) + Bexp(-\omega e) = 0$

These two conditions cannot be achieved simultaneously, so the constant K is negative and equal to $-\omega^2$.

In this case:

$$X = Acos(\omega x) + Bsin(\omega x)$$

$$Y = Ccosh(\omega y) + Dsinh(\omega y)$$

and:

$$T(x, y) = [Acos(\omega x) + Bsin(\omega x)][Ccosh(\omega y) + Dsinh(\omega y)]$$

We first use the homogeneous boundary conditions:

(b): $T(0, y) = 0$ hence: $A = 0$

(d): $T(x, 0) = 0$ hence: $C = 0$

and $T(x, y) = Esin(\omega x)sinh(\omega y)$:

(c): $T(e, y) = 0$ hence: $sin(\omega e) = 0$

All the values ω_n of the form $\omega_n = \frac{n\pi}{e}$ are solutions of this equation; they are called the eigenvalues of the problem. The functions $f_n(x) = sin(\omega_n x)$ are called the eigenfunctions.

The general solution is a superposition of the solutions with separated variables which form a base of the space of the solutions, i.e.:

$$T(x,y) = \sum_{n=1}^{\infty} E_n sin(\omega_n x) sinh(\omega_n y)$$

The method then consists of calculating the integral of the product of the inhomogeneous limit condition (e) by a specific function over its field of application (here between $x = 0$ and $x = e$ for $y = L$). The calculation of this integral which is written here $\int_0^e T(x,L)\, sin(\omega_m x)\, dx$ is carried out in two different ways:

– using the inhomogeneous boundary condition;

– by replacing $T(x,y)$ with its expression obtained by superposition of the particular solutions.

We obtain in the considered example:

$$\int_0^e T(x,L)\, sin(\omega_m x) dx = \int_0^e T_0\, sin(\omega_m x) dx = T_0 \int_0^e sin(\omega_m x) dx$$

$$= \frac{T_0}{\omega_m} cos(\omega_m e)$$

thus:

$$\int_0^e T(x,L)\, sin(\omega_m x) dx = [(-1)^m - 1] \frac{T_0}{\omega_m}$$

We also have:

$$\int_0^e T(x,L)\, sin(\omega_m x) dx = \int_0^e sin(\omega_m x) \sum_{n=1}^{\infty} E_n sinh(\omega_n L) sin(\omega_m x)\, dx$$

The orthogonality theorem of eigenfunctions (see section A.14 of the Appendix) allows us to write:

$$\int_0^e sin(\omega_m x) sin(\omega_n x) dx = 0 \text{ if } m \neq n$$

hence:

$$\int_0^e T(x,L)\,sin(\omega_m x)dx = \int_0^e E_m sinh(\omega_m L) sin^2(\omega_m x)\,dx = \frac{eE_m}{2} sinh(\omega_m L)$$

The equality of the two expressions of $\int_0^L T(x,L)\,sin(\omega_m x)dx$ makes it possible to calculate the value of E_m:

$$E_m = \frac{2\,T_0[(-1)^m - 1]}{e\omega_m sinh(\omega_m L)} = \frac{2\,T_0[(-1)^m - 1]}{m\pi sinh(\omega_m L)}$$

The temperature is finally written as:

$$T(x,y) = \frac{2}{\pi} T_0 \sum_{n=1}^{\infty} \frac{[(-1)^n - 1]}{n} \frac{sinh(\omega_n x) sin(\omega_n y)}{sinh(\omega_n L)} \quad [2.17]$$

REMARK.– In the case of cylindrical coordinates, we calculate the integral of the product of the inhomogeneous limit condition by $r\ f_n(r)$.

2.3.2. *Shape coefficient method*

In two-dimensional or three-dimensional systems with only two limiting temperatures T_1 and T_2, it is shown that the heat flow rate can be written as:

$$\varphi = \lambda F(T_1 - T_2) \quad [2.18]$$

with:

– λ, the thermal conductivity of the medium separating the surfaces S_1 and S_2 ($W\ m^{-1}\ K^{-1}$);

– T_1, the temperature of the surface S_1 (K);

– T_2, the temperature of the surface S_2 (K);

– F, the shape factor (m).

The shape factor F depends only on the shape, dimensions and relative position of the two surfaces S_1 and S_2. The values of F for the most common configurations are presented in section A.3 of the Appendix.

2.3.2.1. *Special case: three-dimensional enclosure (oven, cold room, air-conditioned room, etc.)*

Method: the enclosure is divided into different elements, and the heat flow rate through each of them is calculated according to the representation in Figure 2.15.

If the longitudinal dimensions are large compared to the thickness e of the walls (assumed constant), the shape coefficients of the various elements have the value:

$$F_{wall_i} = \frac{S_i}{L_i}$$

$$F_{edge_i} = 0.54\ D_i$$

$$F_{corner_i} = 0.15\ L_i$$

with:

– S_i, the area of the wall i;

– D_i, the length of the wall or edge i;

– L_i, the wall thickness i.

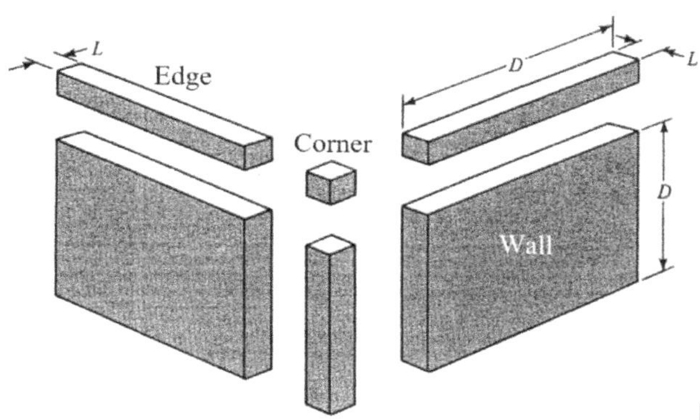

Figure 2.15. *Method of cutting a three-dimensional enclosure*

The heat flow rate crossing the enclosure is then written as:

$$\varphi = \sum_{i=1}^{6} \lambda_i\, F_{wall_i} \Delta T_i + \sum_{i=1}^{12} \lambda_i F_{edge_i}\, \Delta T_i + \sum_{i=1}^{8} \lambda_i F_{corner_i}\, \Delta T_i$$

with:

– λ_i, the thermal conductivity (equivalent if multilayer wall) of the wall i (W m^{-1} K^{-1});

– ΔT_i, the temperature difference between the inner and outer faces of the wall i (°C).

2.3.3. *Numerical methods*

2.3.3.1. *Expression of Laplace's equation in finite differences*

When the shape coefficient method cannot be applied (non-isothermal surfaces, for example), the Laplace equation must be solved numerically. For example, a finite difference method is used by discretizing the domain considered (space or plane). In what follows, the two-dimensional case is solved, the three-dimensional case being deduced by simply adding a dimension of space.

We consider a plane medium meshed with increments Δx and Δy, as represented in Figure 2.16.

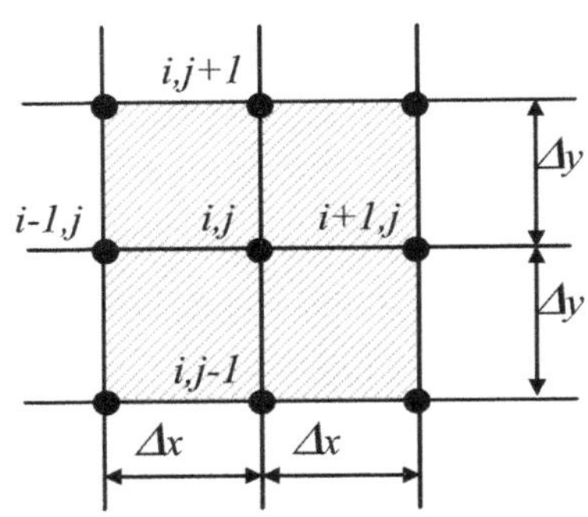

Figure 2.16. *Representation of the surface mesh*

The partial derivatives of the temperature T can be expressed according to the following formulas:

$$\frac{\partial T}{\partial x}\left(i+\frac{1}{2},j\right) \approx \frac{T(i+1,j)-T(i,j)}{\Delta x};\ \frac{\partial T}{\partial x}\left(i-\frac{1}{2},j\right) \approx \frac{T(i,j)-T(i-1,j)}{\Delta x}$$

$$\frac{\partial T}{\partial y}\left(i,j+\frac{1}{2}\right) \approx \frac{T(i,j+1)-T(i,j)}{\Delta y};\ \frac{\partial T}{\partial y}\left(i,j-\frac{1}{2}\right) \approx \frac{T(i,j)-T(i,j-1)}{\Delta y}$$

$$\frac{\partial^2 T}{\partial x^2}(i,j) \approx \frac{\frac{\partial T}{\partial x}\left(i+\frac{1}{2},j\right) - \frac{\partial T}{\partial x}\left(i-\frac{1}{2},j\right)}{\Delta x} = \frac{T(i+1,j) + T(i-1,j) - 2\,T(i,j)}{\Delta x^2}$$

$$\frac{\partial^2 T}{\partial y^2}(i,j) \approx \frac{\frac{\partial T}{\partial y}\left(i,j+\frac{1}{2}\right) - \frac{\partial T}{\partial y}\left(i,j-\frac{1}{2}\right)}{\Delta y} = \frac{T(i,j+1) + T(i,j-1) - 2\,T(i,j)}{\Delta y^2}$$

The two-dimensional Laplace equation $\frac{\partial^2 T}{\partial x^2} + \frac{\partial^2 T}{\partial y^2} = 0$ is then written as:

$$\frac{T(i+1,j) + T(i-1,j) - 2\,T(i,j)}{\Delta x^2} + \frac{T(i,j+1) + T(i,j-1) - 2\,T(i,j)}{\Delta y^2} = 0$$

And if we choose $\Delta x = \Delta y$, we get:

$$T(i,j) = \frac{T(i-1,j) + T(i+1,j) + T(i,j-1) + T(i,j+1)}{4}$$

2.3.3.2. *Expression of boundary conditions in finite differences*

The boundary conditions imposing a surface temperature on an edge are expressed simply by fixing the value of the temperature $T(i,j)$ equal to the imposed value for any pair (i,j) representing a point on this edge. The boundary conditions with convective transfer or imposed heat flow rate are expressed as follows.

2.3.3.2.1. Straight edge

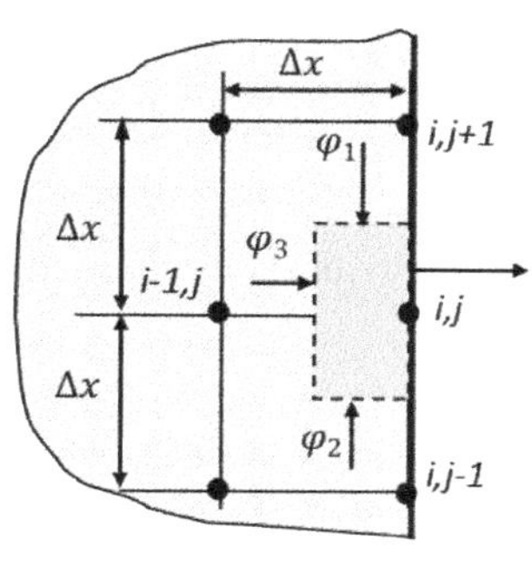

$$\varphi_1 + \varphi_2 + \varphi_3 = h\ell\Delta x\left(T_p - T_\infty\right)$$

or:

$$\varphi_1 + \varphi_2 + \varphi_3 = \phi\ell\Delta x$$

Figure 2.17. *Representation of elementary flows on a straight edge*

A heat balance applied to the gray surface (rectangle with sides $\Delta x/2$ and $\Delta y/2$; see Figure 2.17) leads to the following result taking into account the formulas established previously:

– imposed heat flux ϕ (in W m^{-2}):

$$T(i,j) = \frac{T(i-1,j)}{2} + \frac{T(i,j-1) + T(i,j+1)}{4} - \frac{\phi \Delta x}{2\lambda}$$

– imposed convection coefficient h:

$$T(i,j) = \frac{T(i-1,j) + \frac{T(i,j-1) + T(i,j+1)}{2} + BiT_{\infty}}{2 + Bi}$$

where $Bi = \frac{h\Delta x}{\lambda}$ is the Biot number.

2.3.3.2.2. Outer corner

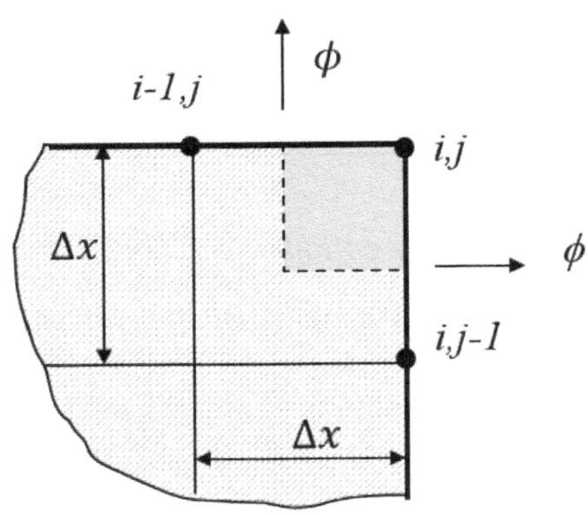

Figure 2.18. *Representation of elementary flows on an outer corner*

A thermal balance applied to the gray surface (see Figure 2.18) leads to the following result taking into account the formulas previously established:

– imposed heat flux ϕ (in W m^{-2}):

$$T(i,j) = \frac{T(i-1,j) + T(i,j-1)}{4} - \frac{\phi \Delta x}{2\lambda}$$

– imposed convection coefficient:

$$T(i,j) = \frac{\frac{T(i-1,j) + T(i,j-1)}{2} + BiT_\infty}{1 + Bi}$$

2.3.3.2.3. Inner corner

A thermal balance applied to the gray surface (see Figure 2.19) leads to the following result taking into account the formulas established previously:

– imposed heat flux ϕ (in Wm^{-2}):

$$T(i,j) = \frac{T(i-1,j) + T(i,j+1)}{3} + \frac{T(i+1,j) + T(i,j-1)}{6} - \frac{\phi \Delta x}{3\lambda}$$

– imposed convection coefficient h:

$$T(i,j) = \frac{T(i-1,j) + T(i,j+1) + \frac{T(i+1,j) + T(i,j-1)}{2} + BiT_\infty}{3 + Bi}$$

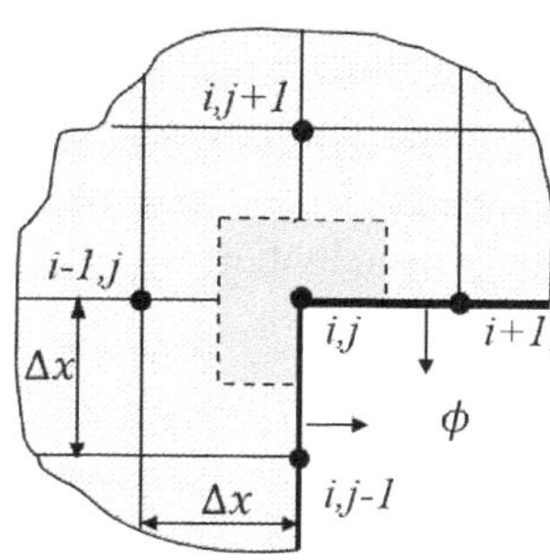

Figure 2.19. *Representation of elementary flows on an inner corner*

2.3.3.3. *Numerical solution method*

Let us solve the Laplace equation on a plane domain (D) bounded by a contour (C). The system is meshed with an increment Δx generally identical in the two directions of the plane. We assign to each point of the field (D) an initial value of the temperature:

– equal to the temperature imposed on the points of the contour where the interface condition imposes a temperature;

– arbitrary elsewhere but as “realistic” as possible.

The resolution uses the iterative method of Gauss–Seidel. Successive iterations are carried out consisting of replacing the value of the temperature at each node of the mesh by the value calculated by the finite difference equation associated with it. Iteration consists of carrying out a complete scan of all the nodes, line after line and from left to right for each line, for example. The recalculated values are immediately taken into account for the calculation of the temperature value at the higher T order points (points located to the right and below in the proposed scanning mode).

2.3.3.4. *Convergence criterion*

For example, the calculation can be stopped as soon as the greatest variation of $T(i,j)$ during iteration remains lower than a given value ε.

REMARKS.–

– No calculation is applied to the points of the contour where the temperature is imposed.

– The value of the temperature will be stored in a table $T(i,j)$; we can use another table $L(i,j)$ whose values will indicate whether the coordinate point $(i\Delta x, j\Delta y)$ belongs to the domain (D) and the type of finite difference equation that applies to it.

It can be noted that the discretization described here amounts exactly to simulating a two-dimensional medium conducting electricity through a network of resistors connecting each node to its neighbors.

2.4. The fins

2.4.1. *The bar equation*

The embedded bar problem schematizes the important practical problem of cooling a solid by fins.

Let us consider a bar of constant section (thickness e and width ℓ) embedded between two surfaces at temperature T_0 and placed in a fluid at temperature T_∞ (see Figure 2.20).

The symmetry of the problem shows the existence of an extremum for the temperature in the middle of the bar. The geometry can be simplified considering only a half-bar with a condition of null flow at the end in contact with a fluid at T_∞ (see Figure 2.20).

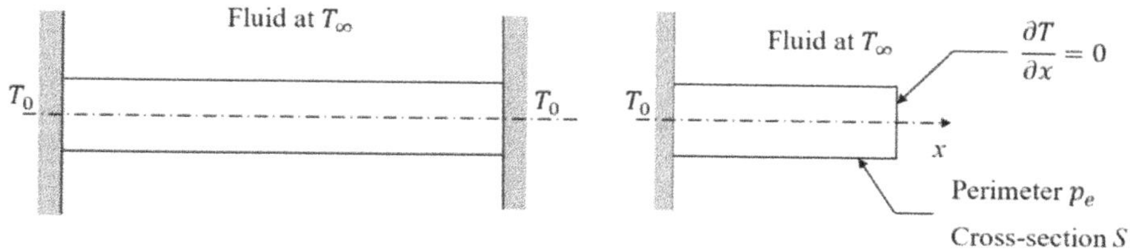

Figure 2.20. *Representation of an embedded bar and simplified diagram*

The cross-section of the bar is assumed to be sufficiently small so that there is no variation of temperature in the same cross-section at a distance x from the embedding in the wall at T_0.

Let us carry out an energy balance (see Figure 2.21) on the system constituted by the portion of the bar included between the abscissas x and $x + dx$ (we retain the hypothesis of the steady state and we neglect radiation).

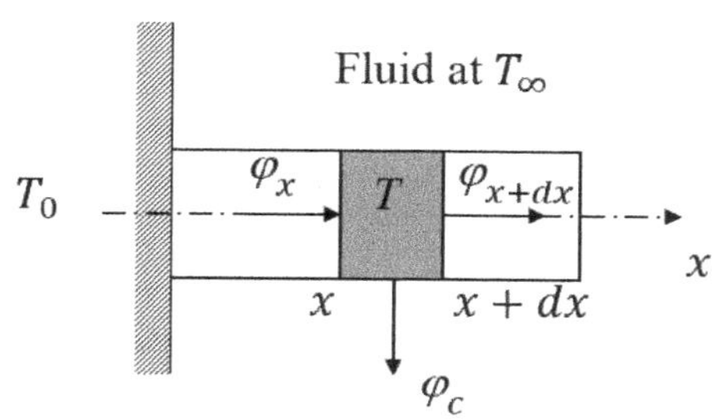

Figure 2.21. *Representation of elementary flows on an embedded bar*

The different flows have the following expression:

– φ_x, heat flow rate transmitted by conduction at the abscissa x; $\varphi_x = -\left(\lambda S \frac{dT}{dx}\right)_x$.

– φ_{x+dx}, heat flow rate transmitted by conduction at the abscissa $x + dx$; $\varphi_{x+dx} = -\left(\lambda S \frac{dT}{dx}\right)_{x+dx}$.

– φ_c, heat flow rate transmitted by convection to the periphery of the bar between x and $x + dx$; $\varphi_c = hp_e dx\,[T(x) - T_\infty]$.

The energy balance is written as $\varphi_x = \varphi_{x+dx} + \varphi_c$:

$$\left(\lambda S\frac{dT}{dx}\right)_{x+dx} - \left(\lambda S\frac{dT}{dx}\right)_x = hp_e dx[T(x) - T_\infty]$$

If λ and S are independent of the abscissa x, we get:

$$\lambda S\frac{\left(\frac{dT}{dx}\right)_{x+dx} - \left(\frac{dT}{dx}\right)_x}{dx} = hp_e[T(x) - T_\infty]$$

So, $T(x)$ is the solution of the following differential equation called equation of the bar or the fin:

$$\frac{d^2T}{dx^2} - \frac{hp_e}{\lambda S}(T - T_\infty) = 0 \qquad [2.19]$$

2.4.2. *Flow extracted by a fin*

A fin is a medium that is a good conductor of heat, one dimension of which is large compared to the others, for example: bar of thickness e and length L, with $e \ll L$. Fins are used when high heat flow rates have to be transmitted in a small space: cooling of electronic components, cooling of a motor by air, etc.

We have established the differential equation [2.19] verified by the temperature $T(x)$ of a fin embedded in a wall at temperature T_0 and immersed in a fluid at temperature T_∞.

By setting $\omega^2 = \frac{hp_e}{\lambda S}$ and $\theta = T - T_\infty$, we can write $\frac{d^2\theta}{dx^2} - \omega^2\theta = 0$.

It is a second-order differential equation with constant coefficients whose general solution is of the form:

$$\theta = Aexp(\omega x) + Bexp(-\omega x)\text{or: } \theta = A_1cosh(\omega x) + B_1sinh(\omega x)$$

2.4.2.1. *Long rectangular fin of constant section*

In the case of the long fin, we assume that: $T(x = L) = T_\infty$, where the length L of the fin tends to infinity.

The boundary conditions are then written as:

– in $x = 0, \theta(0) = T_0 - T_\infty$; (a)

– in $x = L \to \infty, \theta(L) \to 0$. (b)

(b) $\Rightarrow A = 0$

(a) $\Rightarrow B = T_0 - T_\infty$

$$\frac{T(x) - T_\infty}{T_0 - T_\infty} = exp(-\omega x) \qquad [2.20]$$

The heat flow rate dissipated over the entire surface of the fin can be calculated by integrating the local convection flux:

$$\varphi_p = \int_0^L hp_e[T(x) - T_\infty]dx$$

or more easily by noting that in the steady state, this flux is the same as that transmitted by conduction at the base of the fin, i.e. $\varphi_p = \varphi_{cond}(x = 0)$:

$$\varphi_p = -\lambda S \left(\frac{dT}{dx}\right)_{x=0} = -\lambda S(T_0 - T_\infty)(-\omega)exp(-\omega x)$$

with: $\omega = \sqrt{\frac{hp_e}{\lambda S}}$.

$$\varphi_p = \sqrt{hp_e\lambda S}(T_0 - T_\infty) \qquad [2.21]$$

2.4.2.2. *Rectangular fin of constant isolated section at the end*

The general solution obtained is identical to the previous case but the boundary conditions are different:

$$T(x = 0) = T_0$$

$$-\lambda S \left(\frac{dT}{dx}\right)_{x=L} = 0 \text{ (conservation of heat flow rate in } x = L)$$

The solution is written as:

$$\frac{T(x) - T_\infty}{T_0 - T_\infty} = cosh(\omega x) - tanh(\omega L)sinh(\omega x) = \frac{cosh[\omega(L - x)]}{cosh(\omega L)} \quad [2.22]$$

And the total heat flow rate dissipated by the fin has the expression:

$$\varphi_p = \omega\lambda S \, tanh(\omega L)(T_0 - T_\infty) \quad [2.23]$$

REMARK.– If the thickness e of the fin is small compared to its width ℓ, $\omega \approx \sqrt{\frac{2h}{\lambda e}}$.

2.4.2.3. *Rectangular fin of constant section with heat transfer at the end*

The general solution obtained is identical to case 2.4.2.1, but the boundary conditions are different:

$$T(x = 0) = T_0$$

$$-\lambda S \left(\frac{dT}{dx}\right)_{x=L} = hS[T(L) - T_\infty] \text{ (conservation of heat flow rate in } x = L\text{)}$$

The solution is written as:

$$\frac{T(x) - T_\infty}{T_0 - T_\infty} = \frac{cosh[\omega(L - x)] + \frac{h}{\omega\lambda} sinh[\omega(L - x)]}{cosh(\omega L) + \frac{h}{\omega\lambda} sinh(\omega L)} \quad [2.24]$$

And the total flow rate dissipated by the fin has the expression:

$$\varphi_p = \omega\lambda S \, (T_0 - T_\infty) \frac{tanh(\omega L) + \frac{h}{\omega\lambda}}{1 + \frac{h}{\omega\lambda} \, tanh(\omega L)} \quad [2.25]$$

REMARK.– If the thickness e of the fin is small compared to its width ℓ (which is generally verified): $\frac{h}{\omega\lambda} = \sqrt{\frac{he}{2\lambda}}$. The fins being generally made of good conductive material (λ is high) and having a low thickness e, the hypothesis $\sqrt{\frac{he}{2\lambda}} \ll 1$ is most often verified and equations [2.24] and [2.25] are then reduced to the simpler expressions of equations [2.22] and [2.23] which are those used in practice (see section A.4 of the Appendix).

2.4.2.4. *Circular fin of rectangular section*

These fins used to improve heat transfer between the outer wall of a tube and the ambient medium (e.g. automobile radiator tubes) are shown schematically in Figure 2.22.

Let us carry out a heat balance on the element of fin included between the radii r and $r + dr$.

The energy balance is written (see Figure 2.22) $\varphi_r = \varphi_{r+dr} + \varphi_c$ with:

– φ_r, heat flow rate transmitted by conduction on the radius r; $\varphi_r = -2\pi\lambda re\left(\frac{dT}{dr}\right)_r$;

– φ_{r+dr}, heat flow rate transmitted by conduction on the radius $r + dr$; $\varphi_{r+dr} = -2\pi\lambda(r + dr)e\left(\frac{dT}{dr}\right)_{r+dr}$;

– φ_c, heat flow rate transmitted by convection on the surface of the fin between r and $r + dr$; $\varphi_c = 2\,\{2h\,\pi\,rdr[T(r) - T_\infty]\}$.

If λ is independent of radius r, we get:

$$\frac{1}{r}\frac{(r + dr)\left(\frac{dT}{dr}\right)_{r+dr} - r\left(\frac{dT}{dr}\right)_r}{dr} = \frac{2h}{\lambda e}[T(r) - T_\infty]$$

or again: $\frac{d^2\theta}{dr^2} + \frac{1}{r}\frac{d\theta}{dr} - \frac{2h}{\lambda e}\theta = 0$, where: $\theta(r) = T(r) - T_\infty$.

It is a Bessel equation (see section A.5 of the Appendix) which solution is written in the form:

$$\theta = C_1\,I_0(\omega r) + C_2\,K_0(\omega r), \qquad \text{where } \omega = \sqrt{\frac{2h}{\lambda e}}$$

C_1 and C_2 being determined by the boundary conditions:

– in $r = r_0$: $\theta = T_0 - T_\infty$;

– in $r = r_e$: $h\theta = -\lambda\frac{d\theta}{dr}$ (the most general case: heat transfer at the end).

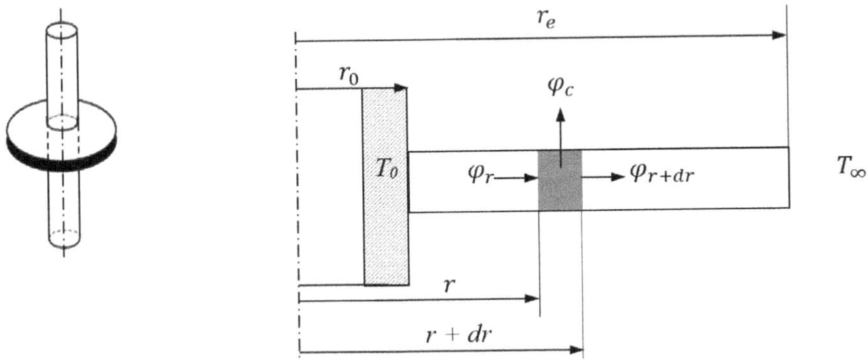

Figure 2.22. *Diagram of a circular fin and elementary fluxes on the fin*

We deduce the values of C_1 and C_2 :

$$C_1 = \frac{K_1(\omega r_e) - \frac{h}{\lambda\omega} K_0(\omega r_e)}{I_1(\omega r_e)K_0(\omega r_0) + I_0(\omega r_0)K_1(\omega r_e) + \frac{h}{\lambda\omega}[I_0(\omega r_e)K_0(\omega r_0) + I_0(\omega r_0)K_0(\omega r_e)]}$$

$$C_2 = \frac{1 - C_1 I_0(\omega r_0)}{K_0(\omega r_0)}$$

If we can make the hypothesis of null heat flow rate at the end: $\sqrt{\frac{he}{2\lambda}} \ll 1$, we end up with the following simplified expression:

$$\frac{T(r) - T_\infty}{T_0 - T_\infty} = \frac{K_1(\omega r_e)I_0(\omega r) + I_1(\omega r_e)K_0(\omega r)}{I_1(\omega r_e)K_0(\omega r_0) + I_0(\omega r_0)K_1(\omega r_e)} \quad [2.26]$$

And the total heat flow rate dissipated by the fin then has the expression:

$$\varphi_p = 2\pi\lambda r_0 e\omega\,(T_0 - T_\infty)\frac{K_1(\omega r_0)I_1(\omega r_e) - I_1(\omega r_0)K_1(\omega r_e)}{I_1(\omega r_e)K_0(\omega r_0) + I_0(\omega r_0)K_1(\omega r_e)} \quad [2.27]$$

2.4.3. *Efficiency of a fin*

The performance of a fin is characterized by comparing the dissipated heat flow rate to the one which would be dissipated in a fin of the same dimensions but whose temperature would be uniform and equal to that of the base (thermal conductivity $\lambda \to \infty$, no thermal conduction resistance, therefore no temperature drop in the fin).

The heat flow rate exchanged by this ideal fin would be:

– $\varphi_{max} = hp_e L(T_0 - T_\infty)$ for a rectangular fin of perimeter p_e and length L;

– $\varphi_{max} = 2\pi h({r_e}^2 - {r_0}^2)(T_0 - T_\infty)$ for a circular fin of base r_0 radius and outer radius r_e.

The efficiency of the fin is therefore written as: $\eta = \frac{\varphi_p}{\varphi_{max}}$.

We deduce the following relations:

– long rectangular fin ($L \to \infty$):

$$\eta = \frac{1}{\omega L} \quad [2.28]$$

– rectangular fin insulated at the end:

$$\eta = \frac{tanh(\omega L)}{\omega L} \quad [2.29]$$

– rectangular fin with heat transfer at the end:

$$\eta = \frac{tanh(\omega L) + \frac{h}{\omega\lambda}}{\omega L + \frac{h}{\omega\lambda}\, tanh(\omega L)} \quad [2.30]$$

with:

$$\omega = \sqrt{\frac{hp_e}{\lambda S}}$$

– fin of rectangular section:

$$\varphi_p = \sqrt{\frac{2\lambda}{hr_0}\frac{e}{r_0}}\frac{1}{\frac{{r_e}^2}{{r_0}^2} - 1}\frac{K_1(\omega r_0)I_1(\omega r_e) - I_1(\omega r_0)K_1(\omega r_e)}{I_1(\omega r_e)K_0(\omega r_0) + I_0(\omega r_0)K_1(\omega r_e)} \quad [2.31]$$

with:

$$\omega = \sqrt{\frac{2h}{\lambda e}}$$

In the case of more complex geometries (variable section fins, needle fins, etc.), there are formulas or charts (see section A.5 of the Appendix) allowing the efficiency of the fins to be determined and then the heat flow rate φ_p extracted by the fin can be calculated thanks to the relationship: $\varphi_p = \eta\varphi_{max}$.

REMARK.– Thermal resistance of a fin.

From the relations $\eta = \frac{\varphi_p}{\varphi_{max}}$ and $\varphi_{max} = hS_e(T_0 - T_\infty)$, we deduce: $\varphi_p = \frac{T_0 - T_\infty}{\frac{1}{\eta h S_e}}$ where S_e is the exchange surface between the fin and the fluid.

The overall thermal resistance between the base of the fin at the temperature T_0 and the fluid at the temperature T_∞ is therefore written as:

$$R_{fin} = \frac{1}{\eta h S_e} \tag{2.32}$$

2.4.4. *Electrical analogy*

2.4.4.1. *Fin in contact with an outside temperature*

The general equation of the fin is written as (see relation [2.19]):

$$\frac{d^2T}{dx^2} - \frac{hp_e}{\lambda S}(T - T_\infty) = 0$$

The solution can take the form of a quadrupole connecting the flux at the fin inlet $\varphi(0)$ at $x = 0$ and the temperature difference $T - T_\infty$ at $x = 0$, to the flux at the outlet $\varphi(L)$ at $x = L$ and the temperature difference $T - T_\infty$ at $x = L$.

Setting $\theta = T - T_\infty$, the equation of the fin becomes:

$$\frac{d^2\theta}{dx^2} - \frac{hp_e}{\lambda S}\theta = 0$$

The solution is of the form:

$$\theta(x) = A\cosh(\omega x) + B\sinh(\omega x)$$

with: $\omega^2 = \frac{hp_e}{\lambda S}$.

The heat flow rate is written as: $\varphi(x) = -\lambda S \frac{d\theta}{dx}$.

Therefore, in $x = 0$:

$$\theta(0) = A\varphi(0) = -\lambda S\omega B$$

And in $x = L$:

$$\theta(L) = A\cosh(\omega L) + B\sinh(\omega L)$$

$$\varphi(L) = -\lambda S\omega[A\sinh(\omega L) + B\cosh(\omega L)]$$

A and B are eliminated between these four relations to obtain:

$$\begin{bmatrix}\theta(0)\\ \varphi(0)\end{bmatrix} = \underbrace{\begin{bmatrix}\cosh(\omega L) & \frac{1}{\lambda\omega S}\sinh(\omega L)\\ \lambda\omega S\sinh(\omega L) & \cosh(\omega L)\end{bmatrix}}_{M\ =\ \text{quadrupole matrix}}\begin{bmatrix}\theta(L)\\ \varphi(L)\end{bmatrix}$$

This passive quadrupole can be represented by an electrical analogy with three resistors (a T-shaped representation has been chosen here), i.e. Figure 2.23.

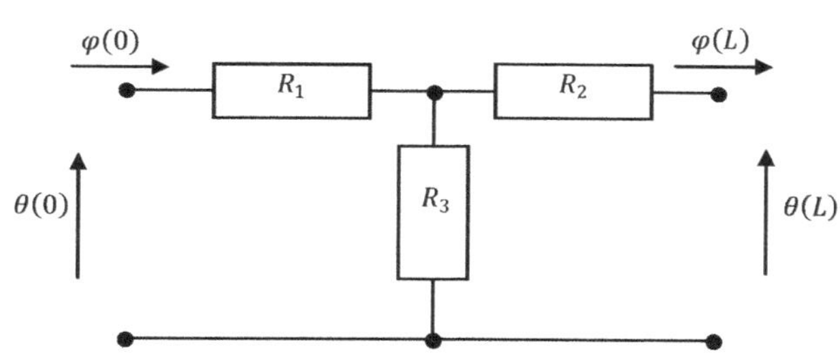

Figure 2.23. *Electrical analogy of the fin at reduced temperature* $\theta = T - T_\infty$

with:

$$R_1 = R_2 = \frac{\cosh(\omega L) - 1}{\lambda\omega S\sinh(\omega L)} \quad \text{and } R_3 = \frac{1}{\lambda\omega S\sinh(\omega L)}$$

Returning to the temperatures T, we obtain the diagram in Figure 2.24, where φ_p is the heat flow rate lost by the fin.

To solve a problem, two boundary conditions are needed, one at $x = 0$ and the other at $x = L$.

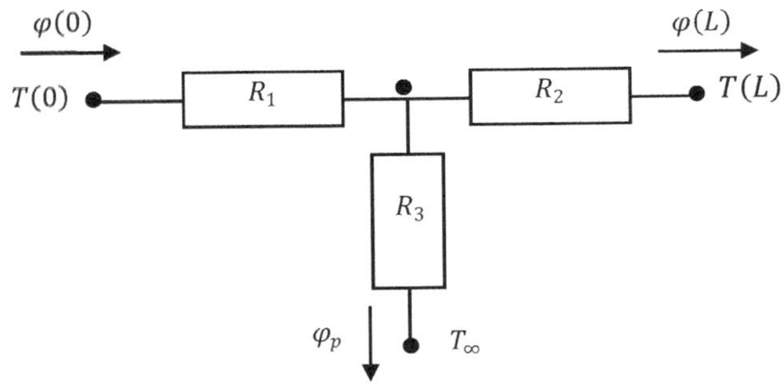

Figure 2.24. *Electrical analogy for fin (where φ_p is the heat flow rate lost by the fin)*

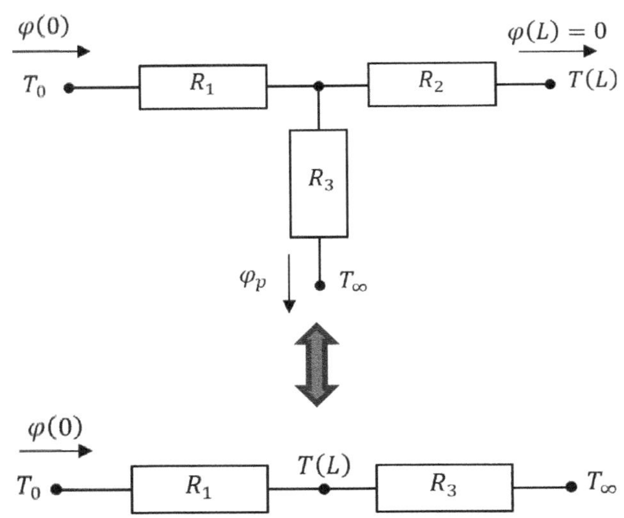

Figure 2.25. *Electrical analogy of a fin with an insulated end. For a color version of this figure, see www.iste.co.uk/jannot/heattransfer1.zip*

For example, if we want to know the temperature at the end of the fin (at $x = L$) assuming a perfect contact of the fin at $x = 0$ and insulated at the extremity $x = L$ using the electrical diagram shown in Figure 2.25, the temperature is given by:

$$T(L) - T_\infty = (T_0 - T_\infty)R_3/(R_1 + R_3)$$

2.4.4.2. *Fin in contact with two outside temperatures*

In a certain number of cases, the heat transfer takes place with two temperatures; in particular, when dealing with convection and radiation heat transfer, the heat flux is then written as:

$$\phi = h_c(T - T_c) + h_r(T - T_r)$$

The fin equation becomes:

$$\frac{d^2T}{dx^2} - \frac{(h_c + h_r)p_e}{\lambda S}\left(T - \frac{h_c T_c + h_r T_r}{h_c + h_r}\right) = 0$$

It is only necessary to replace h by $h_c + h_r$ and T_∞ by $\frac{h_c T_c + h_r T_r}{h_c + h_r}$ in the previous solution.

In some cases, it is more interesting to keep the potentials T_c and T_r in the analogy; the scheme is given in Figure 2.26.

with:

$$R_c = \frac{h_c + h_r}{h_c} R_3;\ R_r = \frac{h_c + h_r}{h_r} R_3 \text{ and } \omega^2 = \frac{(h_c + h_r)p_e}{\lambda S}$$

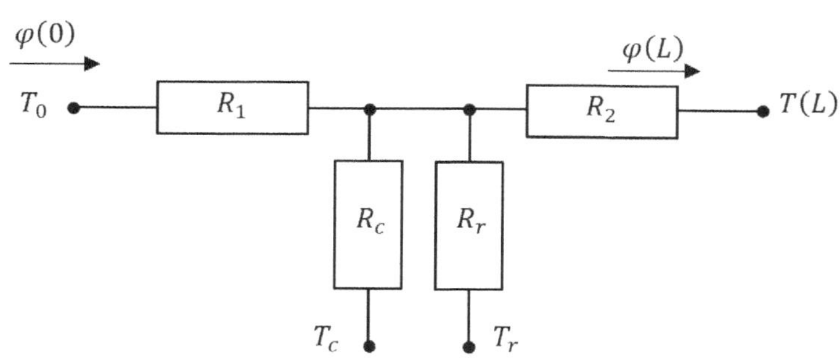

Figure 2.26. *Electrical analogy of a fin in contact with two temperatures*

As an example, in exercise 2.5.11, the boundary conditions are:

– in $x = 0$, $T(0) = T_0$;

– in $x = L$, $\varphi(L) = 0$.

It can be solved by the electric analogy.

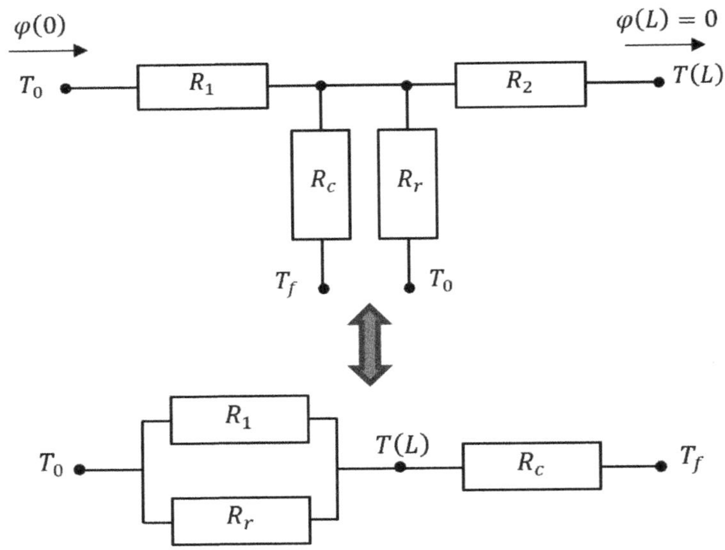

Figure 2.27. *Electrical analogy of a fin with an insulated end and in contact with two temperatures. For a color version of this figure, see www.iste.co.uk/jannot/heattransfer1.zip*

In addition, $T_c = T_f$ (fluid temperature) and $T_r = T_0$ (wall temperature).

The diagram of Figure 2.26 is replaced by the one of Figure 2.27.

Hence: $T(L) - T_f = (T_0 - T_f)R_c/(\frac{R_1R_r}{R_1+R_r} + R_c)$.

All calculation done, we find: $T(L) - T_f = (T_0 - T_f)\left[\frac{h_r}{h_c+h_r} + \frac{h_c}{(h_c+h_r)cosh(\omega L)}\right]$.

(Result identical to exercise 2.5.11).

2.4.5. *Choice of fins*

The fins are used when it is necessary to extract a high flux in a reduced space, for example: automobile radiator, air-cooled engine casing, air conditioner evaporator.

In general, the use of fins is:

– not very useful for liquids because the transfer coefficient by convection h is large;

– useful in the case of gases because h is weak.

Narrow and closely spaced fins are better than large and spaced fins but one is limited by the pressure drop of the fluid with which the fins exchange heat (it increases if the spacing between the fins is reduced too much). The fin is all the more efficient as its thermal conductivity λ is high. The choice of fins is then a compromise between cost, size, mass, pressure drop and heat transfer.

2.5. Corrected exercises

2.5.1. *Heat supply in an air-conditioned room*

The walls of an air-conditioned room at 25°C are made of the following different layers from the outside to the inside:

– 2 cm of cement coating;

– 20 cm of hollow agglomerates;

– 5 cm of polystyrene;

– 1 cm of plaster.

The outside temperature is 35°C; the external and internal convection coefficients have the respective values $h_e = 10\ \mathrm{W\ m^{-2}K^{-1}}$ and $h_i = 5\ \mathrm{W\ m^{-2}K^{-1}}$:

1) Calculate the overall thermal resistance of the wall.

2) Calculate the incoming heat flow rate for an area of 40 m^2.

3) Calculate the daily cost of air conditioning represented by the losses through these walls by considering the following data:

- operation 8 hours per day;
- efficiency of the air conditioning system equal to 2;
- price of electric kWh equal to € 0.15.

4) Calculate the temperature of the inside face of the wall.

5) Repeat questions 3 and 4 in the following two cases: 10 cm of polystyrene and no polystyrene.

6) The consumption of a PC cathode ray screen is 150 W, and that of a flat screen is 75 W. Assuming that it is also left on for 8 hours a day, compare the effect of interposing a polystyrene thickness of 5–10 cm with the effect of a screen change. What annual saving represents a change of screen?

Data: $\lambda_{hollow\ agglomerates} = 0.9\ \text{W m}^{-1}\text{K}^{-1}$; $\lambda_{coating} = 0.95\ \text{W m}^{-1}\text{K}^{-1}$; $\lambda_{plaster} = 1.2\ \text{W m}^{-1}\text{K}^{-1}$; $\lambda_{polystyrene} = 0.035\ \text{W m}^{-1}\text{K}^{-1}$.

Corrected

1) The convection and conduction thermal resistances in the different layers are in series, so the overall thermal resistance between the outside air and the inside air is written as:

$$R = \frac{1}{h_e S} + \frac{e_{coating}}{\lambda_{coating} S} + \frac{e_{agglos}}{\lambda_{agglos} S} + \frac{e_{poly}}{\lambda_{poly} S} + \frac{e_{plaster}}{\lambda_{plaster} S} + \frac{1}{h_i S}$$

$$RS = \frac{1}{10} + \frac{0.02}{0.95} + \frac{0.2}{0.5} + \frac{0.05}{0.035} + \frac{0.01}{1.2} + \frac{1}{5} = 2.15\ \text{K m}^2\text{W}^{-1}$$

2) $\varphi = \frac{\Delta T}{R} = \frac{35-15}{2.15} \times 40 = 185\ \text{W}$.

3) Daily amount of energy consumed: $Q = \varphi t = 0.185\ \text{kW} \times 8\ \text{h} = 1.48\ \text{kWh}$.

Electrical energy consumed: $Q_e = \frac{Q}{2} = 0.74\ \text{kWh}$.

Daily cost of air conditioning: $C = 0.74 \times 0.15 = €0.111$.

4) We have the relation: $\varphi = h_i S(T_{fi} - T_i)$, hence: $T_{fi} = T_i + \frac{\varphi}{h_i S} = 25 + \frac{185}{5 \times 40} = 25.9°\text{C}$.

5) With 10 cm of polystyrene, we obtain: $RS = 3.59\ \text{K m}^2\ \text{W}^{-1}$.

$$\varphi = \frac{\Delta T}{R} = \frac{35 - 15}{3.59} \times 40 = 111.5\ \text{W}$$

Daily quantity of energy consumed: $Q = \varphi t = 0.1115\ \text{kW} \times 8\ \text{h} = 0.89\ \text{kWh}$.

Electrical energy consumed: $Q_e = \frac{Q}{2} = 0.45$ kWh.

Daily cost of air conditioning: $C = 0.45 \times 0.15 =$ €0.067.

We have the relation: $\varphi = h_i S(T_{fi} - T_i)$, hence: $T_{fi} = T_i + \frac{\varphi}{h_i S} = 25 + \frac{111}{5 \times 40} =$ 25.6°C.

Without polystyrene, we obtain: $RS = 0.73$ K m^2 W^{-1}.

$$\varphi = \frac{\Delta T}{R} = \frac{35 - 15}{0.73} \times 40 = 548.4 \text{ W}$$

Daily quantity of energy consumed: $Q = \varphi t = 0.548 \text{ kW} \times 8 \text{ h} =$ 4.39 kWh.

Electrical energy consumed: $Q_e = \frac{Q}{2} = 2.19$ kWh.

Daily cost of air conditioning: $C = 2.19 \times 0.15 =$ €0.329.

We have the relation: $\varphi = h_i S(T_{fi} - T_i)$, hence: $T_{fi} = T_i + \frac{\varphi}{h_i S} = 25 + \frac{548}{5 \times 40} =$ 27.7°C.

Energy consumption is multiplied by nearly 3, and the temperature of the internal walls is 2.7°C higher than the air temperature, which reduces the feeling of thermal comfort due to the radiative exchange between the body and walls.

6) By going from 5 cm to 10 cm of polystyrene, the heat flow rate is reduced by: 185–111= 74 W, which has almost the same effect as changing the screen: 150 – 75 = 75 W.

The annual saving induced by a change of screen is in fact the sum:

– of electrical energy at the level of the screen power supply, i.e.:

$$0.075 \times 8 \times 0.15 \times 365 = €32.85$$

– of electrical energy for air conditioning, i.e.:

$$0.075 \times 8 \times 0.15 \times 365 \times 0.5 = €16.42$$

That is an annual total of €49.27.

2.5.2. *Heat losses from an oil pipeline*

An oil pipeline of diameter 100/108 mm is insulated by a layer of glass wool of thickness $e = 50$ mm and thermal conductivity $\lambda = 0.042\ \text{W m}^{-1}\text{K}^{-1}$. The oil inlet temperature is $T_i = 50°\text{C}$, the outside air temperature is $T_\infty = -15°\text{C}$ and the exchange coefficient with the outside air $h_e = 30\ \text{W m}^{-2}\ \text{K}^{-1}$(strong wind). If the oil flow is $\dot{m} = 6.4\ \text{kg s}^{-1}$, the specific heat of the oil $c = 2{,}100\ \text{J kg}^{-1}\ \text{K}^{-1}$, what is the temperature of the oil at the end of a length $L = 10$ km?

Corrected

By carrying out a heat balance on the portion of fluid located between the abscissas x and $x + dx$, we can write as:

$$\dot{m}cdT = \frac{T_\infty - T}{R}dx \qquad [2.33]$$

where R is the thermal resistance between the fluid at T_∞ and the air at T which is written according to relation [2.15]:

$$R = R_i + R_{tube} + R_{insulator} + R_e = \frac{1}{h_i 2\pi r_1} + \frac{ln\left(\frac{r_2}{r_1}\right)}{2\pi\lambda_{tube}} + \frac{ln\left(\frac{r_3}{r_2}\right)}{2\pi\lambda_{insulator}} + \frac{1}{h_e 2\pi r_3} \qquad [2.34]$$

A liquid/wall convective transfer coefficient is at least 100 times higher than a gas/wall convective coefficient which makes it possible to neglect R_i compared to R_e.

The steel thermal conductivity is larger than 15 W m^{-1} K^{-1}, i.e. 300 times the conductivity of glass wool, the thickness of which is also larger; therefore, it is possible to neglect R_{tube} compared to $R_{insulator}$.

The numerical application then leads to: $R = 2.51\ \text{K W}^{-1}$.

Integration of equation [2.33] gives: $T(x) = T_\infty + (T_i - T_\infty)exp\left(-\frac{x}{R\dot{m}c}\right)$.

Numerical application: $T(L) = -15 + (50 + 15)exp\left(-\frac{10^4}{2.51 \times 6.4 \times 2100}\right) = 33.3°\text{C}$.

2.5.3. *Critical insulation thickness*

Consider a cylindrical tube of internal radius r_1 and external radius r_2 made of a material of thermal conductivity λ_1. Suppose we want to insulate it with a sleeve of external radius r_3 and thermal conductivity λ_2. Let h_i and h_e be the internal and external transfer coefficients respectively.

1) Give the expression of the thermal resistance of the tube alone.

2) Give the expression of the thermal resistance of the tube + sleeve assembly.

3) Determine the conditions for which the addition of a sleeve reduces heat losses.

Data: r_2=1.5 cm; $\lambda = 0.1$ W m^{-1}K^{-1}; $h_e = 6$ W m^{-2}K^{-1}.

Corrected

1) The overall thermal resistance in the case of a non-insulated tube is written as:

$$R_1 = \frac{1}{h_1 2\pi r_1 L} + \frac{ln\left(\frac{r_2}{r_1}\right)}{2\pi\lambda_{tube} L} + \frac{1}{h_2 2\pi r_2 L}$$

2) The overall thermal resistance in the case of the tube + sleeve assembly is written as:

$$R_2 = \frac{1}{h_1 2\pi r_1 L} + \frac{ln\left(\frac{r_2}{r_1}\right)}{2\pi\lambda_{tube} L} + \frac{ln\left(\frac{r_3}{r_2}\right)}{2\pi\lambda_{sleeve}\, L} + \frac{1}{h_2 2\pi r_3 L}$$

3) A decrease in losses results in the condition: $R_1 < R_2$ or $R_2 - R_1 > 0$.

$$R_2 - R_1 > 0 \Rightarrow \frac{ln\left(\frac{r_3}{r_2}\right)}{\lambda_{sleeve}} + \frac{1}{h_2 r_3} - \frac{1}{h_1 r_1} > 0$$

A numerical resolution or plotting of the curve $R_2 - R_1 = f(r_3)$ leads to: $r_3 > 1.86$ cm.

The addition of a sleeve increases the conduction thermal resistance but decreases the external convection thermal resistance by increasing the exchange surface.

2.5.4. *Hot wire anemometry*

Experience shows that a fine filiform conductor, traversed by an electric current and subjected to a transverse flow of a fluid, very quickly reaches a state of thermal equilibrium under the influence, on the one hand, of the Joule effect and, on the other hand, of heat loss by conduction along the wire and, above all, by convection on the lateral surface of the wire immersed in the fluid. If the wire is moderately heated, the effect of thermal radiation is negligible compared to that of convection. This principle is used to measure (from the variation of the exchange coefficient by convection h) the fluid velocity using a device called a "hot-wire anemometer".

Let us consider a circular cylindrical wire of diameter d, of length 2ℓ for which the thermal conductivity λ is constant and the electrical resistivity ρ is a linear function of the temperature T so that $\rho = \rho_e(1 + \beta\theta)$, where $\theta = T - T_e$ and ρ_e is the resistivity of the wire at T_e , the reference temperature. This wire is maintained between two current supply terminals, massive and of large dimensions, so that the temperature at the ends of the wire can be considered as equal to the temperature of the terminals, itself equal to the temperature of the incident fluid T_e. The exchange coefficient by convection h between the incident fluid at the temperature T_e and the lateral surface of the immersed wire will be assumed to be constant even in the vicinity of the extremities.

1) If I is the intensity of the current flowing through the wire in normal operation, i.e. when the thermal equilibrium of the wire is reached, determine the expression of the temperature field along the wire. Calculate the maximum wire temperature if $h = 6000\ \mathrm{W\,m^{-2}\,K^{-1}}$ and $I = 50$ mA.

2) Calculate the conduction losses at the ends of the wire. From what length $2\ell_0$ of the wire can we consider that the heat losses through the ends of the wire are independent of the length?

3) When the wire crossed by a current of intensity I is placed in a flow for which the coefficient of exchange by convection h is known, calculate the expression of the electrical resistance R of the wire and deduce from it the expression of an average temperature $\bar{\theta}$ such that $R = R_e(1 + \beta\bar{\theta})$ with $R_e = \frac{8\rho_e\ell}{\pi d^2}$.

4) In practice, we want to measure the convection heat transfer coefficient h. This is made possible by the simultaneous measurement of the current I and of the resistance R of the wire. Show that we can determine the convection heat transfer coefficient h from the characteristic parameters of the wire at the temperature T_e and from the equilibrium values of the electrical resistance R and the intensity of the current I.

5) For an intensity $I = 50$ mA, we measured $\frac{R}{R_e} = 2.004$. What is the convection heat transfer coefficient h? If we admit that under these experimental conditions the convection heat transfer coefficient h (in W m^{-2} K^{-1}) is given as a function of the velocity v (in m s^{-1}) by the relation $h = 3{,}945\ v^{0.33}$, what is the velocity of the air?

Numerical data: $d = 5 \times 10^{-6}$m; $\ell = 1 \times 10^{-3}$m; $\rho_e = 9.8 \times 10^{-8}\Omega$ m; $\beta = 3.8 \times 10^{-3}$ K^{-1}; $\lambda = 70$ W m^{-1}K^{-1}; $T_e = 20$ °C.

Corrected

1) By carrying out a heat balance on the portion of wire located between the abscissas x and $x + dx$, we can write as: $\varphi_x + \varphi_g = \varphi_{x+dx} + \varphi_c$.

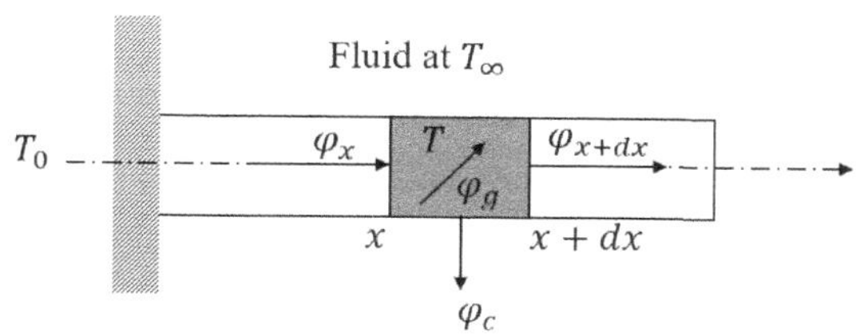

with:

– φ_x, the heat flow rate transmitted by conduction at the abscissa x: $\varphi_x = -\left(\lambda S \frac{dT}{dx}\right)_x$;

– φ_{x+dx}, the heat flow rate transmitted by conduction at the abscissa $x + dx$: $\varphi_{x+dx} = -\left(\lambda S \frac{dT}{dx}\right)_{x+dx}$;

– φ_c, the heat flow rate transmitted by convection at the periphery of the wire between x and $x + dx$: $\varphi_c = hp_e dx\,[T(x) - T_\infty]$;

– φ_g, the heat flow rate generated by the Joule effect between x and $x + dx$: $\varphi_g = \frac{\rho_e[1+\beta(T-T_e)]}{S} I^2 dx$.

Therefore, by setting $\theta = T - T_e$: $\left(\lambda S \frac{d\theta}{dx}\right)_{x+dx} - \left(\lambda S \frac{d\theta}{dx}\right)_x + \frac{\rho_e(1+\beta\theta)}{S} I^2 dx = hp_e dx\theta$.

If λ and S are independent of the abscissa x, we get:

$$\frac{d^2\theta}{dx^2} - \left(\frac{hp_e}{\lambda S} - \frac{\rho_e \beta I^2}{\lambda S^2}\right)\theta = \frac{\rho_e I^2}{\lambda S^2} \qquad [2.35]$$

which is of the form: $\frac{d^2\theta}{dx^2} - \omega^2\theta = \frac{\rho_e I^2}{\lambda S^2}$

with: $\omega^2 = \frac{hp_e}{\lambda S} - \frac{\rho_e \beta I^2}{\lambda S^2} = \frac{4h}{\lambda d} - \frac{\rho_e \beta I^2}{\lambda \pi^2 d^4}$.

It will be verified later with the numerical data that: $\frac{4h}{\lambda d} > \frac{\rho_e \beta I^2}{\lambda \pi^2 d^4}$.

$\theta_0 = \frac{\rho_e I^2}{\lambda S^2 \omega^2}$ is a particular solution of equation [2.35].

The general solution is written as: $\theta(x) = A cosh(\omega x) + \ B sinh(\omega x) + \theta_0$.

The boundary conditions are:

– in $x = 0$: $\theta = 0$;

– in $x = \ell$: $\frac{d\theta}{dx} = 0$.

We can deduce:

$$\theta(x) = \theta_0 \left\{1 - \frac{cosh[\omega(\ell - x)]}{cosh(\omega \ell)}\right\} \quad [2.36]$$

T_{max} is obtained in $x = \ell$, we deduce from equation [2.36]: $T_{max} = T_e + \theta_0 \left[1 - \frac{1}{cosh(\omega \ell)}\right]$.

Numerical application: $\omega^2 = 3.407 \times 10^7$; $\theta_0 = 266.4°C$; $T_{max} = 284.9°C$.

2) The losses by conduction at the extremities are written as: $\varphi = -\left(\lambda S \frac{dT}{dx}\right)_{x=0}$.

$\left(\frac{dT}{dx}\right)_{x=0} = \omega \theta_0 tanh(\omega \ell)$; but $tanh(\omega \ell) \approx 1$ if $\omega \ell > 3$, so the losses at the extremities no longer depend on the length if: $\ell > \frac{3}{\omega}$, where $\omega^2 = \frac{hp_e}{\lambda S} - \frac{\rho_e \beta I^2}{\lambda S^2}$. We find: $\ell_0 = 0.51$ mm.

3) $R = 2 \int_0^\ell \frac{\rho_e(1+\beta\theta)}{S} dx = \frac{2\rho_e \ell}{S} + \frac{2\rho_e \beta}{S} \int_0^\ell \theta(x) dx$.

$\int_0^\ell \theta(x) dx = \theta_0 \int_0^\ell [1 - cosh(\omega x) + tanh(\omega \ell) sinh(\omega x)] dx = \theta_0 \left[\ell - \frac{1}{\omega} tanh(\omega \ell)\right]$

hence:

$$R = \frac{2\rho_e \ell}{S} + \frac{2\rho_e \beta}{S} \theta_0 \left[\ell - \frac{1}{\omega} tanh(\omega \ell)\right] = \frac{8\rho_e \ell}{\pi d^2} + \frac{8\rho_e \beta \ell}{\pi d^2} \theta_0 \left[\ell - \frac{1}{\omega} tanh(\omega \ell)\right]$$

of the form: $R = R_e(1+\beta\bar{\theta})$ with $R_e = \frac{8\rho_e \ell}{\pi d^2}$ and $\bar{\theta} = \theta_0\left[\ell - \frac{1}{\omega}tanh(\omega\ell)\right]$.

$$4)\ \frac{R}{R_e} - 1 = \left[1 - \frac{tanh(\omega\ell)}{\omega\ell}\right]\frac{\beta\rho_e I^2}{\lambda S^2\omega^2}. \qquad [2.37]$$

We can numerically solve this equation where the unknown is ω and then deduce:

$$h = \frac{\lambda S}{p_e}\left(\omega^2 + \frac{\beta\rho_e I^2}{\lambda S^2}\right), \text{ where: } S = \frac{\pi d^2}{4} \text{ and: } p_e = \pi d$$

5) By solving equation [2.37] numerically with the given data, we obtain: $\omega = 5{,}277$.

We deduce: $h = 5{,}455\ \mathrm{W\,m^{-2}\,K^{-1}}$and: $v = \left(\frac{5{,}455}{3{,}945}\right)^{0.33} = 1.11\ \mathrm{m\,s^{-1}}$.

2.5.5. *Calculation of a fin*

An aluminum fin of 5 cm wide, 10 cm long and 3 mm thick is embedded in a wall. The base of the fin is maintained at 300°C, the ambient temperature is 30°C and the transfer coefficient is 10 W $\mathrm{m^{-2}\,K^{-1}}$:

1) Determine the temperature at the end of the fin and the heat flow rate extracted by the fin if the thermal gradients in the width and thickness directions are neglected.

2) Calculate the efficiency of the fin.

3) Calculate the relative errors that would have been made by considering the following two cases:

- the temperature at the extremity of the fin is equal to the ambient temperature;

- the extremity of the fin is insulated.

Corrected

1) The fin temperature is given by the relation [2.24]:

$$\frac{T(x) - T_\infty}{T_0 - T_\infty} = \frac{cosh[\omega(L-x)] + \frac{h}{\omega\lambda}sinh[\omega(L-x)]}{cosh(\omega L)\frac{h}{\omega\lambda}sinh(\omega L)}$$

Either: $T(L) = T_\infty + (T_0 - T_\infty)\frac{1}{cosh(\omega L)\frac{h}{\omega\lambda}sinh(\omega L)}$

with $\omega = \sqrt{\frac{hp_e}{\lambda S}} = \sqrt{\frac{10\times2\times(0.05+0.003)}{204\times0.003\times0.05}} = 5.885$ and $L = 0.1$ m, from which $T(L) =$ 58.1°C.

The total heat flow rate dissipated by the fin is given by the relation [2.25]:

$$\varphi_p = \omega\lambda S(T_0 - T_\infty)\frac{tanh(\omega L) + \frac{h}{\omega\lambda}}{1 + \frac{h}{\omega\lambda}tanh(\omega L)}$$

Hence: $\varphi_p = 26.0$ W.

2) The efficiency of the fin is given by relation [2.30]: $\eta = \frac{tanh(\omega L)+\frac{h}{\omega\lambda}}{\omega L+\frac{h}{\omega\lambda}tanh(\omega L)}$ hence: $\eta = 0.906$.

3) If $T(L) = T_\infty$, the efficiency of the fin is given by relation [2.28]: $\eta = \frac{1}{\omega L}$ which leads to a value greater than 1. The hypothesis of the long fin is not acceptable in this case. If the end of the fin is perfectly insulated, the efficiency of the fin is given by relation [2.29]: $\eta = \frac{tanh(\omega L)}{\omega L}$ hence: $\eta = 0.899$. The 0.8% error is quite acceptable in this case.

2.5.6. *Temperature of teapot handles*

As it can be seen in the photos below, the handles of Moroccan teapots have undergone a modification: they have gone from a solid steel cylinder (case no. 1) to a flat stainless steel cylinder (case no. 2). The purpose of this exercise is to show the interest of this modification.

We are in the presence of two fins; we will call h the exchange coefficient with the air at the temperature T_{ext}, we will call T_0 the temperature assumed to be uniform of the teapot. The handles are assumed to be in perfect contact with the teapot:

1) Establish the system of equations.

2) Give the expression for the temperature in the middle of the loop.

3) Calculate the temperature in the middle of the loop for both cases.

Case 1: solid steel cylinder[1]

Case 2: flat stainless steel cylinder[2]

Let us schematize the handles by:

– case 1: a cylinder of length L, diameter D and thermal conductivity λ_1;

– case 2: a parallelepiped of length L, section $\ell \times e$ and thermal conductivity λ_2.

Data: $T_0 = 70\ °C$; $T_{ext} = 20\ °C$; $h = 15\ \mathrm{W\,m^{-2}K^{-1}}$; $\lambda_1 = 40\ \mathrm{W\,m^{-1}K^{-1}}$; $\lambda_2 = 15\ \mathrm{W\,m^{-1}K^{-1}}$; $D = 10$ mm; $\ell = 15$ mm; $L = 15$ cm; $e = 2$ mm.

Corrected

1) The equation of the fin is written (relation [2.19]) as:

$$\frac{d^2T}{dz^2} - \omega^2(T - T_{ext}) = 0$$

with: $\omega^2 = \frac{hP_e}{\lambda S}$.

where P_e is the perimeter and S is the section.

In $x = 0$: $T = T_0$ (perfect contact).

In $x = \frac{L}{2}$: $\frac{dT}{dx} = 0$ (symmetry).

2) The solution is given by (relation [2.22]):

$$T(x) = T_{ext} + (T_0 - T_{ext})\frac{cosh\left[\omega\left(\frac{L}{2} - x\right)\right]}{cosh\left(\omega\frac{L}{2}\right)}.$$

1 Source: https://artisan-shop.com/theiere-marocaine-ancienne.html.

2 Source: https://www.amazon.fr/Olympia-Th%C3%A9i%C3%A8re-arabe-acier-inoxydable/dp/B00237XOA6/ref=sr_1_5?__mk_fr_FR=%C3%85M%C3%85%C5%BD%C3%95%C3%91&dchild=1&keywords=th%C3%A9i%C3%A8re+morocaine&qid=1626790160&s=kitchen&sr=1-5.

Hence: $T\left(\frac{L}{2}\right) = T_{ext} + (T_0 - T_{ext})\frac{1}{cosh\left(\omega\frac{L}{2}\right)}$.

3) Numerical application:

- case 1:

$$T\left(\frac{L}{2}\right) = T_{ext} + (T_0 - T_{ext})\frac{1}{cosh\left(\sqrt{\frac{hP_e}{\lambda S}}\frac{L}{2}\right)}$$

$$= 20 + (70 - 20)\frac{1}{cosh\left(\sqrt{\frac{15 \times \pi \times 0.01}{40 \times \pi \times 0.005^2}} \times 0.075\right)} = 54.5°C$$

- case 2:

$$T\left(\frac{L}{2}\right) = T_{ext} + (T_0 - T_{ext})\frac{1}{cosh\left(\sqrt{\frac{hP_e}{\lambda S}}\frac{L}{2}\right)}$$

$$= 20 + (70 - 20)\frac{1}{cosh\left(\sqrt{\frac{15 \times 2 \times (0.002 + 0.01)}{15 \times 0.002 \times 0.01}} \times 0.075\right)} = 27.9°C$$

The handle, which was "hot" in case 1 because $T\left(\frac{L}{2}\right) \gg T_{skin}$, is no longer hot in case 2 because $T\left(\frac{L}{2}\right) < T_{skin}$ justifying the modification.

2.5.7. *Thermal resistance of a finned tube*

Consider a circular cylindrical steel tube of thermal conductivity $\lambda = 40\ \mathrm{W\,m^{-1}K^{-1}}$, length $\ell = 3$ m, inner radius, $R_i = 40$ mm and outer radius $R_e = 50$ mm. Hot water at temperature T_i flows through the tube and the internal interfacial exchange coefficient, on the water side, is equal to $h_i = 250\ \mathrm{W\,m^{-2}K^{-1}}$. Fresh air at the temperature T_i circulates axially on the outer surface and the outer interfacial exchange coefficient, on the air side, is equal to $h_e = 10\ \mathrm{W\,m^{-2}K^{-1}}$.

In order to increase the heat transfer between the water and the air, it was decided to set longitudinal steel fins ($\lambda = 40\ \mathrm{W\,m^{-1}K^{-1}}$) on the tube.

1) Should the fins be placed on the water side or on the air side? Justify their location. For aerodynamic reasons, their number is fixed at 20. They are welded along the generatrices of the tube over the entire length ℓ and regularly spaced on the directing circle. Let L be their height and e their constant thickness (rectangular fin). In order to minimize the weight of raw material, the value of the product eL is imposed: $eL = 128\ \text{mm}^2$.

2) Determine the optimal geometric dimensions of a fin to evacuate a maximum flow.

3) What is the efficiency η_t of the fins?

4) What is the total thermal resistance between the water at the temperature T_i and the outside air at the temperature T_e? Give the equivalent electrical diagram and discuss.

REMARK.– In the calculation of the interfacial thermal resistance of a wall with fins, the surface which must be taken into account is the surface $\eta_t S$, where η_t is the total efficiency and S *is* the surface in contact with the fluid.

5) What is the heat flow rate exchanged between water and air if $T_i - T_e = 60°\text{C}$? Compare this heat flow rate to the one which would be exchanged by the same device without fins.

Corrected

1) The fins are placed on the side where the convective transfer coefficient is the lowest, i.e. on the air side.

2) The heat flow rate dissipated on the surface of the 20 fins with transfer at the end is written as:

$$\varphi_{p_1} = 20\omega\lambda S \frac{tanh(\omega L) + \dfrac{h}{\omega\lambda}}{1 + \dfrac{h}{\omega\lambda} tanh(\omega L)} [T(R_e) - T_e]$$

where: $S = e\ell$; $P = 2(e + \ell)$; $\omega = \sqrt{\dfrac{h_e P}{\lambda S}}$.

The heat flow rate dissipated on the outer surface without fin is written as:

$$\varphi_{p_2} = h_e(2\pi R_e - 20e)\,\ell\,[T(R_e) - T_e]$$

The total heat flow rate dissipated is written as:

$$\varphi_t = \varphi_{p_1} + \varphi_{p_2} = \left[20\omega\lambda S \frac{tanh(\omega L)+\frac{h}{\omega\lambda}}{1+\frac{h}{\omega\lambda}tanh(\omega L)} + h_e(2\pi R_e - 20e)\,\ell\right][T(R_e) - T_e]$$

The parameters e and L being linked by the relation $eL = constant$, the various flows depend in fact only on the single parameter e. We thus return to the problem of finding the extremum of the function φ_t of a single variable. Several methods are applicable:

– Derive the function and find the value that cancels the derivative. In this specific case, the expression of the heat flow rate as a function of e is relatively complex, which makes this method complicated. However, it could be applied by considering the simplified expression corresponding to the hypothesis of a zero flow at the end of the fin leading to: $\varphi_{p_1} = 20\omega\lambda S\, tanh(\omega L)[T(R_e) - T_e]$ which derivation is easier.

– Drawing the curve $\varphi_t(e)$ and determining the extremum, very easily achievable in a spreadsheet.

– Using a spreadsheet solver to directly find the value of e which maximizes φ_t.

This last method led us to: $e = 2.53$ mm and $L = 5.05$ mm. It should be noted that these values do not maximize the flux dissipated on the fin but on the finned surface + surface without fin assembly. The extrema are slightly different because increasing the thickness of the fins decreases the exchange surface without fins. The values that maximize the flux dissipated by the fins are: $e = 2.61$ mm and $L = 4.90$ mm.

3) In the case $e = 2.53$ mm and $L = 5.05$ mm, the efficiency of the fin which is written $\eta = \frac{tanh(\omega L)+\frac{h}{\omega\lambda}}{\omega L+\frac{h}{\omega\lambda}tanh(\omega L)}$ and has the value $\eta = 0.707$.

4) The equivalent electrical diagram is as follows:

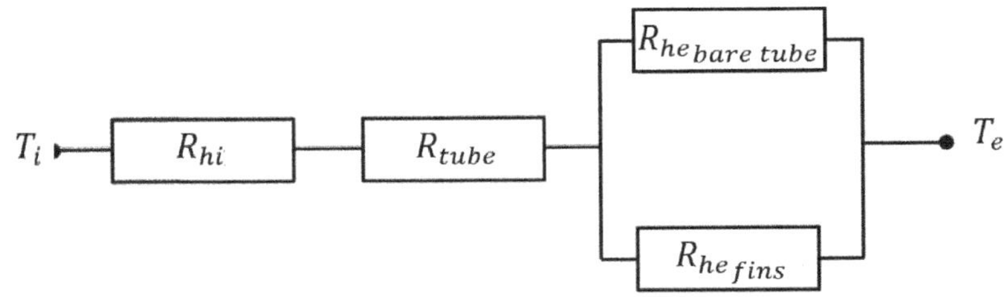

where: $R_{h_i} = \frac{1}{h_i 2\pi R_i \ell}$; $R_{tube} = \frac{ln\left(\frac{R_e}{R_i}\right)}{2\pi\lambda\ell}$; $R_{he\,baretube} = \frac{1}{h_e(2\pi R_e - 20e)\ell}$; $R_{he\,fins} = \frac{1}{20\,P\ell h_e \eta}$.

The global resistance is written as: $R = R_{h_i} + R_{tube} + \frac{1}{\frac{1}{R_{hebaretube}} + \frac{1}{R_{hefins}}}$

or: $R = \frac{1}{h_i 2\pi R_i \ell} + \frac{ln\left(\frac{R_e}{R_i}\right)}{2\pi\lambda\ell} + \frac{1}{h_e(2\pi R_e - 20e)\ell + 20\,P\ell h_e \eta}$.

Numerical application: $R = 0.00331\ \mathrm{K\,W^{-1}}$.

Without fins: $R_0 = R_{h_i} + R_{tube} + R_{h_e}$ with $R_{h_e} = \frac{1}{h_e 2\pi R_e \ell}$.

Numerical application: $R_0 = 0.0198\ \mathrm{K\,W^{-1}}$.

5) The heat flow rate is written as: $\varphi = \frac{T_i - T_e}{R}$; with fins: $\varphi = \frac{T_i - T_e}{R} = \frac{60}{0.00331} = 18.1$ kW; without fins: $\varphi = \frac{T_i - T_e}{R_0} = \frac{60}{0.0198} = 3.03$ kW.

The presence of fins multiplies the dissipated heat flow rate by 6.

2.5.8. *Heat input in a cold room*

We consider a cubic cold room with an inner side of $\ell = 3$ m and at an inner temperature of –18°C. The walls are made of 15 cm of polystyrene and 10 cm of concrete. The outside temperature is 35°C, and the internal and external transfer coefficients are respectively equal to $h_e = 15\ \mathrm{W\,m^{-2}\,K^{-1}}$ and $h_i = 10\ \mathrm{W\,m^{-2}\,K^{-1}}$. Assuming that the floor is perfectly insulated, calculate the heat flow rate entering the enclosure.

Corrected

As a first approximation, we can neglect the heat flow rate crossing the "edges" and the "corners" of the cold room and consider only the 1D transfer through the plane walls.

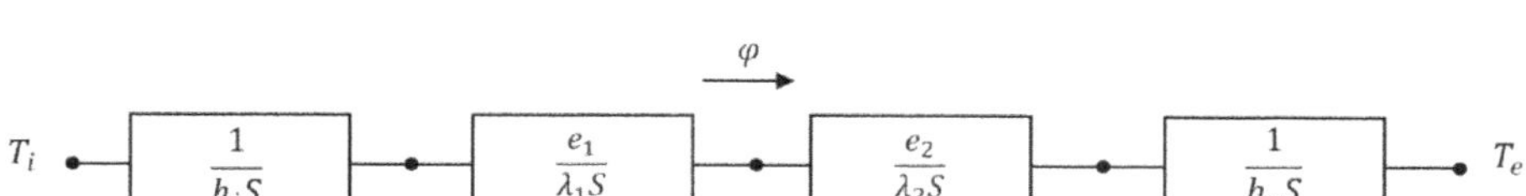

with: $S = 5\ell^2$.

According to the appendices:

$$\lambda_1 = 0.043 \text{ W m}^{-1}\text{K}^{-1};\ e_1 = 0.15 \text{ m (polystyrene)}$$

$$\lambda_2 = 1.75 \text{ W m}^{-1}\text{K}^{-1};\ e_2 = 0.10 \text{ m (concrete)}$$

A first default value can be calculated by:

$$\varphi = \frac{T_i - T_e}{\frac{1}{h_i S} + \frac{e_1}{\lambda_1 S} + \frac{e_2}{\lambda_2 S} + \frac{1}{h_e S}}$$

We deduce a first approximate value: $\varphi = 642$ W.

The calculation can be refined by considering the losses through the edges and the corners. Since the floor is perfectly insulated, only eight edges and four corners should be considered. The problem that arises is that the wall is two-layered; in this case, we consider an equivalent thermal conductivity λ such that the wall of thickness $e = e_1 + e_2$ and thermal conductivity has λ the same thermal resistance as the real wall, i.e.:

$$\frac{e}{\lambda} = \frac{e_1}{\lambda_1} + \frac{e_2}{\lambda_2}$$

hence: $\lambda = 0.0705 \text{ W m}^{-1}\text{K}^{-1}$.

The heat flow rate crossing the walls by conduction is written as:

$$\varphi = \lambda\left[\sum_{i=1}^{5} F_{wall\,i} + \sum_{i=1}^{8} F_{edge_i} + \sum_{i=1}^{5} F_{corner\,i}\right]\Delta T$$

or again: $\varphi = \frac{\Delta T}{R_{walls}}$ with: $R_{walls} = \frac{1}{\lambda\left[\sum_{i=1}^{5} F_{wall\,i} + \sum_{i=1}^{8} F_{edge_i} + \sum_{i=1}^{5} F_{corner\,i}\right]}$.

And: $F_{wall\,i} = F_{wall} = \frac{\ell^2}{e}$; $F_{edge_i} = F_{edge} = 0.54\ell$; $F_{corner\,i} = F_{corner} = 0.15e$ because the enclosure is cubic.

From this, we deduce: $R_{walls} = 0.0734 \text{ W K}^{-1}$.

Hence: $\varphi = \frac{\Delta T}{\frac{1}{h_i S} + R_{walls} + \frac{1}{h_e S}} = 687$ W.

The difference between the two calculated values is only 6.5%.

2.5.9. *Pipe insulation*

Ice water at 8°C circulates at a velocity of 0.5 m s^{-1} inside a steel tube with inside and outside diameters of 20/27 mm. This pipe is insulated with a layer of insulation of 10 mm thick. The ambient air temperature is equal to 25°C. The length of the pipe is 30 meters.

1) Calculate the heat flow rate gained by the tube.

2) Calculate the internal temperature of the tube.

3) Calculate the heating of the water as it flows through the tube.

Data: external convection coefficient: $h_e = 10 \text{ W m}^{-2}\text{ K}^{-1}$; internal convection coefficient: $h_e = 2027 \text{ W m}^{-2}\text{ K}^{-1}$; $\lambda_{insulator} = 0.35 \text{ W m}^{-1}\text{K}^{-1}$; water velocity $v = 1 \text{ m s}^{-1}$.

Corrected

1) The heat flow rate transferred is calculated by:

$$\varphi = L \frac{T_{air} - T_{water}}{\frac{1}{h_i \pi D_i} + \frac{ln\left(\frac{D_e}{D_i}\right)}{2\pi\lambda_{tube}} + \frac{ln\left(\frac{D_{insulator}}{D_e}\right)}{2\pi\lambda_{insulator}} + \frac{1}{h_e \pi D_e}}$$

Numerical application: $\varphi = 159.0$ W.

REMARK.– This calculation is approximate because the difference $T_{air} - T_{water}$ is considered to be constant all along the tube while the temperature of the water rises between the inlet and the outlet.

2) The heat flow rate exchanged can also be expressed by:

$$\varphi = L \frac{T_{tubeint.} - T_{water}}{\frac{1}{h_i \pi D_i}}$$

hence: $T_{tubeint.} = T_{water} + \frac{\varphi}{h_i \pi D_i L} = 8.04$°C.

The physical properties of water should therefore have been calculated for a temperature $T_{film} = 8°C$; they were calculated at 12°C, which is, however, very close. It is therefore not necessary to repeat the calculation.

3) A heat balance on the water between the inlet and the outlet of the tube leads to:

$$\varphi = \dot{m}_{water} c_{water} \Delta T_{water}$$

From this, we deduce: $\Delta T_{water} = \dfrac{\varphi}{\dot{m}_{water} c_{water}} = 0.12°C.$

This low difference validates a posteriori the approximate calculation of the heat flux.

2.5.10. *Heat losses from a pipe*

In an individual heating installation, water from a boiler at a temperature of 50°C is distributed to the radiators located in the rooms to be heated by a network of copper tubes with a diameter of 12/14 mm. A 5 m length of pipe is located in an uninsulated attic where the average air temperature is 10°C. The water flows in the tubes with a velocity $v = 0.5 \text{ m s}^{-1}$. The natural convection coefficient between the outer surface of the tube and the air is $h_e = 10 \text{ W m}^{-2}\text{K}^{-1}$, and the emissivity of copper is 0.4. It will be assumed that the temperature drop in water is small compared to the difference between the water temperature and the air temperature.

1) Calculate the radiation heat loss coefficient on the outer surface of the tube.

2) Represent the equivalent electrical diagram of the system. What simplifying assumptions can be made? Justify them.

3) Calculate the heat flow rate lost over the 5 m length and the corresponding water temperature drop.

4) We want to insulate the tube with a foam sleeve of conductivity of 0.045 W m^{-1} K^{-1}. Represent the new equivalent electrical diagram and give the expression of the overall thermal resistance between water and air.

5) What sleeve thickness should be used to reduce the wasted heat flow rate by 80%? The emissivity of the sleeve is 0.9.

Corrected

1) $\phi_r = \sigma\varepsilon(T^4 - T_a{}^4) = 5.67 \times 10^{-8} \times 0.4 \times (323^4 - 283^4) = 101.4 \text{ W m}^{-2}$.

$$h_r = \frac{\phi_r}{T - T_a} = \frac{101.4}{323 - 283} = 2.53 \text{ W m}^{-2} \text{ K}^{-1}$$

2)

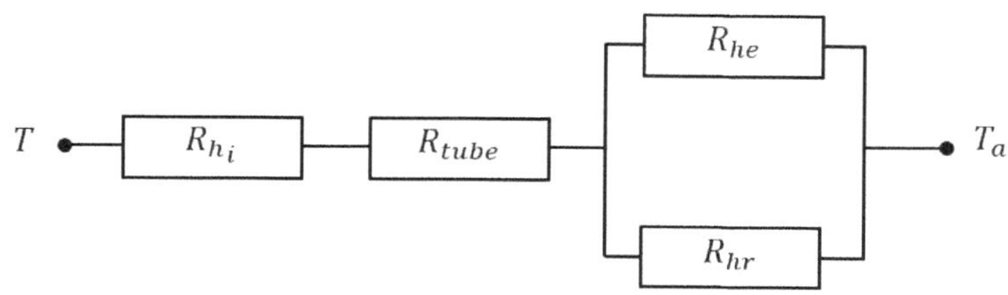

A forced convection coefficient with a liquid is of the order of 100–1,000 times larger than a forced convection coefficient with a gas in natural convection, so we can neglect R_{h_i}compared to R_{h_e}.

The conductivity of copper is of the order of 400 W m^{-1} K^{-1} and its thickness is 1 mm, i.e. the thermal resistance R_{h_e} is completely negligible compared to $R_{tube} = \frac{e}{\lambda} \approx 2.5 \times 10^{-6} \text{ K m}^2 \text{ W}^{-1}$.

3) Neglecting the internal convection and conduction resistances in the copper, the equivalent thermal resistance is written as:

$$R_{eq} = \frac{1}{\frac{1}{R_{h_e}} + \frac{1}{R_{h_r}}} = \frac{1}{(h_e + h_r)\pi D_2 L}$$

and:

$$\varphi = \frac{T - T_a}{R_{eq}} = (h_e + h_r)\pi D_2 L(T - T_a)$$

$$\varphi = (10 + 2.53) \times \pi \times 0.014 \times 5 \times (323 - 283) = 110.3 \text{ W}$$

and: $\Delta T = \frac{\varphi}{\rho v \frac{\pi D_1{}^2}{4} c} = \frac{110.3}{1000 \times 0.5 \times \frac{\pi \times 0.012^2}{4} \times 4180} = 0.47°\text{C}.$

4) $R_{eq}{}' = R_{sleeve} + \frac{1}{\frac{1}{R_{he}}+\frac{1}{R_{hr}{}'}} = \frac{ln\left(\frac{D_3}{D_2}\right)}{2\pi\lambda_m L} + \frac{1}{(h_e+h_r{}')\pi D_3 L}$

with: $h_r{}' = \frac{\sigma\varepsilon_m(T_e{}^4-T_a{}^4)}{T_e-T_a} \approx 4\sigma\varepsilon_m T_a{}^3$ because the difference $T_e - T_a$ will be small.

Hence: $h_r{}' = 4 \times 5.67 \times 10^{-8} \times 0.9 \times 283^3 = 4.62 \text{ W m}^{-2}\text{ K}^{-1}$.

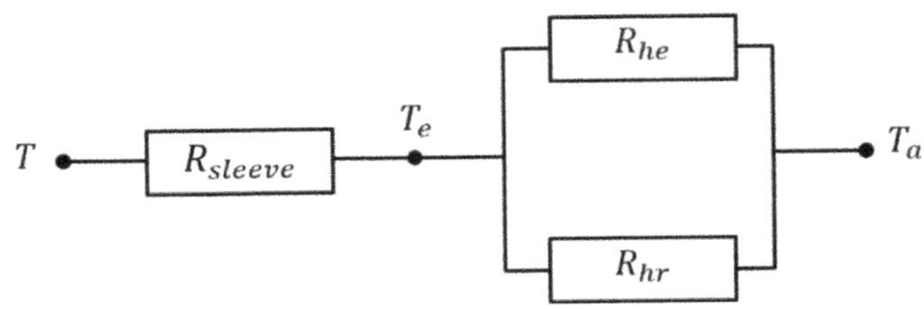

5) We want to get $\varphi' = 0.2\varphi$ or still $R_{eq}{}' = 5R_{eq}$.

Hence: $\frac{ln\left(\frac{D_3}{D_2}\right)}{2\pi\lambda_m L} + \frac{1}{(h_e+h_r{}')\pi D_3 L} = \frac{5}{(h_e+h_r)\pi D_2 L}$.

The resolution of this equation leads to $D_3 = 4.73$ cm or an insulation thickness of 1.67 cm.

2.5.11. *Effect of a fin and radiation on a thermocouple*

The temperature of a moving fluid in a pipe is measured with a thermocouple passing through the wall. The temperature θ measured is that of the junction located at the end $x = l$ of the thermocouple. The thermocouple is assumed to be made of a conductive material λ and its diameter d is small compared to its length l; the approximation of the fin in the steady state can be used to model the thermocouple. It will be assumed that the wall imposes its temperature T_p on the thermocouple at $x = 0$.

1) Describe the exchanges between the thermocouple and its environment.

2) Write and solve the system of equations giving the temperature of the thermocouple assuming that it is welded to the wall.

3) Calculate the measurement error and discuss the influence of the different parameters.

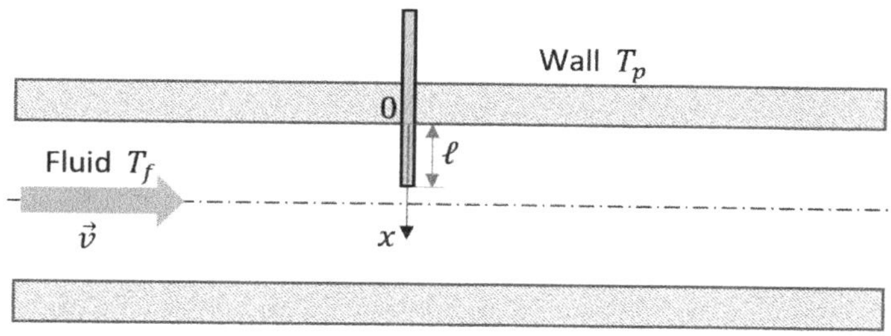

Data: $d = 1$ mm; $\lambda = 20$ W m^{-1}K^{-1}; $l = 10$ cm; emissivity of the thermocouple $\varepsilon_t = 0.9$; wire/fluid convection coefficient: $h_c = 100$ W m^{-2} K^{-1}; $T_f = 20$°C; $T_p = 40$°C.

4) What would happen to this error with $T_f = 800$°C and $T_p = 820$°C, the other parameters being unchanged?

Corrected

1) The heat exchanges on the thermocouple between the crossing of the wall $x = 0$ and its end $x = l$ are as follows:

- conduction in the section of the wire: $\varphi_{cond} = -\lambda S \frac{\partial T}{\partial x}$;
- convection between the wire and the fluid at T_f: $\varphi_{conv} = h_c S_l [T(x) - T_f]$;
- radiation between the wire and the wall at T_p: $\varphi_{rad} = \sigma \varepsilon_t S_l [T(x)^4 - T_p{}^4]$.

We can linearize the radiation heat flow rate in the form: $\varphi_{rad} = h_r S_l [T(x) - T_p]$

with: $h_r = \sigma \varepsilon_t [T(x)^2 + T_p{}^2][T(x) + T_p]$ assumed to be constant in the first approximation.

It is necessary to solve the heat equation with the boundary conditions between $x = 0$ and $x = l$ to be able to calculate the temperature $T(l)$ indicated by the thermocouple.

2) The wire diameter is small, so the radial temperature variation is neglected; the temperature only depends on x.

The fin equation [2.19] only takes convection into account: $\frac{d^2T}{dx^2} - \frac{h_c P}{\lambda S}(T - T_f) = 0$.

Where P is the perimeter of the wire, S is its section and λ is its thermal conductivity.

With radiation, it becomes $\frac{d^2T}{dx^2} - \frac{h_c P}{\lambda S}(T - T_f) - \frac{h_r P}{\lambda S}(T - T_p) = 0.$

$$\frac{d^2T}{dx^2} - \frac{h_c P}{\lambda S}(T - T_f) - \frac{h_r P}{\lambda S}(T - T_f + T_f - T_p) = 0$$

$$\frac{d^2T}{dx^2} - \frac{hP}{\lambda S}(T - T_f) = \frac{h_r P}{\lambda S}(T_f - T_p) \quad \text{(a)}$$

with: $h = h_c + h_r$.

The boundary conditions are:

$$T(x = 0) = T_p \quad \text{(b)}$$

$$\frac{dT}{dx}(x = l) = 0 \quad \text{(c)}$$

The general solution of the homogeneous equation (without second member) is given by:

$$T(x) - T_f = A\cosh(\omega x) + \mathrm{B}\sinh(\omega x)$$

with: $\omega = \sqrt{\frac{hP}{\lambda S}} = \sqrt{\frac{4h}{\lambda d}}$.

$T - T_f = -\frac{h_r}{h}(T_f - T_p)$ is a particular solution of equation (a).

The general solution is therefore written as:

$$T - T_f = Acosh(\omega x) + \mathrm{B}\, sinh(\omega x) + \frac{h_r}{h}(T_p - T_f)$$

with:

– in $x = 0$: $T_p - T_f = A + \frac{h_r}{h}(T_p - T_f)$ from where: $A = \left(1 - \frac{h_r}{h}\right)(T_p - T_f)$;

– in $x = l$: $\frac{dT}{dx} = 0$ hence: $A\omega sinh(\omega l) + B\omega cosh(\omega l) = 0$ from where: $B = -A\frac{sinh(\omega l)}{cosh(\omega l)}$.

$$T - T_f = \left(1 - \frac{h_r}{h}\right)(T_p - T_f)\left[cosh(\omega x) - \frac{sinh(\omega l)}{cosh(\omega l)} sinh(\omega x)\right] + \frac{h_r}{h}(T_p - T_f)$$

$$T - T_f = (T_p - T_f)\left[\left(1 - \frac{h_r}{h}\right)\frac{cosh[\omega(l - x)]}{cosh(\omega l)} + \frac{h_r}{h}\right]$$

And in particular in $(x = l)$: $T(l) - T_f = (T_p - T_f)\left[\left(1 - \frac{h_r}{h}\right)\frac{1}{cosh(\omega l)} + \frac{h_r}{h}\right]$.

3) The measurement error is equal to $T(l) - T_f$, so we can immediately note that it is proportional to $(T_p - T_f)$:

$$h_r \approx \sigma\varepsilon_t({T_f}^2 + {T_p}^2)(T_f + T_p)$$
$$= 5.67 \times 10^{-8} \times 0.9 \times (293^2 + 313^2)(293 + 313)$$

$$h_r \approx 5.68 \text{ W m}^{-2}\text{K}^{-1}$$

$$h = h_c + h_r = 100 + 5.68 = 105.68 \text{ W m}^{-2}\text{K}^{-1}$$

$$\omega = \sqrt{\frac{4h}{\lambda d}} = \sqrt{\frac{4 \times 105.68}{20 \times 0.001}} = 145.4$$

$$T(l) - T_f = (40 - 20)\left[\frac{5.68}{105.68} + \left(1 - \frac{5.68}{105.68}\right)\frac{1}{cosh(145.4 \times 0.1)}\right]$$
$$= (40 - 20)[0.0537 + 1.5 \times 10^{-6}] = 1.08°\text{C}$$

The radiation effect (term $\frac{h_r}{h}$) is much more important than the fin effect (term $\frac{1}{cosh(\omega l)}$) in this case.

By taking $h_c = 20 \text{ W m}^{-2}\text{K}^{-1}$, the error is 4.51°C, so it is interesting to increase convection and reduce radiation.

By taking a length of thermocouple $l = 3$ cm, the error is 1.68°C, the fin effect is no longer negligible. However, by decreasing the diameter of the thermocouple to $d = 0.1$ mm, the error is 1.08°C. It is therefore advantageous to use a fine thermocouple to reduce the fin effect, but the diameter has no influence on the radiative effect.

By taking a thermal conductivity thermocouple wire $\lambda = 200 \text{ W m}^{-1}\text{K}^{-1}$, the error is 1.56°C, so there is no point in using a type T thermocouple where one of the two wires is made of thermal conductivity copper $\lambda = 386 \text{ W m}^{-1}\text{K}^{-1}$.

4) The value of h_r becomes:

$$h_r \approx \sigma \varepsilon_t \left(T_f{}^2 + T_p{}^2\right)\left(T_f + T_p\right)$$
$$= 5.67 \times 10^{-8} \times 0.9 \times (1073^2 + 1093^2)(1073 + 1093)$$

$$h_r \approx 259.3 \text{ W m}^{-2}\text{K}^{-1}$$

$$h = h_c + h_r = 100 + 259.3 = 359.3 \text{ W m}^{-2}\text{K}^{-1}$$

$$\omega = \sqrt{\frac{4h}{\lambda d}} = \sqrt{\frac{4 \times 359.3}{20 \times 0.001}} = 268.1$$

$$T(l) - T_f = 20 \times \left[\frac{259.3}{359.3} + \left(1 - \frac{259.3}{359.3}\right)\frac{1}{cosh(268.1 \times 0.1)}\right]$$
$$= 20 \times [0.722 + 7.8.10^{-12}] = 14.4°\text{C}$$

The measurement error, which is due only to radiation in the case studied, increases very strongly at high temperature.

2.5.12. *Internal heat transfer in a pipe*

The hydrogen supply pipe of a space engine (diameter $D = 5$ cm, thickness, $e = 5$ mm thermal conductivity $\lambda = 25 \text{ W m}^{-1}\text{K}^{-1}$) is as shown in the figure, with its lower part occupied by liquid H_2 with a convection exchange coefficient $h_\ell = 3{,}000 \text{ W m}^{-2}\text{K}^{-1}$and its upper part occupied by gaseous H_2 with an exchange coefficient by convection $h_g = 30 \text{ W m}^{-2}\text{K}^{-1}$. The internal temperature of the pipe (liquid–gas balance of H_2) is $T_i = 20$ K. The temperature outside the pipe is $T_e = 250$ K, and the external exchange coefficient is $h_e = 20 \text{ W m}^{-2} \text{ K}^{-1}$. The objective of the exercise is to predict the heat losses per unit length of the pipe.

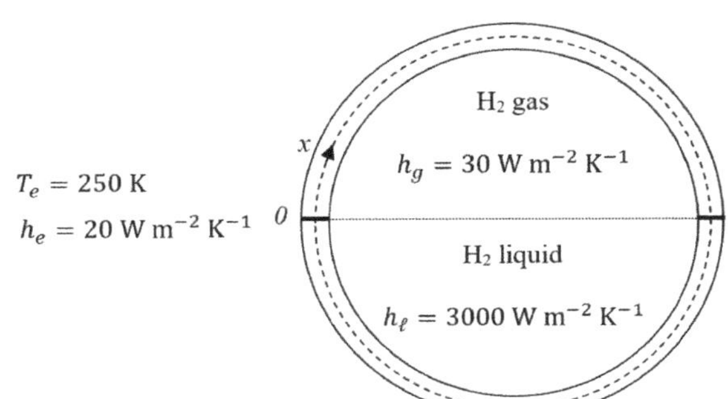

1) Justify that the temperature of the pipe can be considered as independent on the radius r. In the following, we will assume a temperature field $T(x)$, where x designates the curvilinear abscissa from the liquid–vapor interface ($x = 0$) assuming that we can treat the problem in a one-dimensional mode.

2) Justify that the temperature of the part of the pipe in contact with the liquid is close to the temperature of the liquid.

3) Assuming the two previous approximations, calculate the temperature $T(x)$ in the upper part of the pipe.

4) What is the heat flux gained per unit length of the pipe?

5) Justify the fact that heat transfer by radiation have not been taken into account.

Corrected

1) The thickness e is small compared to the radius R, so as a first approximation, the problem can be treated like that of a plate of thickness e. There are three thermal resistances in series in the direction of the thickness:

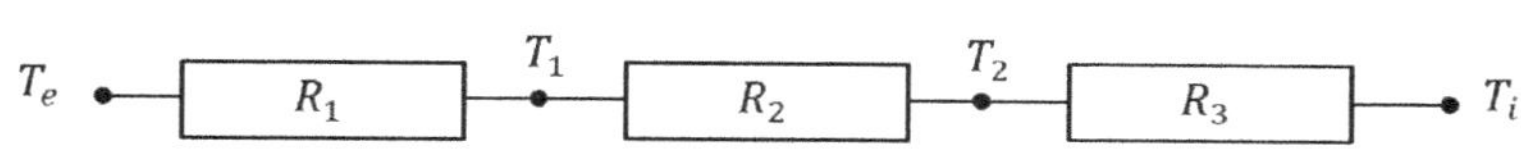

$$R_1 = \frac{1}{h_e} = \frac{1}{20} = 5 \times 10^{-2}\ \mathrm{K\,m^2\,W^{-1}}$$

$$R_2 = \frac{e}{\lambda} = \frac{5.10^{-3}}{15} = 3.3 \times 10^{-4}\ \mathrm{K\,m^2\,W^{-1}}$$

$$R_{3g} = \frac{1}{h_g} = \frac{1}{30} = 3.3 \times 10^{-2}\ \mathrm{K\,m^2\,W^{-1}};\ R_{3\ell} = \frac{1}{h_\ell} = \frac{1}{3000} = 3.3 \times 10^{-4}\ \mathrm{K\,m^2\,W^{-1}}$$

so R_2 is very small compared to R_1 in both cases, and the temperature variation in the pipe as a function of r can be neglected.

2) In the lower part of the pipe: $R_{3\ell} = 3.3 \times 10^{-4}\mathrm{K\,m^2\,W^{-1}}$, this value is negligible compared to the value of R_1; therefore, the tube is at a temperature very close to that of liquid hydrogen.

3) At $x = \frac{\pi D}{4}$, the flux along Ox is zero, so the heat balance is written as: $h_g(T_{eq} - T_i) = h_e(T_e - T_{eq})$, we deduce: $T_{eq} = \frac{h_e T_e + h_g T_i}{h_e + h_g} = \frac{20\times 250 + 30\times 20}{20+30} = 112$ K.

By taking the origin of the curvilinear abscissa x at the interface between the liquid and the gas, the equation of the heat in the upper wall of the tube is written as:

$$\lambda e \frac{d^2T}{dx^2} - h_g(T - T_g) - h_e(T - T_e) = 0 \text{ for } 0 \leq x \leq \frac{\pi D}{4}$$

$$T = T_i \text{ for } x = 0$$

$$\frac{dT}{dx} = 0 \text{ for } x = \frac{\pi D}{4}$$

By setting: $\omega = \sqrt{\frac{h_g + h_e}{\lambda e}}$, $T_{eq} = \frac{h_e T_e + h_g T_i}{h_e + h_g}$ and $\ell = \frac{\pi D}{4}$, the solution is written as:

$$\frac{T(x) - T_{eq}}{T_i - T_{eq}} = \frac{cosh[\omega(\ell - x)]}{cosh(\omega\ell)}$$

4) $\varphi_{inf} = 2\pi\ell D h_e(T_e - T_i) = 2\pi \times 0.005 \times 0.05 \times 20 \times (T_e - T_i) = 56.8$ W.

$$\omega = \sqrt{\frac{h_g + h_e}{\lambda e}} = \sqrt{\frac{30 + 20}{25 \times 0.005}} = 20$$

$$\varphi_{sup} = 2\pi D \int_0^\ell h_e[T_e - T(x)]dx = 2\pi\ell D h_e(T_e - T_i)\frac{h_g + h_e \frac{tanh(\omega\ell)}{\omega\ell}}{h_g + h_e}$$

$$= \varphi_{inf}\frac{h_g + h_e \frac{tanh(\omega\ell)}{\omega\ell}}{h_g + h_e}$$

$$\varphi_{sup} = 56.8 \times \frac{30 + 20 \times \frac{tanh(20 \times 0.0393)}{20 \times 0.0393}}{30 + 20} = 53.0 \text{ W}$$

$$\varphi = \varphi_{inf} + \varphi_{sup} = 56.8 + 53.0 = 109.8 \text{ W}$$

5) A heat transfer coefficient by radiation is of the form: $h_r = 4\sigma\varepsilon T_{average}{}^3$.

In the upper part of the tube: $T_{average} \approx \frac{(250+20)}{2} \approx 135$ K.

Hence, the order of magnitude is: $h_r \approx 0.55\varepsilon$ which is quite negligible compared to 20 W m^{-2} K^{-1}.

2.5.13. *Buried pipes*

1) What is the expression of the temperature field around a horizontal linear source delivering a heat flow rate φ per unit length and buried at a depth ℓ in a ground whose surface is isothermal?

2) Deduce the conduction factor between a cylindrical pipeline of radius R buried at a depth ℓ and the ground surface assumed to be isothermal.

3) Two tubes are buried in the ground and maintained at 300°C and 125°C respectively. Their diameters are 8 cm and 16 cm, and the distance between their axes is 40 cm. Calculate the heat flow rate exchanged per meter of length if the thermal conductivity of the soil is 0.7 W m^{-1} K^{-1}.

Corrected

The field created by a linear intensity source φ satisfies the equation: $\varphi = -\lambda 2\pi r \frac{dT}{dr}$ and is therefore written as: $T(r) = \frac{\varphi}{2\pi\lambda} ln(r) + constant.$

The image method is used: to obtain a plane isotherm; a second source with intensity $-\varphi$ is placed symmetrical to the first source with respect to the surface of the ground. For reasons of symmetry, the points of the surface of the ground located at equal distance from the two sources (of equal intensity in modulus and opposite in sign) undergo an equal and opposite influence on the part of these two sources and are located on an isotherm. The boundary conditions are thus respected.

Due to the linearity of the Laplace equation (steady-state heat equation), the sum of two solutions is a solution, so in the configuration with two sources, the temperature is written as:

$$T = \frac{\varphi}{2\pi\lambda} ln\left(\frac{r_1}{r_2}\right) + constant$$

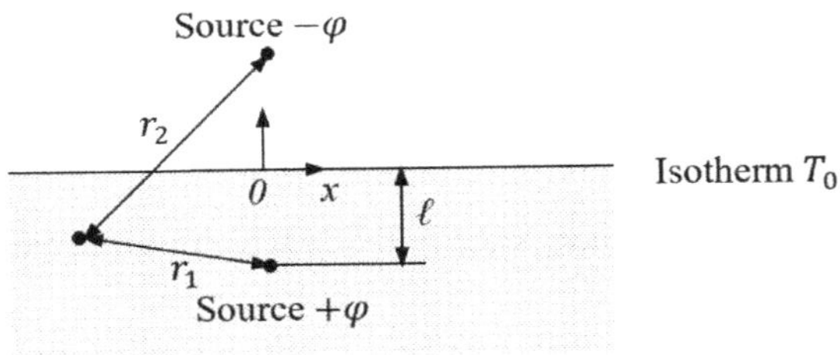

If $r_2 = r_1$: $T = T_0$, hence: $T = T_0 + \frac{\varphi}{2\pi\lambda} ln\left(\frac{r_1}{r_2}\right)$.

2) The points on the isotherm T verify:

$$\left(\frac{r_1}{r_2}\right)^2 = \frac{x^2 + (y+\ell)^2}{x^2 + (y-\ell)^2} = exp\left[\frac{4\pi\lambda}{\varphi}(T - T_0)\right] = C$$

hence: $x^2 + (y+\ell)^2 = C[x^2 + (y-\ell)^2]$.

Or again: $x^2 + \left(y + \ell\frac{1+C}{1-C}\right)^2 = \frac{4\ell^2 C}{(1-C)^2}$.

They are circles with center $(0, -h)$ and radius R with:

$h = \ell\frac{1+C}{1-C}$ and $R^2 = \frac{4\ell^2 C}{(1-C)^2}$

From this, we deduce: $\frac{h^2}{R^2} = \frac{(1+C)^2}{4C}$.

C can be expressed as a function of $\frac{h^2}{R^2}$ by solving this quadratic equation, we get:

$$C = -1 + 2\frac{h^2}{R^2} \pm \sqrt{\left(1 - 2\frac{h^2}{R^2}\right)^2 - 1} = -1 + 2\frac{h^2}{R^2} + \sqrt{\left(1 - 2\frac{h^2}{R^2}\right)^2 - 1}$$

because C must be greater than 1.

Using the relationship: $C = exp\left[-\frac{4\pi\lambda}{\varphi}(T - T_0)\right]$, the heat flow rate can be calculated by:

$$\varphi = \frac{4\pi\lambda L(T - T_0)}{ln\left[-1 + 2\frac{h^2}{R^2} + \sqrt{\left(1 - 2\frac{h^2}{R^2}\right)^2 - 1}\right]} = \frac{4\pi\lambda L(T - T_0)}{ln\left(-1 + 2\frac{h^2}{R^2} + \frac{2h}{R}\sqrt{\frac{h^2}{R^2} - 1}\right)}$$

$$= \frac{4\pi\lambda L(T - T_0)}{ln\left[\left(\frac{h}{R} + \sqrt{\frac{h^2}{R^2} - 1}\right)^2\right]}$$

or else: $\varphi = \frac{2\pi\lambda L(T-T_0)}{cosh^{-1}\left(\frac{h}{R}\right)}$.

The conduction form factor therefore has the value: $F = \frac{2\pi L}{cosh^{-1}\left(\frac{h}{R}\right)}$.

We find the value given in the appendices.

3) In the case of two buried tubes, the value of the conduction form factor is found in the Appendix:

$$F = \frac{2\pi L}{cosh^{-1}\left(\frac{D^2 - r_1{}^2 - r_2{}^2}{2r_1 r_2}\right)} = \frac{2\pi \times 0.4}{cosh^{-1}\left(\frac{0.4 - 0.08^2 - 0.16^2}{2 \times 0.08 \times 0.16}\right)} = 1.096 \text{ m}$$

We deduce the heat flow rate exchanged by: $\varphi = \lambda F(T_1 - T_2) = 0.7 \times 1.096 \times (300 - 125) = 134.3$ W.

2.5.14. *Measurement of the thermal conductivity of a rock*

In order to measure the thermal conductivity λ of a rock, two cylindrical blocks of diameter d and height e ($d = e = 5$ cm) are placed symmetrically on either side of a thin heating element dissipating a heat flow rate $\varphi_0 = 2$ W. The temperature of the other face of the blocks is maintained equal to the ambient temperature T_a.

After a time long enough for the indications to be stable, the temperature in the center of the heater is $T = 30$°C. The scrupulous experimenter also measures with a fluxmeter the heat flux $\phi_1 = 330 \text{ W m}^{-2}$ on the two extremities of the blocks maintained at $T_a = 20$°C.

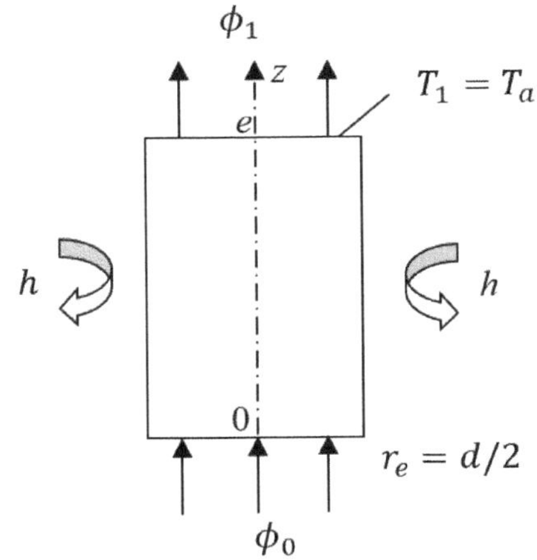

1) Explain why the heat flow rate measurements are inconsistent.

2) Write the system equations assuming a 1D temperature field. What assumption should be considered to assume a 1D temperature field?

3) Give the expression for temperature and heat flow rate as a function of z.

4) Give the expression for the temperature of the heating element and the flux ϕ_1 on the unheated side.

5) Deduce from the experimental data the values of the unknown parameters, in particular that of the thermal conductivity λ of the rock. Do these values seem realistic to you? Was the assumption of a 1D temperature field on the blocks justified?

6) Establish the expression of the 2D temperature field using the method of resolution by separation of variables.

7) Propose a method for numerical resolution of the 2D problem.

Corrected

1) $\phi_0 = \frac{\varphi}{2\pi\frac{d^2}{4}} = \frac{2}{2\times\pi\times\frac{0.05^2}{4}} = 509.3 \text{ W m}^{-2}$ thus $\phi_1 = 330 \text{ W m}^{-2}$.

$\phi_0 \neq \phi_1$ due to convective lateral losses.

2) If the condition $Bi < 0.1$ is respected either if $\frac{hr_e}{\lambda} < 0.1$, then we can neglect the temperature gradient along the radius and consider that the temperature field in the block is 1D.

3) In this case, we can write the heat equation as for a fin:

$$\frac{d^2T}{dz^2} - m^2(T - T_a) = 0 \qquad \text{(a)}$$

with: $m^2 = \frac{hP}{\lambda S} = \frac{2h}{\lambda r_e}$.

Boundary conditions:

$$T(z = e) = T_a \qquad \text{(b)}$$

$$\phi_0 = -\lambda \left(\frac{dT}{dz}\right)_{z=0} \qquad \text{(c)}$$

We let: $\bar{T}(z) = T(z) - T_a$.

The general solution is written as: $\bar{T}(z) = Acosh(mz) + Bsinh(mz)$.

The boundary condition (b) leads to: $A = -Btanh(me)$.

Hence: $\bar{T}(z) = B[-tanh(me)cosh(mz) + sinh(mz)]$.

The boundary condition (c) leads to: $\phi_0 = -\lambda mB$.

Hence finally: $\bar{T}(z) = \frac{\phi_0}{\lambda m}[tanh(me)cosh(mz) - sinh(mz)]$.

And: $\phi(z) = -\lambda\frac{d\bar{T}}{dz} = \phi_0[cosh(mz) - tanh(me)sinh(mz)]$.

4) The temperature reached by the heating element is calculated at $z = 0$:

$$\bar{T}(0) = \frac{\phi_0 tanh(me)}{\lambda m}$$

The heat flux ϕ_1 has the value:

$$\phi_1 = \phi_0[cosh(me) - tanh(me)sinh(me)] = \frac{\phi_0}{cosh(me)}$$

5) We can write: $\frac{\phi_1}{\phi_0} = \frac{1}{cosh(me)}$ hence: $me = cosh^{-1}\left(\frac{\phi_1}{\phi_0}\right) = 1.0$.

Now: $\bar{T}(0) = \frac{\phi_0 tanh(me)}{\lambda m}$, we deduce: $\lambda = \frac{e\phi_0 tanh(me)}{me\bar{T}(0)} = \frac{0.05 \times 509.3 \times tanh(1.0)}{1.0 \times 10}$.

Or finally: $\lambda = 1.94 \text{ W m}^{-1}\text{K}^{-1}$.

The order of magnitude of this value complies with the values given in the Appendix:

$$m^2 = \frac{2h}{\lambda r_e} \text{ hence: } h = \frac{\lambda r_e}{2e^2}(me)^2 = \frac{1.94\times 0.0025\times 1.0^2}{2\times 0.05^2}$$

or finally: $h = 9.7\ \mathrm{W\,m^{-2}K^{-1}}$ which corresponds well to the usual values in natural convection.

Calculation of the Biot number: $Bi = \frac{hR}{\lambda} = \frac{9.7\times 0.025}{1.94} = 0.125.$

This value is slightly higher than the limit of 0.1; it can therefore be expected that the estimate of the thermal conductivity is slightly biased.

6) The problem has cylindrical symmetry, so we use the heat equation in cylindrical coordinates:

$$\frac{\partial^2 T}{\partial r^2} + \frac{1}{r}\frac{\partial T}{\partial r} + \frac{\partial^2 T}{\partial z^2} = 0 \tag{a}$$

Boundary conditions:

$$-\lambda\left(\frac{\partial T}{\partial r}\right)_{(r,0)} = \phi_0 \tag{b}$$

$$-\lambda\left(\frac{\partial T}{\partial r}\right)_{(r_e,z)} = h[T(r_e,z) - T_a] \tag{c}$$

$$T(r,e) = T_a \tag{d}$$

The method of resolution used is the separation of the variables; we write the temperature in the following form: $T(r,z) - T_a = R(r)\times Z(z)$. Equation (a) is then written as:

$$\frac{\partial^2(RZ)}{\partial r^2} + \frac{1}{r}\frac{\partial(RZ)}{\partial r} + \frac{\partial^2(RZ)}{\partial z^2} = 0$$

or again: $\frac{1}{R}\left[\frac{\partial^2 R}{\partial r^2} + \frac{1}{r}\frac{\partial R}{\partial r}\right] + \frac{1}{Z}\frac{\partial^2 Z}{\partial z^2} = 0.$

We deduce that the two terms of the sum are constants with the same absolute value and opposite sign.

The problem posed is to determine the sign of each of them. We use the boundary conditions for this.

Suppose that: $\frac{1}{R}\left[\frac{\partial^2 R}{\partial r^2}+\frac{1}{r}\frac{\partial R}{\partial r}\right]=+\alpha^2$and: $\frac{1}{Z}\frac{\partial^2 Z}{\partial z^2}=-\alpha^2$.

The solutions of the above equations are:

$$R(r)=AI_0(\alpha r)+BK_0(\alpha r)$$
$$Z(z)=Csinh(\beta z)+Dcosh(\beta z)$$

Application of the homogeneous boundary conditions: $K_0(\alpha r)\to\infty$ when $r\to 0$, but the temperature must remain finite; therefore, $B=0$ and $R(r)=AI_0(\alpha r)$:

In $r=r_e$: $-\lambda\left(\frac{\partial T}{\partial r}\right)_{(r_e,z)}=h[T(r_e,z)-T_a]\Rightarrow-\lambda\left(\frac{\partial R}{\partial r}\right)_{(r_e)}=hR(r_e)$.

$\frac{\partial R}{\partial r}=A\alpha I_1(\alpha r)$; therefore:$-\lambda A\alpha I_1(\alpha r_e)=hAI_0(\alpha r_e)$

Both functions I_0 and I_1 are positive, so this equality cannot be true.

We therefore have: $\frac{1}{R}\left[\frac{\partial^2 R}{\partial r^2}+\frac{1}{r}\frac{\partial R}{\partial r}\right]=-\alpha^2$ and: $\frac{1}{Z}\frac{\partial^2 Z}{\partial z^2}=+\alpha^2$.

The solutions of the above equations are:

$$R(r)=AJ_0(\alpha r)+BY_0(\alpha r)$$

$$Z(z)=Csinh(\beta z)+Dcosh(\beta z)$$

Application of the homogeneous boundary conditions: $Y_0(\alpha r)\to-\infty$ when $r\to 0$, but the temperature must remain finite; therefore, $B=0$ and $R(r)=AJ_0(\alpha r)$:

At $r=r_e$: $-\lambda\left(\frac{\partial T}{\partial r}\right)_{(r_e,z)}=h[T(r_e,z)-T_a]\Rightarrow-\lambda\left(\frac{\partial R}{\partial r}\right)_{(r_e)}=hR(r_e)$.

$\frac{\partial R}{\partial r}=-A\alpha J_1(\alpha r)$; therefore:$-\lambda A\alpha J_1(\alpha r_e)=hAJ_0(\alpha r_e)$

We set: $H=\frac{hr_e}{\lambda}$ and $\omega=\alpha r_e$.

The eigenvalues ω_n are solutions of the transcendent equation:

$$\omega J_1(\omega)=HJ_0(\omega)$$

In $z=e$: $T(r,e)=T_a$.

So: $Csinh(\beta e) + Dcosh(\beta e) = 0$ and: $D = -Ctanh(\beta e)$.

Finally: $T(r,z) = \sum_{n=1}^{\infty} B_n J_0(\alpha_n r) sinh[\alpha_n(e-z)]$.

From this, we deduce: $\phi(r,z) = -\lambda \frac{\partial T}{\partial r} = \lambda \sum_{n=1}^{\infty} \alpha_n B_n J_0(\alpha_n r) cosh[\alpha_n(e-z)]$.

We now use the inhomogeneous condition (b) to determine the coefficients B_n by calculating in two different ways the integral between 0 and r_e of the heat flux multiplied by rdr and by the eigenfunction $J_0(\alpha_n r)$: $\int_0^{r_e} r J_0(\alpha_n r)\phi(r)dr = \phi_0 \int_0^{r_e} r J_0(\alpha_n r)dr = \phi_0 \frac{1}{\alpha_n} r_e J_1(\alpha_n r_e)$.

We can also write:

$$\int_0^{r_e} r J_0(\alpha_n r)\phi(r)dr = \int_0^{r_e} r J_0(\alpha_m r)\lambda \sum_{m=1}^{\infty} \alpha_m B_m J_0(\alpha_m r) cosh[\alpha_m(e-z)]\, dr$$

By using the orthogonality property of eigenfunctions, we have:

$$\int_0^{r_e} r J_0(\alpha_n r)\phi(r)dr = \lambda \int_0^{r_e} \alpha_n B_n r\, [J_0(\alpha_n r)]^2 cosh(\alpha_n e)dr$$

$$\int_0^{r_e} r J_0(\alpha_n r)\phi(r)dr = \lambda \alpha_n cosh(\alpha_n e) B_n \int_0^{r_e} r\, [J_0(\alpha_n r)]^2 dr$$

$$\int_0^{r_e} r J_0(\alpha_n r)\phi(r)dr = \lambda \alpha_n cosh(\alpha_n e) B_n \frac{r_e^2}{2} \{[J_0(\alpha_n r_e)]^2 + [J_1(\alpha_n r_e)]^2\}$$

(see section A.5 of the Appendix).

From this, we deduce: $B_n = \frac{2\, J_1(\alpha_n r_e)\phi_0}{\lambda \alpha_n^2 cosh(\alpha_n e) \frac{r_e}{2} \{[J_0(\alpha_n r_e)]^2 + [J_1(\alpha_n r_e)]^2\}}$.

And finally:

$$T(r,z) = \sum_{n=1}^{\infty} \frac{2\, J_1(\alpha_n r_e)\phi_0}{\lambda \alpha_n^2 cosh(\alpha_n e) \frac{r_e}{2} \{[J_0(\alpha_n r_e)]^2 + [J_1(\alpha_n r_e)]^2\}} J_0(\alpha_n r) sinh[\alpha_n(e-z)]$$

7) We can also set up a numerical resolution by finite differences of the heat equation. For this, the system is meshed; six different types of elementary surfaces are identified on this mesh on which the general thermal balance represented in the following figure will be carried out.

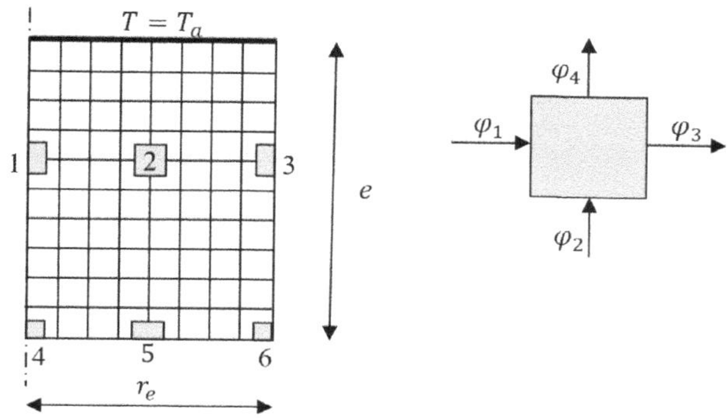

Schematization of the mesh and the thermal balance carried out on each mesh.

The thermal balance of the elementary surface is written as: $\varphi_1 + \varphi_2 = \varphi_3 + \varphi_4$.

Let us write, for example, the balances on the surfaces of types 1, 2 and 5:

– *type* 1:

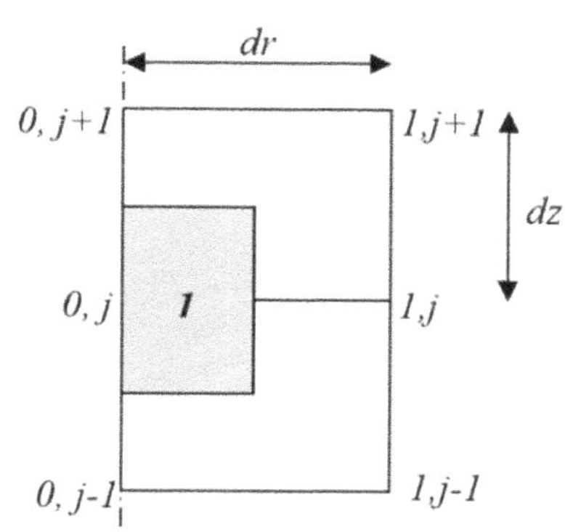

$$\varphi_1 = 0$$

$$\varphi_2 = -\lambda \frac{\pi\, dr^2}{4} [T(0, j) - T(0, j - 1)]$$

$$\varphi_3 = -\lambda\, \pi\, dr\, dz[T(1,j) - T(0,j)]$$

$$\varphi_4 = -\lambda\, \frac{\pi\, dr^2}{4}[T(0,j+1) - T(0,j)]$$

$$T(0,j) = \frac{\frac{dr}{4}T(0,j-1) + dzT(1,j) + \frac{dr}{4}T(0,j+1)}{\frac{dr}{2} + dz}$$

– *type 2*:

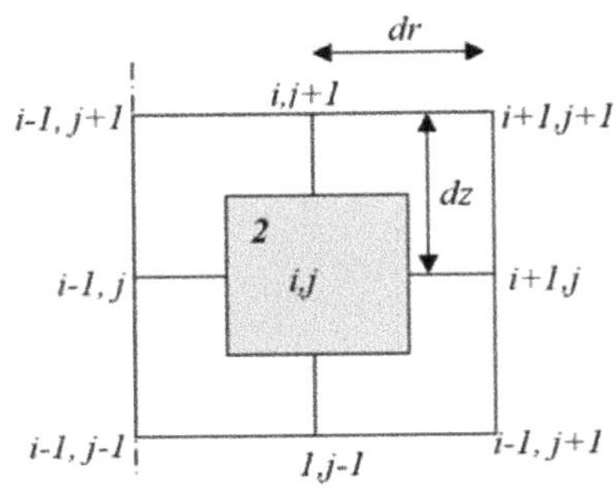

$$\varphi_1 = -\lambda\, 2\pi\left(i - \frac{3}{2}\right)drdz[T(i,j) - T(i-1,j)]$$

$$\varphi_2 = -\lambda\pi\left\{\left[\left(i - \frac{1}{2}\right)dr\right]^2 - \left[\left(i - \frac{3}{2}\right)dr\right]^2\right\}[T(i,j) - T(i,j-1)]$$

$$\varphi_2 = -2\,\lambda\pi(i-1)\quad dr^2[T(i,j) - T(i,j-1)]$$

$$\varphi_3 = -\lambda\, 2\pi\left(i - \frac{1}{2}\right)drdz[T(i+1,j) - T(i,j)]$$

$$\varphi_4 = -\lambda\pi\left\{\left[\left(i - \frac{1}{2}\right)dr\right]^2 - \left[\left(i - \frac{3}{2}\right)dr\right]^2\right\}[T(i,j+1) - T(i,j)]$$

$$\varphi_4 = -2\,\lambda\pi(i-1)\quad dr^2[T(i,j+1) - T(i,j)]$$

$$T(i,j) = \frac{2\left(i - \frac{3}{2}\right)drdzT(i-1,j) + 2(i-1)\quad dr^2[T(i,j-1) + T(i,j+1)] + 2\left(i - \frac{1}{2}\right)drdzT(i+1,j)}{4(i-1)(2dr + 4dz)}$$

– *type 5*:

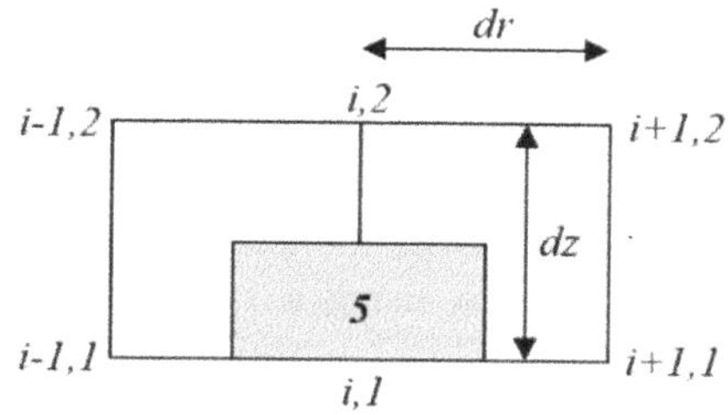

$$\varphi_1 = -\lambda\,\pi\left(i-\frac{3}{2}\right)dr\;dz[T(i,1)-T(i-1,1)]$$

$$\varphi_2 = \pi\left\{\left[\left(i-\frac{1}{2}\right)dr\right]^2-\left[\left(i-\frac{3}{2}\right)dr\right]^2\right\}\phi_0$$

$$\varphi_3 = -\lambda\,\pi\left(i-\frac{1}{2}\right)dr\;dz[T(i+1,1)-T(i,1)]$$

$$\varphi_4 = -\lambda\,\pi\left\{\left[\left(i-\frac{1}{2}\right)dr\right]^2-\left[\left(i-\frac{3}{2}\right)dr\right]^2\right\}[T(i,2)-T(i,1)]$$

$$T(i,1) = \frac{\left(i-\frac{3}{2}\right)drdz\;T(i-1,1)+2(i-1)\;\;dr^2\frac{\phi_0}{\lambda}+\left(i-\frac{1}{2}\right)drdz\;T(i+1,1)+2(i-1)\;\;dr^2\;\;T(i,2)}{2(i-1)(dr+dz)}$$

By carrying out in the same way the heat balance of the elements of types 3, 4 and 6, we obtain for all the points of the mesh a relation allowing us to express the temperature of the point according to the temperature of the surrounding points. The result is a linear system of $N \times M$ equations with as many unknowns, with $N = \frac{r_e}{dr}$ and $M = \frac{e}{dz}$. It can be solved by several methods; the iterative method whose principle is given below permits a solution to be obtained easily:

– initialize the temperature of all the points of the mesh with a "realistic" value, for example, 25°C here;

– as long as the convergence criterion is not reached:

- do from $i = 1$ to N and from $j = 1$ to M,

- recalculate $T(i,j)$ from relations resulting from elementary heat balances,

- compare the recalculated values with the values obtained in the previous calculation step.

If the maximum difference is lower than the convergence criterion, stop computation, if not continue.

By imposing $\phi_0 = 509.3\ \text{W m}^{-2}$, $\lambda = 1.94\ \text{W m}^{-1}\text{K}^{-1}$ and $T(r, e) = 20°\text{C}$, the calculation leads to $T(0,0) = 30.3°\text{C}$. To find the experimental value of 30°C with the 2D model, it is necessary to consider a thermal conductivity $\lambda = 2.01\ \text{W m}^{-1}\text{K}^{-1}$ which is therefore the exact value of the conductivity of the rock. The estimation error made by using a 1D "fin" type model to exploit the measurement would therefore be: $\frac{2.01-1.94}{2.01} \times 100\% = 3.5\%$.

3

Heat Transfer by Conduction in Transient Regime

3.1. Unidirectional conduction in transient regime without change of state

3.1.1. *Uniform temperature medium*

We are going to study the transfer of heat towards a medium at uniform temperature, which is a priori contradictory, because there is no heat transfer if the thermal gradient is null. This approximation of the medium at uniform temperature can nevertheless be justified in certain cases which will be specified. Take for example the quenching of a metal ball which consists of immersing a ball initially at temperature T_i in a bath at temperature T_0 kept constant. If the temperature inside the ball is assumed to be uniform, which will be all the more true as its size is small and its thermal conductivity high, the heat balance of this ball between two instants t and $t + dt$ is written as:

$$-hS(T - T_0) = \rho c V \frac{dT}{dt} \text{ or: } \frac{dT}{T - T_0} = -\frac{hS}{\rho c V} dt$$

hence:

$$\frac{T - T_0}{T_i - T_0} = exp\left(-\frac{hS}{\rho c V} t\right) \qquad [3.1]$$

We note that the grouping $\frac{\rho c V}{hS}$ is homogeneous to a time that we will call τ and which is the time constant of the system:

$$\tau = \frac{\rho c V}{hS} \qquad [3.2]$$

This quantity is fundamental insofar as it gives the time order of magnitude of the physical phenomenon, we have in fact: $\frac{T-T_0}{T_i-T_0} = exp\left(-\frac{t}{\tau}\right)$.

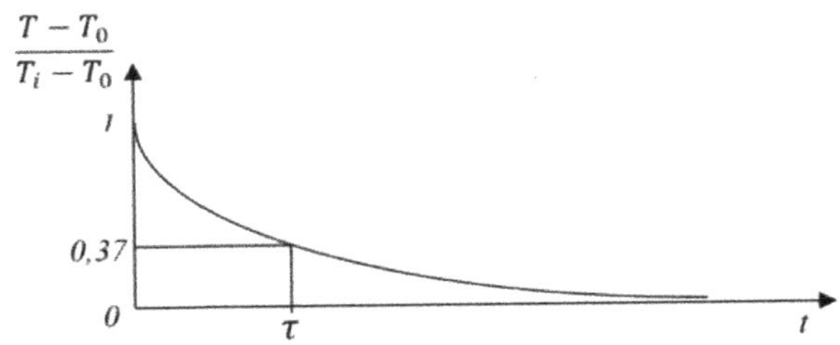

Figure 3.1. *Evolution of the temperature of a medium at uniform temperature*

It is always interesting in physics to present the results in dimensionless form, two dimensionless numbers are particularly important in transient regime:

– the Biot number: $Bi = \frac{internal\ thermal\ resistance}{external\ thermal\ resistance} = \frac{\frac{\ell}{\lambda S}}{\frac{1}{hS}}$, ℓ is the characteristic dimension of the medium, $\ell = \frac{R}{3}$ for a sphere:

$$Bi = \frac{h\ell}{\lambda} \quad [3.3]$$

– the Fourier number:

$$Fo = \frac{at}{\ell^2} \quad [3.4]$$

The Fourier number characterizes the penetration of heat in transient regime.

The definition of these two dimensionless numbers allows us to write the temperature of the ball in the form:

$$\frac{T-T_0}{T_i-T_0} = exp(-Bi\ Fo) \quad [3.5]$$

The knowledge of the product of the Biot and Fourier numbers allows the temperature evolution of the sphere to be determined. It is generally accepted that a system can be considered at uniform temperature if $Bi < 0.1$, the criterion $Bi < 0.1$ is called the "thermal accommodation" criterion.

3.1.2. *Semi-infinite medium*

A semi-infinite medium is a wall with a thickness large enough so that the disturbance applied to one face is not felt by the other face. Such a system represents the evolution of a wall of finite thickness for a sufficiently short time so that the disturbance created on one face has not reached the other face (true as long as the temperature of the other face has not changed).

3.1.2.1. *Constant temperature imposed on the surface*

Method: Laplace integral transform over time and inversion by tables.

The semi-infinite medium is initially at uniform temperature T_i. The temperature T_0 is suddenly imposed on its surface; this boundary condition is called a Dirichlet condition.

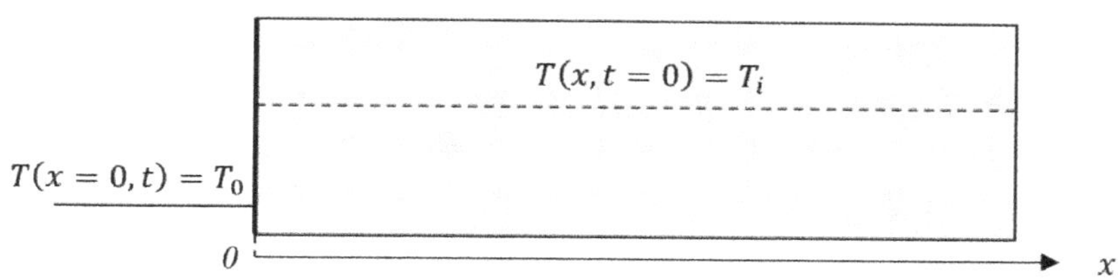

Figure 3.2. *Diagram of the semi-infinite medium with imposed surface temperature. For a color version of this figure, see www.iste.co.uk/jannot/heattransfer1.zip*

The heat equation is written as:

$$\frac{\partial^2 T}{\partial x^2}=\frac{1}{a}\frac{\partial T}{\partial t} \qquad \text{(a)}$$

With the boundary conditions:

$$T(x.0)=T_i \qquad \text{(b)}$$

$$T(0,t)=T_0 \qquad \text{(c)}$$

$$T(x,t)\to T_i \text{ when } x\to\infty \qquad \text{(d)}$$

We perform the following change of variable: $\bar{T}=T-T_i$.

Hence: $\frac{\partial \bar{T}}{\partial x}=\frac{\partial T}{\partial x}$, $\frac{\partial^2 \bar{T}}{\partial x^2}=\frac{\partial^2 T}{\partial x^2}$ and $\frac{\partial \bar{T}}{\partial t}=\frac{\partial T}{\partial t}$.

Equation (a) can then be written as: $\frac{\partial^2 \bar{T}}{\partial x^2} = \frac{1}{a}\frac{\partial \bar{T}}{\partial t}$.

The boundary conditions become:

$$\bar{T}(x,0) = 0 \quad \text{(b)}$$

$$\bar{T}(0,t) = T_0 - T_i \quad \text{(c)}$$

$$\bar{T}(x,t) \to 0 \text{ when } x \to \infty \quad \text{(d)}$$

The Laplace transform of $\bar{T}(x,t)$ with respect to time is written as (see section A.6 of the Appendix on integral transforms):

$$\theta(x,p) = L[\bar{T}(x,t)] = \int_0^\infty exp(-pt)\bar{T}(x,t)dt$$

The Laplace transform of equation (a) leads to:

$$\frac{d^2\theta}{dx^2} - \frac{1}{a}[p\theta - \bar{T}(x,0)] = 0$$

with $\bar{T}(x,0) = 0$ (which justifies the change of variable).

This equation is therefore of the form: $\frac{d^2\theta}{dx^2} - q^2\theta = 0$ with $q = \sqrt{\frac{p}{a}}$.

Hence: $\theta(x,p) = Aexp(-qx) + Bexp(qx)$.

The temperature remains finite when $x \to \infty$ therefore $B = 0$ and: $\theta(x,p) = Aexp(-qx)$.

The Laplace transform of equation (c) leads to: $\theta(0,p) = \frac{T_0 - T_i}{p}$

and: $\theta(x,p) = \frac{T_0 - T_i}{p} exp(-qx)$.

The use of the inverse Laplace transform tables presented in section A.7 of the Appendix leads to the following result:

$$\frac{T(x,t) - T_0}{T_i - T_0} = erf\left(\frac{x}{2\sqrt{at}}\right) \quad [3.6]$$

with: $erf(u) = \frac{2}{\sqrt{\pi}}\int_0^u exp(-t^2)dt$.

The function erf is called the error function (see values in section A.9 of the Appendix).

3.1.2.2. *Imposed heat flux*

Method: Laplace integral transform over time and inversion by tables.

A heat flux ϕ_0 is imposed on the surface of the semi-infinite medium; this boundary condition is called a Neumann condition.

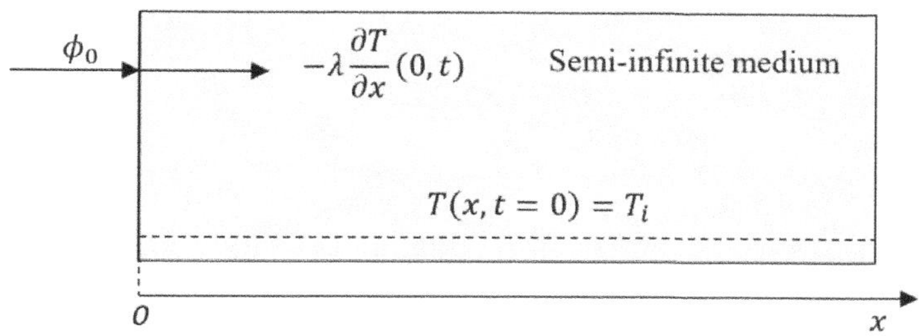

Figure 3.3. *Diagram of the semi-infinite medium with imposed surface flux. For a color version of this figure, see www.iste.co.uk/jannot/heattransfer1.zip*

The heat equation is written as:

$$\frac{\partial^2 T}{\partial x^2} = \frac{1}{a}\frac{\partial T}{\partial t} \quad \text{(a)}$$

With the boundary conditions:

$$T(x,0) = T_i \quad \text{(b)}$$

$$-\lambda\frac{\partial T}{\partial x}(0,t) = \phi_0 \quad \text{(c)}$$

$$T(x,t) \to T_i \text{ when } x \to \infty \quad \text{(d)}$$

Condition (c) reflects the conservation of the heat flux at the surface of the semi-infinite medium.

We perform the following change of variable: $\bar{T} = T - T_i$.

Equation (a) can then be written as: $\frac{\partial^2 \bar{T}}{\partial x^2} = \frac{1}{a}\frac{\partial \bar{T}}{\partial t}$.

The boundary conditions become:

$$\bar{T}(x.0) = 0 \quad \text{(b)}$$

$$-\lambda \frac{\partial \bar{T}}{\partial x}(0,t) = \phi_0 \quad \text{(c)}$$

$$\bar{T}(x,t) \to 0 \text{ when } x \to \infty \quad \text{(d)}$$

The Laplace transform of equation (a) leads to: $\frac{d^2\theta}{dx^2} - q^2\theta = 0$ with $q = \sqrt{\frac{p}{a}}$.

Hence: $\theta(x,p) = Aexp(-qx) + Bexp(qx)$.

The temperature remains finite when $x \to \infty$ and $B = 0$: $\theta(x,p) = Aexp(-qx)$.

The Laplace transform of equation (c) leads to: $\frac{\phi_0}{p} = -\lambda \frac{d\theta}{dx}(0,p)$.

Hence: $A = \frac{\phi_0}{\lambda pq}$ and $\theta(x,p) = \frac{\phi_0}{\lambda}\frac{exp(-qx)}{pq}$.

The use of the inverse Laplace transform tables presented in section A.7 of the Appendix leads to the following result:

$$T(x,t) - T_i = \frac{2\phi_0}{\lambda}\sqrt{at}\; ierfc\left(\frac{x}{2\sqrt{at}}\right) \quad [3.7]$$

With: $ierfc(u) = \frac{1}{\sqrt{\pi}}exp(-u^2) - u[1 - erf(u)]$, this function is tabulated in section A.9 of the Appendix.

3.1.2.3. *Imposed transfer coefficient*

Method: Laplace integral transform over time and inversion by tables.

Consider the case where the coefficient of heat transfer by convection h between the semi-infinite medium and the ambient medium is imposed; this boundary condition is called a Newton condition.

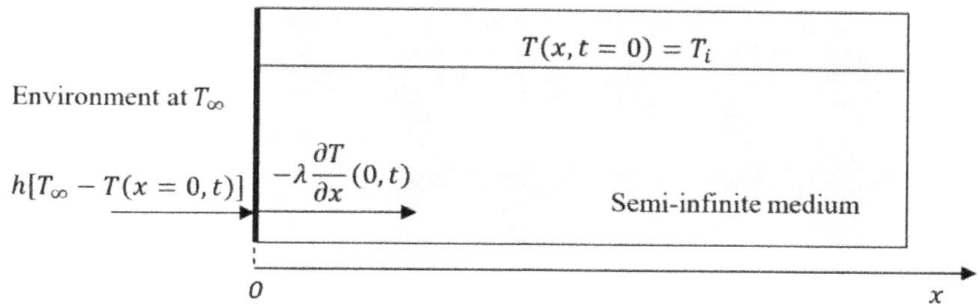

Figure 3.4. *Diagram of the semi-infinite medium with imposed convective transfer coefficient. For a color version of this figure, see www.iste.co.uk/jannot/heattransfer1.zip*

The heat equation is written as:

$$\frac{\partial^2 T}{\partial x^2} = \frac{1}{a}\frac{\partial T}{\partial t} \quad \text{(a)}$$

With the boundary conditions:

$$T(x, 0) = T_i \quad \text{(b)}$$

$$-\lambda \frac{\partial T}{\partial x}(0, t) = h[T_\infty - T(x = 0, t)] \quad \text{(c)}$$

$T(x, t) \to T_i$ when $x \to \infty$(d).

We perform the following change of variable: $\bar{T} = T - T_i$.

Equation (a) can then be written as: $\frac{\partial^2 \bar{T}}{\partial x^2} = \frac{1}{a}\frac{\partial \bar{T}}{\partial t}$.

The boundary conditions become:

$$\bar{T}(x, 0) = 0 \quad \text{(b)}$$

$$\lambda \frac{\partial \bar{T}}{\partial x}(0, t) = h[\bar{T}(x = 0, t) - (T_\infty - T_i)] \quad \text{(c)}$$

$$\bar{T}(x, t) \to 0 \text{ when } x \to \infty \quad \text{(d)}$$

The Laplace transform of equation (a) leads to: $\frac{d^2\theta}{dx^2} - q^2\theta = 0$ with $q = \sqrt{\frac{p}{a}}$.

Hence: $\theta(x, p) = A exp(-qx) + B exp(qx)$.

The temperature remains finite when $x \to \infty$; therefore, $B = 0$ and $\theta(x,p) = A exp(-qx)$.

The Laplace transform of equation (c) is written as: $\lambda \frac{d\theta}{dx}(0,p) = h\theta(0,p) + h\frac{T_i - T_\infty}{p}$.

Therefore: $-\lambda A q = hA + h\frac{T_i - T_\infty}{p}$ hence: $A = \frac{\frac{h}{\lambda}(T_\infty - T_i)}{p\left(\frac{h}{\lambda}+q\right)}$.

And: $\theta(x,p) = \frac{h}{\lambda}(T_\infty - T_i)\frac{exp(-qx)}{p\left(q+\frac{h}{\lambda}\right)}$.

The use of the inverse Laplace transform tables presented in section A.7 of the Appendix leads to the following result:

$$\frac{T(x,t) - T_\infty}{T_i - T_\infty} = erf\left(\frac{x}{2\sqrt{at}}\right) + exp\left(\frac{hx}{\lambda} + \frac{ah^2 t}{\lambda^2}\right) erfc\left(\frac{x}{2\sqrt{at}} + \frac{h\sqrt{at}}{\lambda}\right) \quad [3.8]$$

For an approximate calculation, section A.10 of the Appendix presents an abacus representing this formula graphically.

3.1.2.4. *Sinusoidal temperature imposed on the surface, established periodic state*

Method: search for a solution with the same frequency as the excitation.

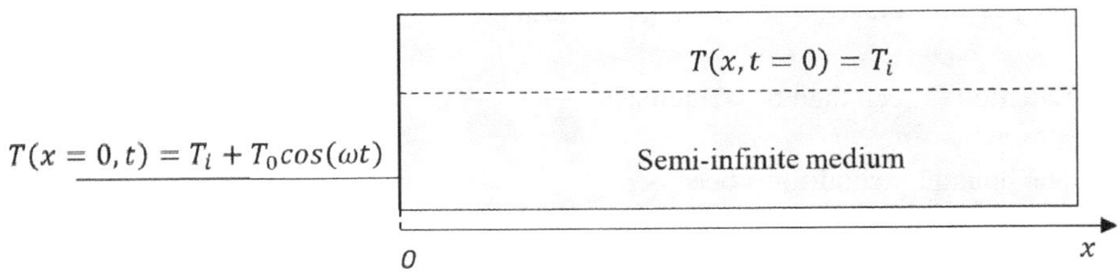

Figure 3.5. *Diagram of the semi-infinite medium with sinusoidal temperature imposed on the surface. For a color version of this figure, see www.iste.co.uk/jannot/heattransfer1.zip*

The heat equation is written as:

$$\frac{\partial^2 T}{\partial x^2} = \frac{1}{a}\frac{\partial T}{\partial t} \quad (a)$$

With the boundary conditions:

$$T(x = 0, t) = T_i + T_0 \cos(\omega t) \tag{b}$$

$$T(x, t) \to T_i \text{ when } x \to \infty. \tag{c}$$

We seek a solution in the steady state for which the medium temperature field evolves like:

$$T(x, t) = T_i + f(x) exp(i\omega t)$$

The problem being linear, we consider either the real part or the imaginary part of the solution according to whether the temperature varies like $\cos(\omega t)$ or $\sin(\omega t)$. The complex function f is a solution of:

$$\frac{d^2 f}{dx^2} - i\frac{\omega}{a} f = 0$$

with: $f(0) = T_0$.

$$f(x) = A exp\left(-\sqrt{\frac{i\omega}{a}} x\right) + B exp\left(\sqrt{\frac{i\omega}{a}} x\right)$$

with: $\sqrt{\frac{i\omega}{a}} = \sqrt{\frac{\omega}{2a}}(1 + i)$.

The function f must remain finite when $x \to \infty$ thus $B = 0$ and $f(0) = T_0$ thus $A = T_0$.

Hence:

$$T(x, t) - T_i = T_0 exp\left[-\sqrt{\frac{\omega}{2a}}(1 + i)x + i\omega t\right]$$

$$= T_0 exp\left(-\sqrt{\frac{\omega}{2a}} x\right) exp\left[i\left(\omega t - \sqrt{\frac{\omega}{2a}} x\right)\right]$$

Therefore, by taking the real part of the solution:

$$T(x, t) = T_i + T_0\, exp\left(-\sqrt{\frac{\omega}{2a}} x\right) \cos\left(\omega t - \sqrt{\frac{\omega}{2a}} x\right) \tag{3.9}$$

REMARK.–

– The amplitude of the oscillations decreases rapidly when moving away from the interface.

– The amplitude of the oscillations also decreases rapidly when the frequency of the excitation increases: a high-frequency excitation applied to the surface of a solid modifies the temperature only over a small depth.

– Between the temperatures T_1 and T_2 of two points at respective distances x_1 and x_2 from the surface, there is a phase shift equal to $\sqrt{\frac{\omega}{2a}}(x_1 - x_2)$. The knowledge of ω and the measurement of the temperature within the medium at two points located at known distances x_1 and x_2 from the surface enables evaluating the thermal diffusivity a.

3.1.2.5. *Sudden contact between two semi-infinite media*

Method: Laplace integral transform over time and inversion by tables.

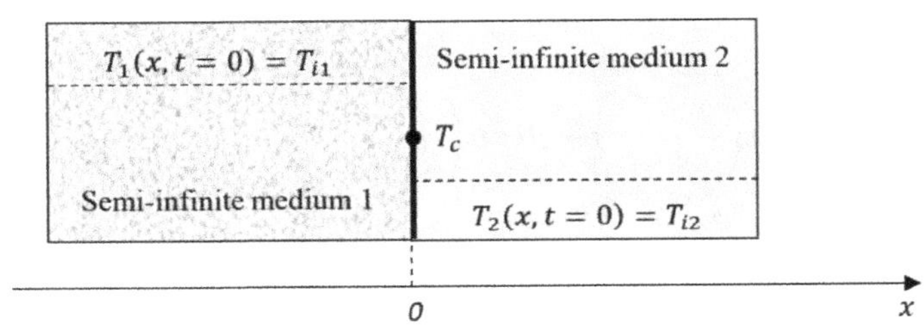

Figure 3.6. *Diagram of the sudden contact between two semi-infinite media. For a color version of this figure, see www.iste.co.uk/jannot/heattransfer1.zip*

We consider two semi-infinite media initially at two different uniform temperatures T_{i1} and T_{i2}. At the initial time, the two media are brought into contact and we seek the temperature evolution within both media.

The heat equation is written for each of the two media as:

$$\frac{\partial^2 T_1}{\partial x^2} = \frac{1}{a_1}\frac{\partial T_1}{\partial t} \qquad \text{(a)}$$

and

$$\frac{\partial^2 T_2}{\partial x^2} = \frac{1}{a_2}\frac{\partial T_2}{\partial t} \qquad \text{(b)}$$

The origin of the abscissa is taken at the contact point between the two media. The boundary conditions are written as:

$$T_1(x,0) = T_{i1} \tag{c}$$

$$T_2(x,0) = T_{i2} \tag{d}$$

$$\lambda_1 \left(\frac{\partial T_1}{\partial x}\right)_{x=0} = \lambda_2 \left(\frac{\partial T_2}{\partial x}\right)_{x=0} \tag{e}$$

$$T_1(0,t) = T_2(0,t) \tag{f}$$

$$T_1(x,t) \to T_{i1} \text{ when } x \to -\infty \tag{g}$$

$$T_2(x,t) \to T_{i2} \text{ when } x \to \infty \tag{h}$$

We perform the following changes of variable: $\bar{T}_1 = T_1 - T_{i1}$ and $\bar{T}_2 = T_2 - T_{i2}$.

Equations (a) and (b) can then be written as: $\frac{\partial^2 \bar{T}_1}{\partial x^2} = \frac{1}{a_1}\frac{\partial \bar{T}_1}{\partial t}$ and $\frac{\partial^2 \bar{T}_2}{\partial x^2} = \frac{1}{a_2}\frac{\partial \bar{T}_2}{\partial t}$.

And the boundary conditions become:

$$\bar{T}_1(x,0) = 0 \tag{c}$$

$$\bar{T}_2(x,0) = 0 \tag{d}$$

$$\lambda_1 \left(\frac{\partial \bar{T}_1}{\partial x}\right)_{x=0} = \lambda_2 \left(\frac{\partial \bar{T}_2}{\partial x}\right)_{x=0} \tag{e}$$

$$\bar{T}_1(0,t) = \bar{T}_2(0,t) + T_{i2} - T_{i1} \tag{f}$$

$$\bar{T}_1(x,t) \to 0 \text{ when } x \to -\infty \tag{g}$$

$$\bar{T}_2(x,t) \to 0 \text{ when } x \to \infty \tag{h}$$

The Laplace transforms of equations (a) and (b) lead, as in the previous cases, to solutions of the type:

$$\theta_1(x,p) = A_1 exp(-q_1 x) + B_1 exp(+q_1 x) \text{ and } \theta_2(x,p) = A_2 exp(-q_2 x) + B_2 exp(+q_2 x)$$

with: $q_1 = \sqrt{\frac{p}{a_1}}$ and $q_2 = \sqrt{\frac{p}{a_2}}$.

The temperature remains finite when $x \to \pm\infty$; therefore, $A_1 = 0$ and $B_2 = 0$, we deduce that: $\theta_1(x,p) = B_1 exp(+q_1 x)$ and $\theta_2(x,p) = A_2 exp(-q_2 x)$.

The Laplace transforms of equations (e) and (f) are then written as:

$$B_1 \lambda_1 q_1 = -A_2 \lambda_2 q_2 \tag{e}$$

$$B_1 = A_2 + \frac{T_{i2} - T_{i1}}{p} \tag{f}$$

The resolution of this linear system allows us to calculate the values of B_1 and A_2:

$$B_1 = \frac{E_2}{E_1 + E_2}(T_{i2} - T_{i1})\frac{1}{p}$$

$$A_2 = \frac{E_1}{E_1 + E_2}(T_{i1} - T_{i2})\frac{1}{p}$$

where $E_i = \sqrt{\lambda_i \rho_i c_i}$ is the thermal effusivity of the medium i.

We deduce the values of θ_1 and θ_2:

$$\theta_1(x,p) = \frac{E_2(T_{i2} - T_{i1})}{E_1 + E_2}\frac{exp(q_1 x)}{p}$$

$$\theta_2(x,p) = \frac{E_1(T_{i1} - T_{i2})}{E_1 + E_2}\frac{exp(-q_2 x)}{p}$$

The use of the inverse Laplace transform tables presented in section A.7 of the Appendix leads to the following result:

$$\frac{T_1(x,t) - T_{i1}}{T_{i2} - T_{i1}} = \frac{E_2}{E_1 + E_2} erfc\left(\frac{|x|}{2\sqrt{a_1 t}}\right) \tag{3.10}$$

$$\frac{T_2(x,t) - T_{i2}}{T_{i1} - T_{i2}} = \frac{E_1}{E_1 + E_2} erfc\left(\frac{x}{2\sqrt{a_2 t}}\right) \tag{3.11}$$

Contact temperature T_c: the contact temperature is calculated by $T_c(t) = T_1(0,t) = T_2(0,t)$ knowing that $erfc(0) = 1$.

Hence:

$$T_c = \frac{E_1 T_{i1} + E_2 T_{i2}}{E_1 + E_2} \qquad [3.12]$$

It can be noted that the contact temperature between the two media remains constant throughout the duration of the heat transfer. The medium with the greatest thermal effusivity imposes the contact temperature.

Application: thermal sensation during contact of the skin with a metal (very high effusivity) or insulation (very low effusivity), choice of materials improving thermal comfort, possibility of walking over embers, etc.

3.1.3. ***Unidirectional transfer in limited media: plate, cylinder, sphere***

3.1.3.1. *Infinite plate*

We consider the case of a plate with a thickness $2L$ and lateral dimensions large enough to consider that the heat transfer is unidirectional. The study of this case illustrates the different methods used to solve the one-dimensional heat equation in transient regime.

3.1.3.1.1. First case: plate with constant temperature imposed on the surface

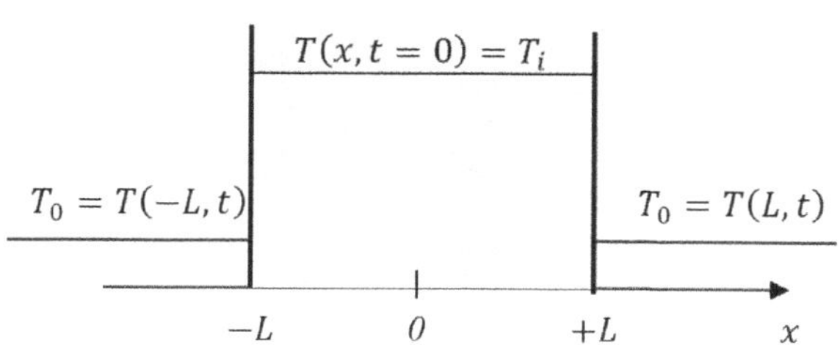

Figure 3.7. *Diagram of a plate with imposed surface temperature. For a color version of this figure, see www.iste.co.uk/jannot/heattransfer1.zip*

First method: Laplace transform, series expansion and term-to-term inversion using tables.

The heat equation is written as:

$$\frac{\partial^2 T}{\partial x^2} = \frac{1}{a}\frac{\partial T}{\partial t} \qquad \text{(a)}$$

With the boundary conditions:

$$T(x.0) = T_i \qquad \text{(b)}$$

$$T(L,t) = T_0 \qquad \text{(c)}$$

$$\left(\frac{\partial T}{\partial x}\right)_{x=0} = 0. \qquad \text{(d)}$$

We perform the following change of variable: $\bar{T} = T - T_i$.

Equation (a) can then be written as: $\frac{\partial^2 \bar{T}}{\partial x^2} = \frac{1}{a}\frac{\partial \bar{T}}{\partial t}$.

And the boundary conditions become:

$$\bar{T}(x,0) = 0 \qquad \text{(b)}$$

$$\bar{T}(L,t) = T_0 - T_i \qquad \text{(c)}$$

$$\left(\frac{\partial \bar{T}}{\partial x}\right)_{x=0} = 0 \qquad \text{(d)}$$

The Laplace transform $\theta(x,p)$ of $\bar{T}(x,t)$ with respect to time is written as:

$$\theta(x,p) = L[\bar{T}(x,t)] = \int_0^\infty exp(-pt)\bar{T}(x,t)dt$$

The Laplace transform of equation (a) leads to:

$$\frac{d^2\theta}{dx^2} - \frac{1}{a}[p\theta - \bar{T}(x,0)] = 0$$

with $\bar{T}(x,0) = 0$ (which justifies the change of variable).

This equation is of the form: $\frac{d^2\theta}{dx^2} - q^2\theta = 0$ with $q = \sqrt{\frac{p}{a}}$.

Hence: $\theta(x,p) = Acosh(qx) + Bsinh(qx)$.

The Laplace transform of equation (d) leads to: $\left(\frac{d\theta}{dx}\right)_{x=0} = 0$.

From where $B = 0$ and $\theta(x,p) = Acosh(qx)$.

The Laplace transform of equation (c) leads to: $\theta(L,p)=\frac{T_0-T_i}{p}$, hence $A=\frac{T_0-T_i}{p\cosh(qL)}$.

And: $\theta(x,p)=\frac{T_0-T_i}{p\cosh(qL)}\cosh(qx)=\frac{\Delta T\cosh(qx)}{p\cosh(qL)}$.

We can use a series expansion of $\frac{1}{1+e^{-2qL}}$ to write $\theta(x,p)$ in the form:

$$\theta(x,p)=\frac{\Delta T}{p}\frac{e^{qx}+e^{-qx}}{e^{qL}(1+e^{-2qL})}=\frac{\Delta T}{p}\left[e^{-q(L-x)}+e^{-q(L+x)}\right]\sum_{n=0}^{\infty}e^{-2nqL}$$

$$\theta(x,p)=\frac{\Delta T}{p}\sum_{n=0}^{\infty}(-1)^n e^{-q[(2n+1)L-x]}+\frac{\Delta T}{p}\sum_{n=0}^{\infty}(-1)^n e^{-q[(2n+1)L+x]}$$

The inverse Laplace transform term by term (linearity property) leads to:

$$\frac{T(x,t)-T_i}{T_0-T_i}=\sum_{n=0}^{\infty}(-1)^n erfc\left[\frac{(2n+1)L-x}{2\sqrt{at}}\right]+\sum_{n=0}^{\infty}(-1)^n erfc\left[\frac{(2n+1)L+x}{2\sqrt{at}}\right] \quad [3.13]$$

REMARKS.–

– This solution converges quickly for low values of t.

– The reduced temperature $\frac{T(x,t)-T_i}{T_0-T_i}$ is only a function of two dimensionless quantities because: $\frac{(2n+1)L\pm x}{\sqrt{at}}=\frac{(2n+1)\pm\frac{x}{L}}{\frac{\sqrt{at}}{L}}=f\left(\frac{x}{L},Fo\right)$ with $Fo=\frac{at}{L^2}$ (Fourier number) which allows a graphic representation to be established (see section A.11 of the Appendix).

Second method: resolution by the method of separation of variables.

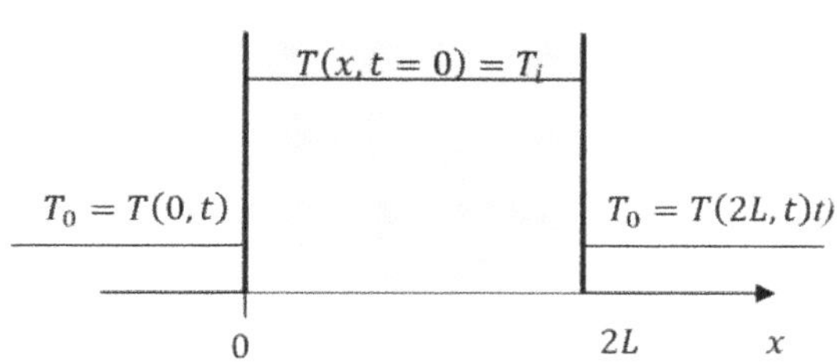

Figure 3.8. *Diagram of a plate with imposed surface temperature. For a color version of this figure, see www.iste.co.uk/jannot/heattransfer1.zip*

The heat equation is written as:

$$\frac{\partial^2 T}{\partial x^2} = \frac{1}{a}\frac{\partial T}{\partial t} \tag{a}$$

With the boundary conditions:

$$T(x, 0) = T_i \tag{b}$$

$$T(0, t) = T(2L, t) = T_0 \tag{c}$$

We perform the following change of variable: $\bar{T} = T - T_0$.

Equation (a) can then be written as: $\frac{\partial^2 \bar{T}}{\partial x^2} = \frac{1}{a}\frac{\partial \bar{T}}{\partial t}$.

And the boundary conditions become:

$$\bar{T}(x, 0) = T_i - T_0 \tag{b}$$

$$\bar{T}(0, t) = \bar{T}(2L, t) = 0 \tag{c}$$

We can also consider because of the symmetry a plate of thickness L with a condition of null heat flow rate at $x = L$; therefore, the second boundary condition is:

$$\left(\frac{\partial T}{\partial x}\right)_{x=L} = 0 \tag{d}$$

We carry out a decomposition of the temperature into a product of functions in the form: $T(x, t) = X(x)Y(t)$. The heat equation leads to the following equation:

$$X''Y = \frac{1}{a}XY' \text{ or } \frac{X''}{X} = \frac{1}{a}\frac{Y'}{Y} = K$$

where K is a constant because the two functions X and Y depend on x and t respectively. We deduce:

$$X'' - KX = 0$$

$$Y' + aKY = 0 \Rightarrow Y = C\, e^{aKt}$$

The temperature must remain finite when $t \to \infty$; therefore, $K < 0$ can be written as: $K = -\omega^2$.

Hence:

$$X = Acos(\omega x) + Bsin(\omega x)\text{and } Y = Ce^{-a\omega^2 t}$$

and $\bar{T}(x,t) = Ce^{-a\omega^2 t}[Acos(\omega x) + Bsin(\omega x)]$

The boundary condition $\bar{T}(0,t) = 0$ is then written as: $A = 0$.

Hence: $\bar{T}(x,t) = De^{-a\omega^2 t}sin(\omega x)$, the functions $\psi_n(x) = sin(\omega_n x)$ are the eigenfunctions of the system.

The boundary condition $\left(\frac{\partial T}{\partial x}\right)_{x=L} = 0$ for all t is then written as: $cos(\omega L) = 0$.

This equation admits an infinite number of solutions that are called eigenvalues: $\omega_n = (2n+1)\frac{\pi}{2L}$ with n varying from 0 to infinity.

The solution superposition theorem allows us to write the general solution of (a) in the form:

$$\bar{T}(x,t) = \sum_{n=0}^{\infty} D_n \, sin(\omega_n x) exp(-a\omega_n{}^2 t)$$

The general solution method is as follows:

– the terms D_n are determined by calculating $\int_0^L \bar{T}(x,0)\psi_m(x)dx$ in two ways:

- by replacing $\bar{T}(x,0)$ with its expression deduced from the data of the system in the initial state: $\bar{T}(x,0) = T_i - T_0$ (e),

- by replacing $\bar{T}(x,0)$ with $\bar{T}(x,0) = \sum_{n=0}^{\infty} D_n \, sin(\omega_n x)$, we obtain the infinite sum: $\int_0^L \bar{T}(x,0)\psi_m(x)dx = \int_0^L [\sum_{n=0}^{\infty} D_n \, \psi_n(x)]\psi_m(x)dx$;

– we show that if $n \neq m$, then $\int_0^L \psi_n(x)\psi_m(x)dx = 0$ (orthogonality of eigenfunctions, see section A.14 of the Appendix) and then: $\int_0^L \bar{T}(x,0)\psi_m(x)dx = D_m \int_0^L \psi_m{}^2(x)dx$ (f);

– the value of the constants D_m is determined by equalizing the expressions (e) and (f).

Let us apply this method to the example considered, we have: $\int_0^L \bar{T}(x,0)\psi_m(x)dx = \int_0^L (T_i - T_0)sin(\omega_m x)dx = \frac{T_i - T_0}{\omega_n}$ and: $\int_0^L \bar{T}(x,0)\psi_m(x)dx = D_m \int_0^L sin^2(\omega_m x)dx = D_m \frac{L}{2}$.

We deduce: $D_m = \frac{2(T_i - T_0)}{\omega_m L} = \frac{4(T_i - T_0)}{(2m+1)\pi}$.

And finally:

$$\frac{T(x,t) - T_0}{T_i - T_0} = \frac{4}{\pi}\sum_{n=0}^{\infty}\frac{1}{2n+1} sin\left[(2n+1)\frac{\pi}{2}\frac{x}{L}\right] exp\left[-(2n+1)^2\frac{\pi^2}{4}\frac{at}{L^2}\right] \qquad [3.14]$$

This solution converges with a small number of terms for large values of t (the first term may be sufficient for large values of t).

Third method: use of an integral transformation on the space variable.

Principle of the use of an integral transform to the resolution of the heat equation: an integral transform is applied to the heat equation and to the boundary conditions allowing us to obtain a new differential equations system whose resolution (easier) leads to the expression of the temperature θ in the transformed space. The inverse transform is then applied to θ to obtain the expression for the temperature T in the real space.

The choice of the most suitable integral transformation depends on the configuration and on the boundary conditions. If the temperature depends on the space variable r, we choose a transformation of the following type:

$$\theta(\omega) = \int_D w(r)\, T_\omega(r) T(r,t) dr$$

where D is the domain of definition of the temperature and $T_\omega(r)$ is an eigenfunction solution of the system formed by the heat equation and the boundary conditions for an infinite number of values ω_n $(n = 1, 2, \dots)$. The equation for which ω_n are solutions is called the transcendent equation. The function $w(r)$ is chosen constant and equal to 1 in rectangular geometry and equal to r in cylindrical geometry. The general inversion formula is then:

$$T(r,t) = \sum_{n=1}^{\infty}\frac{T_{\omega n}(r)}{N(\omega_n)}\theta(\omega_n) \text{ with: } N(\omega_n) = \int_D \left[T_{\omega_n}(r)\right]^2 w(r) dr$$

$N(\omega_n)$ is called the norm of the eigenfunction $T_\omega(r)$.

The definition and properties of the most commonly used integral transforms on the space variable, namely the Fourier and Hankel transforms, are given in section A.6 of the Appendix. A table giving the eigenfunctions and their norms, transcendental equations and eigenvalues for the most common cases is also given in section A.8 of the Appendix.

Let us apply this method to the system schematized in Figure 3.8.

The heat equation is written as:

$$\frac{\partial^2 T}{\partial x^2} = \frac{1}{a}\frac{\partial T}{\partial t} \qquad \text{(a)}$$

With the boundary conditions:

$$T(x,0) = T_i \qquad \text{(b)}$$

$$T(0,t) = T(2L,t) = T_0 \qquad \text{(c)}$$

We make the following change of variable: $\bar{T} = T - T_0$.

According to section A.8 of the Appendix, the eigenfunction is $T_\omega(x) = sin\left(\frac{n\pi x}{e}\right)$ with $e = 2L$; we therefore apply a Fourier in sine transformation (finite because the medium is finite) (see section A.6 of the Appendix) to equation (a):

$$F_s[(a)] \Rightarrow \frac{n\pi}{e}[\bar{T}(0) - (-1)^n \bar{T}(e)] - \frac{n^2\pi^2}{e^2}\theta_s(n,t) = \frac{1}{a}\frac{d\theta_s}{dt}(n,t)$$

with: $\bar{T}(0,t) = \bar{T}(e,t) = 0$.

Hence: $-\frac{n^2\pi^2}{e^2}\theta_s(n,t) = \frac{1}{a}\frac{d\theta_s}{dt}(n,t)$.

The general solution of this equation is written as: $\theta_s(n,t) = A_n exp\left(-\frac{n^2\pi^2 at}{e^2}\right)$.

The boundary condition $\bar{T}(x,0) = T_i - T_0$ leads to:

$$\theta_s(n,t=0) = F_s(T_i - T_0) = \int_0^L (T_i - T_0) sin\left(\frac{n\pi x}{e}\right) dx$$

$$= (T_i - T_0)\frac{e}{n\pi}[1 - (-1)^n]$$

We deduce: $A_n = (T_i - T_0)\frac{e}{n\pi}[1 - (-1)^n]$ and:

$$\theta_s(n,t) = (T_i - T_0)\frac{e}{n\pi}[1-(-1)^n]exp\left(-\frac{n^2\pi^2at}{e^2}\right)$$

The inverse transform makes it possible to calculate $T(x,t)$:

$$T(x,t) - T_0 = \frac{2}{e}(T_i - T_0)\sum_{n=1}^{\infty}\frac{e}{n\pi}[1-(-1)^n]\,exp\left(-\frac{n^2\pi^2at}{e^2}\right)sin\left(\frac{n\pi x}{e}\right)$$

with: $e = 2L$.

Or finally:

$$T(x,t) - T_0 = \frac{4\,(T_i - T_0)}{\pi}\sum_{n=0}^{\infty}\frac{1}{2n+1}sin\left[(2n+1)\frac{\pi x}{2L}\right]exp\left[-(2n+1)^2\frac{\pi^2at}{4L^2}\right]$$

We find the relation [3.14].

Fourth method: Laplace transform, resolution and inversion by a numerical method.

We have shown by applying the first method that the Laplace transform $\theta(x,p)$ of the temperature $T(x,t) - T_i$ is written as:

$$\theta(x,p) = \frac{(T_0 - T_i)cosh(qx)}{cosh(ql)} = \frac{\Delta T cosh(qx)}{cosh(ql)} \text{ with } q = \sqrt{\frac{p}{a}}$$

The temperature $T(x,t)$can be deduced by applying a numerical method to find the inverse Laplace transform of $\theta(x,p)$. We can, for example, use the Stehfest method to calculate $T(x,t)$ by:

$$T(x,t) = T_i + \frac{ln(2)}{t}\sum_{j=1}^{N}V_j\,\theta\left[x,\frac{j\,ln(2)}{t}\right] \quad [3.15]$$

A number of terms $N = 10$ is sufficient to obtain a satisfactory precision. The values of the corresponding coefficients V_j are given in section A.7 of the Appendix.

Comparison of the methods: the simplest method to calculate $T(x,t)$ is the fourth method, which however only provides an approximate numerical solution of the solution and which is not free from numerical instabilities in some very particular cases. The number of terms $N = 10$ in formula [3.15] enables satisfactory precision to be obtained. Then come, in order of increasing difficulty, the first method and the second and third methods.

The first term of formula [3.13] represents well the temperature at short times, while the first term of formula [3.14] represents well the temperature at long times.

As an example, Figure 3.9 shows the representation of the reduced temperature $\frac{T(x,t)-T_i}{T_0-T_i}$ at 2.5 cm from the edge of a plate 10 cm thick for a material with a diffusivity $a = 10^{-6}\mathrm{m}^2 s^{-1}$.

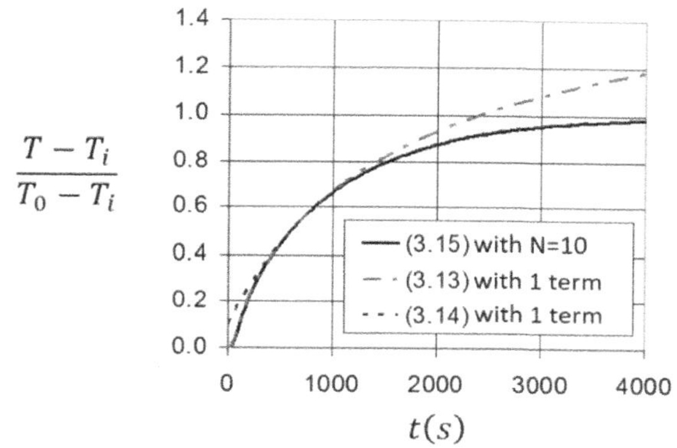

Figure 3.9. *Reduced temperature in a plate calculated by the different relations. For a color version of this figure, see www.iste.co.uk/jannot/heattransfer1.zip*

3.1.3.1.2. Second case: plate with imposed flux

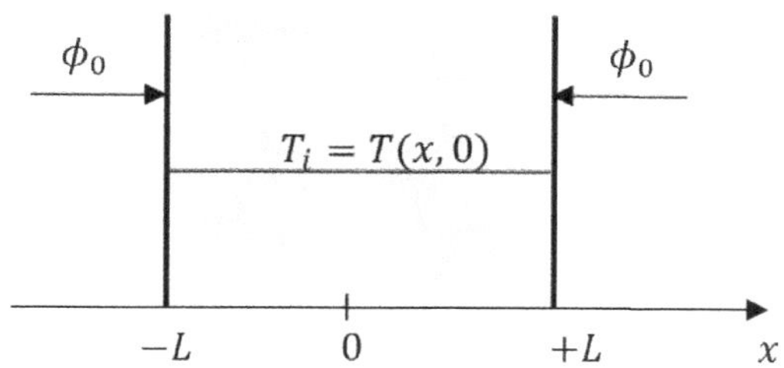

Figure 3.10. *Schematization of a plate with heat flux imposed on the surface. For a color version of this figure, see www.iste.co.uk/jannot/heattransfer1.zip*

The heat equation is written as:

$$\frac{\partial^2 T}{\partial x^2} = \frac{1}{a}\frac{\partial T}{\partial t} \tag{a}$$

With the boundary conditions:

$$T(x,0) = T_i \tag{b}$$

$$\left(\frac{\partial T}{\partial x}\right)_{x=0} = 0 \tag{c}$$

$$\lambda\left(\frac{\partial T}{\partial x}\right)_{x=L} = \phi_0 \tag{d}$$

Using the first two methods of the previous section, we obtain the following results:

$$T(x,t) = T_i + \frac{\phi_0 t}{\rho c L} + \frac{\phi_0 L}{\lambda}$$
$$\left[\frac{3x^2 - L^2}{6L^2} - \frac{2}{\pi^2}\sum_{n=1}^{\infty}\frac{(-1)^n}{n^2}\, exp\left(-\frac{an^2\pi^2 t}{L^2}\right)\cos\left(\frac{n\pi x}{L}\right)\right] \tag{[3.16]}$$

And:

$$T(x,t) = T_i + \frac{2\phi_0\sqrt{at}}{\lambda}$$
$$\sum_{n=0}^{\infty} ierfc\left[\frac{(2n+1)L - x}{2\sqrt{at}}\right] + ierfc\left[\frac{(2n+1)L + x}{2\sqrt{at}}\right] \tag{[3.17]}$$

These formulas are complex to calculate because they include an infinite sum of terms. Relation [3.16] however converges quickly for a small number of terms at long times where the first term provides a good estimate.

Relation [3.16] can also be written as:

$$T_r(x,t) = \frac{T - T_i}{\frac{\phi_0 L}{\lambda}} = \frac{at}{L^2} + \frac{3\left(\frac{x}{L}\right)^2 - 1}{6} - \frac{2}{\pi^2}\sum_{n=1}^{\infty} exp\left(-n^2\pi^2\frac{at}{L^2}\right)\cos\left(n\pi\frac{x}{L}\right)$$

The reduced temperature $T_r(x,t)$ is only a function of the two dimensionless quantities $\frac{x}{L}$ and $Fo = \frac{at}{L^2}$, which allows us to establish a graphic representation.

Application of the first method to a plate with imposed heat flow rate shows that the Laplace transform of the temperature $T(x,t) - T_i$ is: $\theta(x,p) = A\cosh(qx)$.

The boundary condition at $x = L$ is written as: $\lambda\left(\frac{\partial T}{\partial x}\right)_{x=L} = \phi_0$.

The Laplace transform of this equation leads to: $A\lambda q\sinh(qL) = \frac{\phi_0}{p}$, hence: $A = \frac{\phi_0}{p}\frac{1}{\lambda q\sinh(qL)}$.

And:

$$\theta(x,p) = \frac{\phi_0}{p}\frac{\cosh(qx)}{\lambda q\sinh(qL)} \quad [3.18]$$

The temperature $T(x,t)$ can be deduced easily by applying a numerical method (Stehfest or other) to find the inverse Laplace transform of $\theta(x,p)$.

This method easily leads to a solution that is much simpler to calculate than that given by formula [3.16] or [3.17].

3.1.3.1.3. Third case: plate with imposed transfer coefficient

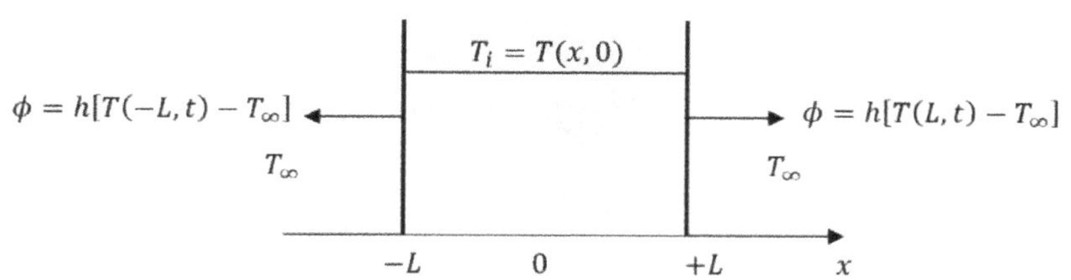

Figure 3.11. *Schematization of a plate with transfer coefficient imposed on the surface. For a color version of this figure, see www.iste.co.uk/jannot/heattransfer1.zip*

The heat equation is written as:

$$\frac{\partial^2 T}{\partial x^2} = \frac{1}{a}\frac{\partial T}{\partial t} \quad (a)$$

With the boundary conditions:

$$T(x,0) = T_i \quad (b)$$

$$\left(\frac{\partial T}{\partial x}\right)_{x=0} = 0 \quad \text{(c)}$$

$$-\lambda\left(\frac{\partial T}{\partial x}\right)_{x=L} = h[T(L,t) - T_\infty] \quad \text{(d)}$$

The temperature is decomposed into a product of functions in the form: $\bar{T}(x,t) = T(x,t) - T_i = X(x)Y(t)$. The heat equation leads to the following equation:

$$X''Y = \frac{1}{a}XY' \text{ or } \frac{X''}{X} = \frac{1}{a}\frac{Y'}{Y} = -\omega^2$$

where ω is a constant because the two functions X and Y depend one on x and the other on t. We deduce:

$$X = A_1 cos(\omega x) + B_1 sin(\omega x)$$
$$Y = Ce^{-a\omega^2 t}$$

and: $\bar{T}(x,t) = e^{-a\omega^2 t}[Acos(\omega x) + Bsin(\omega x)]$. The exponent of the exponential must be negative so that the temperature remains finite at long times, which justifies a posteriori the choice of the sign of the constant $-\omega^2$.

The boundary condition $\left(\frac{\partial \bar{T}}{\partial x}\right)_{x=0} = 0$ is then written as: $B = 0$.

Hence: $\bar{T}(x,t) = Ae^{-a\omega^2 t}cos(\omega x)$, the eigenfunctions of the system are: $\psi_n(x) = cos(\omega_n x)$.

The boundary condition $-\lambda\left(\frac{\partial \bar{T}}{\partial x}\right)_{x=L} = h\bar{T}(L,t)$ for all t is then written as: $\omega tan(\omega L) = \frac{h}{\lambda}$.

This equation admits an infinity of solutions (eigenvalues) ω_n.

The solutions superposition theorem allows the general solution of (a) to be written in the form:

$$\bar{T}(x,t) = \sum_{n=1}^{\infty} D_n\, cos(\omega_n x) exp(-a\omega_n{}^2 t)$$

We have on one hand:

$$\int_0^L \bar{T}(x,0)\psi_m(x)dx = \int_0^L (T_i - T_0)cos(\omega_m x)dx = \frac{T_i - T_0}{\omega_m} sin(\omega_m L)$$

And on the other hand:

$$\int_0^L \bar{T}(x,0)\psi_m(x)dx = D_m \int_0^L cos^2(\omega_m x)dx = D_m \int_0^L \frac{1+cos(2\omega_m x)}{2}dx$$

hence: $\int_0^L \bar{T}(x,0)\psi_m(x)dx = D_m\left[\frac{L}{2}+\frac{1}{4\omega_m}\frac{2tan(\omega_m L)}{1+tan^2(\omega_m L)}\right] = \frac{D_m}{2}\left[L+\frac{1}{\omega_m}\frac{\frac{h}{\lambda\omega_m}}{1+\left(\frac{h}{\lambda\omega_m}\right)^2}\right].$

We deduce: $D_m = 2\,(T_i - T_0)\frac{\omega_m{}^2+\frac{h^2}{\lambda^2}}{L\left(\omega_m{}^2+\frac{h^2}{\lambda^2}\right)+\frac{h}{\lambda}}\frac{sin(\omega_m L)}{\omega_m}$ and finally:

$$\frac{T(x,t)-T_0}{T_i-T_0} = 2\sum_{n=1}^{\infty}\frac{\omega_n{}^2+\frac{h^2}{\lambda^2}}{L\left(\omega_n{}^2+\frac{h^2}{\lambda^2}\right)+\frac{h}{\lambda}}\frac{sin(\omega_n L)}{\omega_n}exp(-a\omega_n{}^2 t)\,cos(\omega_n x) \qquad [3.19]$$

where $\omega_n (n = 1.2, \dots)$ are the solutions of the equation $\omega tan(\omega L) = \frac{h}{\lambda}$, which is also written as: $\omega L tan(\omega L) = \frac{hL}{\lambda}$. The solutions $\omega_n L$ of this equation (shown in Figure 3.12) are therefore only functions of $Bi = \frac{hL}{\lambda}$.

Equation [3.19] can also be written as:

$$T_r(x,t) = \frac{T(x,t)-T_0}{T_i-T_0} = 2\sum_{n=1}^{\infty}\frac{(\omega_n L)^2+\left(\frac{hL}{\lambda}\right)^2}{(\omega_n L)^2+\left(\frac{hL}{\lambda}\right)^2+\frac{hL}{\lambda}}\frac{sin(\omega_n L)}{\omega_n L}exp(-a\omega_n{}^2 t)cos(\omega_n x)$$

The dimensionless temperature $T_r(x,t)$ is only a function of the dimensionless numbers $Bi = \frac{hL}{\lambda}$, $Fo = \frac{at}{L^2}$ and $\frac{x}{L}$. We can therefore construct for a given value of $\frac{x}{L}$ an abacus giving the value of $T_r(x,t)$ as a function of Fo for several values of Bi. Two examples can be found for $\frac{x}{L} = 0$ and $\frac{x}{L} = 1$ in section A.12 of the Appendix.

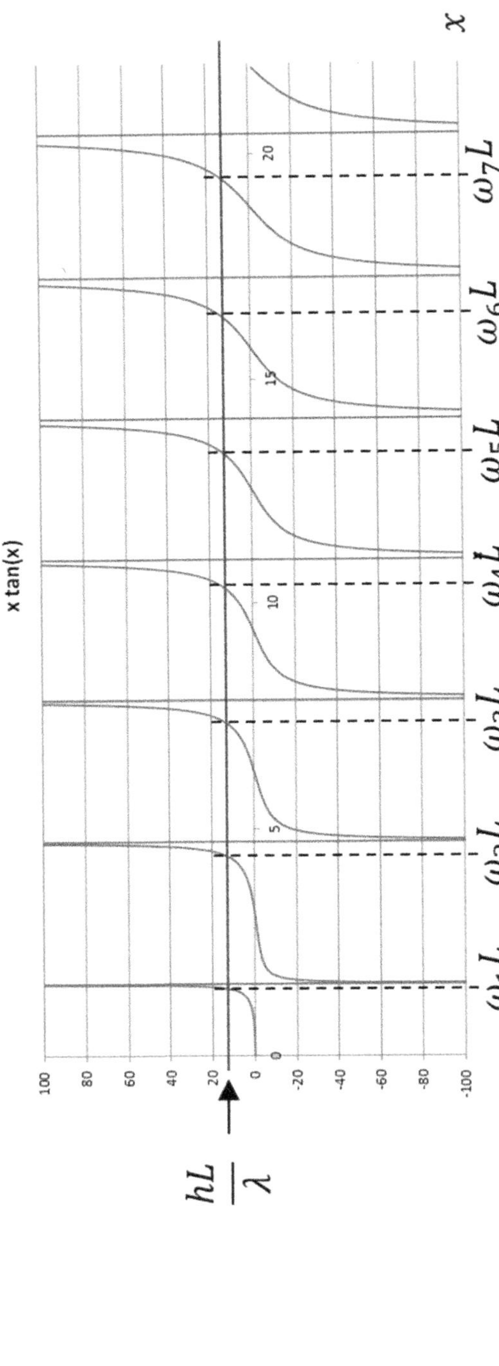

Figure 3.12. *Schematization of eigenvalues ω_n. For a color version of this figure, see www.iste.co.uk/jannot/heattransfer1.zip*

Moreover, at "long" times, the first term of the sum is sufficient to give a good estimate of $T_r(x,t)$. In this case, the relationship between $T_r(x,t)$ and $T_r(0,t)$ is simply written as: $\frac{T_r(x,t)}{T_r(0,t)} = cos(\omega_1 x)$, where $\omega_1 L$ is the first root of the equation $\omega L tan(\omega L) = \frac{hL}{\lambda}$. This ratio has the particularity of not depending on Fo; it is therefore easy to construct an abacus giving the value at long times of the ratio $\frac{T_r(x,t)}{T_r(0,t)}$ as a function of the Biot number for several values of $\frac{x}{L}$ (see section A.12 of the Appendix).

Application of the first method to the case of a plate with imposed transfer coefficient shows that the Laplace transform of the temperature $T(x,t) - T_i$ can be written as: $\theta(x,p) = Acosh(qx)$.

We have the boundary condition at $x = L$: $\lambda\left(\frac{\partial T}{\partial x}\right)_{x=L} = h[T(L,t) - T_\infty] = h[T(L,t) - T_i] + h[T_i - T_\infty]$.

The Laplace transform of this equation leads to: $-A\lambda q sinh(qL) = hAcosh(qL) + h\frac{T_i - T_\infty}{p}$, whence: $A = \frac{T_\infty - T_i}{p}\frac{1}{cosh(qL) + \frac{\lambda q}{h} sinh(qL)}$.

And:

$$\theta(x,p) = \frac{T_\infty - T_i}{p}\frac{cosh(qx)}{cosh(qL) + \frac{\lambda q}{h}\, sinh(qL)} \quad [3.20]$$

The temperature $T(x,t)$ can be deduced easily by applying a numerical method (Stehfest or other) to find the inverse Laplace transform of $\theta(x,p)$.

3.1.3.2. *Infinite cylinder*

We consider here an infinite cylinder (very large length compared to the diameter) of radius R initially at temperature T_i. It can be assumed in this case that the heat transfer is only radial.

3.1.3.2.1. First case: infinite cylinder with imposed surface temperature

A temperature T_0 is suddenly imposed on the surface of the cylinder initially at the uniform temperature T_i (see Figure 3.13).

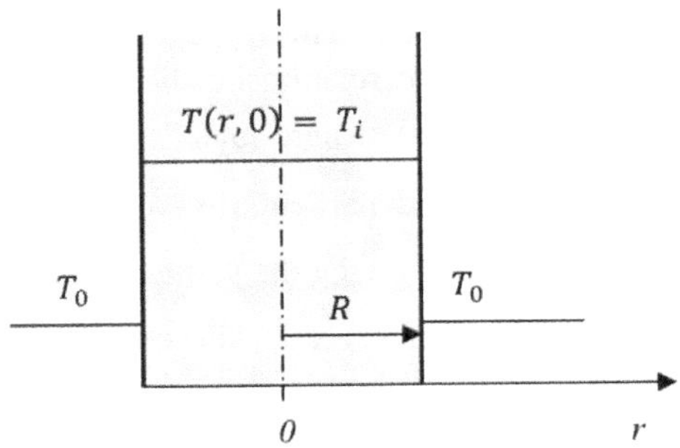

Figure 3.13. *Schematization of an infinite cylinder with imposed surface temperature. For a color version of this figure, see www.iste.co.uk/jannot/heattransfer1.zip*

First method: decomposition of the temperature into a product of functions and Hankel transformation.

The heat equation is written as:

$$\frac{\partial^2 T}{\partial r^2} + \frac{1}{r}\frac{\partial T}{\partial r} = \frac{1}{a}\frac{\partial T}{\partial t} \qquad \text{(a)}$$

With the boundary conditions:

$$T(r,0) = T_i \qquad \text{(b)}$$

$$T(R,t) = T_0 \qquad \text{(c)}$$

We perform the following change of variable: $\bar{T} = T - T_0$.

The boundary conditions become:

$$\bar{T}(r,0) = T_i - T_0 \qquad \text{(b)}$$

$$\bar{T}(R,t) = 0 \qquad \text{(c)}$$

The temperature is decomposed into a product of functions in the form: $\bar{T}(r,t) = X(r)Y(t)$. The heat equation leads to the following equation:

$$X''Y + \frac{1}{r}X'Y = \frac{1}{a}XY' \text{ or } \frac{X'' + \frac{1}{r}X'}{X} = \frac{1}{a}\frac{Y'}{Y} = -\omega^2$$

where ω is a constant because the two functions X and Y are independent. We deduce:

$$X'' + \frac{1}{r}X' + \omega^2 X = 0 \Rightarrow X = AJ_0(\omega r) + BY_0(\omega r)$$

$$Y' + a\omega^2 Y = 0 \Rightarrow Y = Ce^{-a\omega^2 t}$$

where J_0 is the unmodified first kind Bessel function of order 0 and Y_0 is the unmodified second kind Bessel function of order 0. Section A.5 of the Appendix provides the definition and main properties of the Bessel functions.

We deduce that the solutions of (a) are of the form: $\bar{T}(r,t) = Ce^{-a\omega^2 t}[AJ_0(\omega r) + BY_0(\omega r)]$.

Moreover, we know that $Y_0(0) = -\infty$, which imposes $B = 0$ from where $\bar{T}(r,t) = De^{-a\omega^2 t}J_0(\omega r)$.

The homogeneous boundary condition $\bar{T}(R,t) = 0$ is then written as: $De^{-a\omega^2 t}J_0(\omega R) = 0$, which imposes that ω_n is a solution of the equation $J_0(\omega R) = 0$.

The solution superposition theorem allows the general solution of (a) to be written in the form:

$$T(r,t) = \sum_{n=1}^{\infty} D_n e^{-a\omega_n{}^2 t} J_0(\omega_n r)$$

The inhomogeneous initial condition $\bar{T}(r,0) = T_i - T_0$ is then written as:

$$T_i - T_0 = \textstyle\sum_{n=1}^{\infty} D_n J_0(\omega_n r) \qquad \text{(d)}$$

The eigenfunction is $J_0(\omega_n r)$, which leads us to apply the Hankel transform to the boundary condition (d), i.e. to multiply each side of equation (d) by $rJ_0(\omega_m r)$ and to integrate between 0 and R:

$$\int_0^R rJ_0(\omega_m r)(T_i - T_0)dr = \int_0^R rJ_0(\omega_m r)\sum_{n=1}^{\infty} D_n J_0(\omega_n r)\,dr = \int_0^R D_m r[J_0(\omega_m r)]^2 dr$$

because we know that: $\int_0^R rJ_0(\omega_m r)J_0(\omega_n r)dr = 0$ if $m \neq n$.

$$\int_0^R rJ_0(\omega_m r)(T_i - T_0)dr = D_m \int_0^R r[J_0(\omega_m r)]^2 dr = D_m \frac{R^2}{2}[J'_0(\omega_m R)]^2$$
$$= D_m \frac{R^2}{2}[J_1(\omega_m R)]^2$$

because the functions J_n satisfy the relations (see section A.5 of the Appendix):

$\int_0^R r[J_n(\omega r)]^2 dr = \frac{R^2}{2}[J'_n(\omega R)]^2$ and $J'_n(\omega r) = -J_{n+1}(\omega r) + \frac{n}{\omega r}J_n(\omega r)$

We finally deduce:

$$\frac{T(x,t) - T_0}{T_i - T_0} = \frac{2}{R}\sum_{n=1}^{\infty} \frac{J_0(\omega_n r)}{\omega_n J_1(\omega_n R)} exp(-a\omega_n{}^2 t) \qquad [3.21]$$

where $\omega_n (n = 1.2, ...)$ are the solutions of the equation: $J_0(\omega_n R) = 0$.

Second method: Laplace transformation, resolution and numerical inversion.

The heat equation is written as:

$$\frac{\partial^2 T}{\partial r^2} + \frac{1}{r}\frac{\partial T}{\partial r} = \frac{1}{a}\frac{\partial T}{\partial t} \qquad (a)$$

With the boundary conditions:

$$T(r,0) = T_i \qquad (b)$$

$$T(R,t) = T_0 \qquad (c)$$

We perform the following change of variable: $\bar{T} = T - T_i$, and we set $\theta(r,p) = L[T(r,t)]$.

The Laplace transform of equation (a) is written as:

$$\frac{d^2\theta}{dr^2} + \frac{1}{r}\frac{d\theta}{dr} = \frac{p}{a}\theta \qquad (d)$$

We perform the change of variable: $u = \sqrt{\frac{p}{a}}r = qr$.

Equation (d) is then written as: $\frac{d^2\theta}{du^2} + \frac{1}{r}\frac{d\theta}{du} - \theta = 0$.

The general solution of this Bessel equation is written as (see section A.5 of the Appendix): $\theta(qr,p) = AI_0(qr) + BK_0(qr)$.

We have the condition: $\left(\frac{\partial T}{\partial r}\right)_{r=0} = 0$, hence: $\left(\frac{d\theta}{dr}\right)_{r=0} = 0$ and: $\left(\frac{d\theta}{du}\right)_{r=0} = 0$.

So: $AI_1(0) - BK_1(0) = 0$ with: $K_1(0) \to \infty$, hence: $B = 0$.

The second boundary condition is written as: $T(R,t) = T_0$, thus $\bar{T}(R,t) = T_0 - T_i$ and $\theta(R,p) = \frac{T_0 - T_i}{p}$.

We deduce: $A = \frac{T_0 - T_i}{p}\frac{1}{I_0(qR)}$.

And:

$$\theta(r,p) = \frac{T_0 - T_i}{p}\frac{I_0(qr)}{I_0(qR)} \qquad [3.22]$$

The temperature $T(r,t)$ can be deduced easily by applying a numerical method (Stehfest or other) to find the inverse Laplace transform of $\theta(r,p)$.

3.1.3.2.2. Second case: imposed heat flux

A heat flux ϕ_0 is suddenly imposed on the surface of the cylinder initially at the uniform temperature T_i.

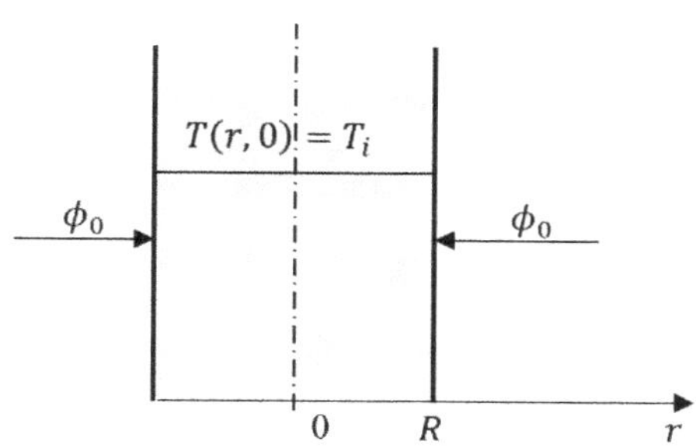

Figure 3.14. *Schematization of an infinite cylinder with imposed heat flux. For a color version of this figure, see www.iste.co.uk/jannot/heattransfer1.zip*

The heat equation is written as:

$$\frac{\partial^2 T}{\partial r^2} + \frac{1}{r}\frac{\partial T}{\partial r} = \frac{1}{a}\frac{\partial T}{\partial t} \qquad (a)$$

With the boundary conditions:

$$T(r,0) = T_i \tag{b}$$

$$\lambda\left(\frac{\partial T}{\partial r}\right)_{r=R} = \phi_0 \tag{c}$$

We perform the following change of variable: $\bar{T} = T - T_0$ Where $T_0 = T_i + \frac{2\phi_0 at}{\lambda R} + \frac{\phi_0 R}{\lambda}\left(\frac{r^2}{2R^2} - \frac{1}{4}\right)$

The boundary conditions become:

$$\bar{T}(r,0) = T_i - T_0 \tag{b}$$

$$\lambda\left(\frac{\partial \bar{T}}{\partial r}\right)_{r=R} = 0 \tag{c}$$

The temperature is decomposed into a product of functions in the form: $\bar{T}(r,t) = X(r)Y(t)$. The heat equation leads to the following equation:

$$X''Y + \frac{1}{r}X'Y = \frac{1}{a}XY' \text{ or: } \frac{X'' + \frac{1}{r}X'}{X} = \frac{1}{a}\frac{Y'}{Y} = -\omega^2$$

whose solution leads to the following result:

$$T(r,t) = T_i + \frac{2\phi_0 at}{\lambda R} + \frac{\phi_0 R}{\lambda}\left[\frac{r^2}{2R^2} - \frac{1}{4} - 2\sum_{n=1}^{\infty} exp(-a\omega_n{}^2 t)\frac{J_0(\omega_n r)}{\omega_n{}^2\, R^2 J_0(\omega_n R)}\right] \tag{3.23}$$

Where ω_n are the solutions of the equation $J_1(\omega R) = 0$

We can also solve the problem by a Laplace transform as in the case of an imposed surface temperature.

It was shown in the previous section that the temperature is of the form: $\theta(qr,p) = AI_0(qr)$ with $q = \sqrt{\frac{p}{a}}$.

The boundary condition $r = R$ is written as: $\lambda\left(\frac{\partial \bar{T}}{\partial r}\right)_{r=R} = \phi_0$.

Therefore: $\left(\frac{d\theta}{dr}\right)_{r=R} = \frac{\phi_0}{p}$, where: $A\lambda q I_1(qR) = \frac{\phi_0}{p}$ and $A = \frac{\phi_0}{p}\frac{1}{\lambda q I_1(qR)}$.

Hence:

$$\theta(r,p) = \frac{\phi_0}{p}\frac{I_0(qr)}{\lambda q I_1(qR)} \qquad [3.24]$$

The temperature $T(r,t)$ can be deduced easily by applying a numerical method (Stehfest or other) to find the inverse Laplace transform of $\theta(r,p)$.

3.1.3.2.3. Third case: imposed transfer coefficient

A heat exchange by convection with a transfer coefficient h and a fluid at temperature T_∞ is suddenly imposed on the surface of the cylinder initially at uniform temperature.

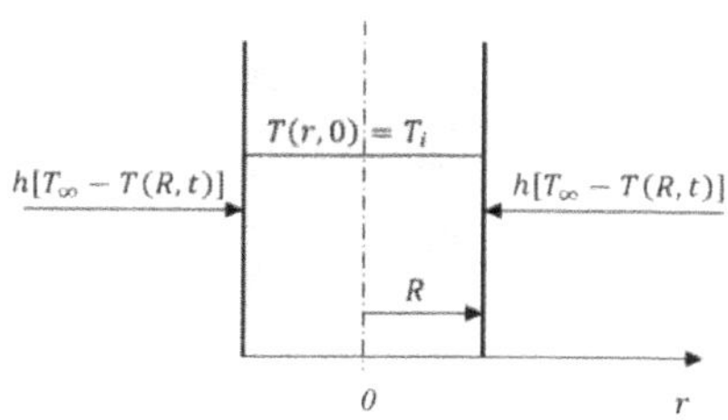

Figure 3.15. *Schematization of an infinite cylinder with imposed convective transfer coefficient. For a color version of this figure, see www.iste.co.uk/jannot/heattransfer1.zip*

The heat equation is written as:

$$\frac{\partial^2 T}{\partial r^2} + \frac{1}{r}\frac{\partial T}{\partial r} = \frac{1}{a}\frac{\partial T}{\partial t} \qquad (a)$$

With the boundary conditions:

$$T(r,0) = T_i \qquad (b)$$

$$\lambda\left(\frac{\partial T}{\partial r}\right)_{r=R} = h[T_\infty - T(R,t)] \qquad (c)$$

The solution can be written as:

$$T(r,T) = T_\infty + \frac{2\frac{h}{\lambda}(T_i - T_\infty)}{R}\sum_{n=1}^{\infty} exp(-a\omega_n{}^2 t)\frac{J_0(\omega_n r)}{\left(\frac{h^2}{\lambda^2} + \omega_n{}^2\right)J_0(\omega_n R)} \qquad [3.25]$$

where $\omega_n (n = 1.2.3 \ldots)$ are the eigenvalues, roots of the equation $\omega_n R\, J_1(\omega_n R) = \frac{hR}{\lambda}\, J_1(\omega_n R)$.

We can also solve the problem by a Laplace transform as in the case of an imposed surface temperature.

It has been shown that the temperature is of the form: $\theta(qr, p) = A I_0(qr)$ with $q = \sqrt{\frac{p}{a}}$.

The boundary condition $r = R$ is written as:

$$-\lambda\left(\frac{\partial T}{\partial r}\right)_{r=R} = h[T(R,t) - T_\infty] = h[T(R,t) - T_i] + h(T_i - T_\infty)$$

Therefore: $-\lambda\left(\frac{d\theta}{dr}\right)_{r=R} = h\theta(r = R) + \frac{h(T_i - T_\infty)}{p}$, where: $-\lambda q A I_1(qR) = hAI_0(qR) + \frac{h(T_i - T_\infty)}{p}$ and: $A = \frac{T_\infty - T_i}{p} \frac{1}{I_0(qR) + \frac{\lambda q}{h} I_1(qR)}$.

Hence:

$$\theta(r,p) = \frac{T_\infty - T_i}{p} \frac{I_0(qr)}{I_0(qR) + \frac{\lambda q}{h} I_1(qR)} \qquad [3.26]$$

The temperature $T(r, t)$ can be deduced easily by applying a numerical method (Stehfest or other) to find the inverse Laplace transform of $\theta(r, p)$.

3.1.3.3. *Sphere*

3.1.3.3.1. First case: imposed surface temperature

We consider here a sphere of radius R at the uniform initial temperature T_i to which a surface temperature T_0 is suddenly imposed.

The heat equation is written as:

$$\frac{1}{r}\frac{\partial^2(r\bar{T})}{\partial r^2} = \frac{1}{a}\frac{\partial \bar{T}}{\partial t} \qquad (a)$$

where: $\bar{T} = T - T_0$.

With the boundary conditions:

$$\bar{T}(r,0) = T_i - T_0 \quad \text{(b)}$$

$$\bar{T}(R,t) = 0 \quad \text{(c)}$$

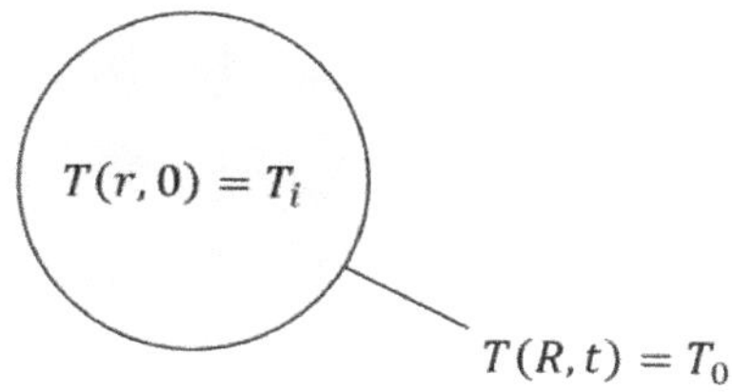

Figure 3.16. *Schematization of a sphere with imposed surface temperature. For a color version of this figure, see www.iste.co.uk/jannot/heattransfer1.zip*

Performing the following change of variable: $U(r,t) = r\bar{T}(r,t)$, equation (a) becomes: $\frac{\partial^2 U}{\partial r^2} = \frac{1}{a}\frac{\partial U}{\partial t}$.

With the boundary conditions:

$$U(r,0) = r(T_i - T_0) \quad \text{(b)}$$

$$U(R,t) = 0 \quad \text{(c)}$$

We find the system of equations of the infinitely thick plate $2L$ (section 3.1.3.1) with the following changes:

$$x \rightarrow r - R$$

$$T \rightarrow U$$

$$T_i - T_0 \rightarrow r(T_i - T_0)$$

which leads us to obtain:

$$T(r,T) = T_0 + \frac{2R(T_0 - T_i)}{\pi r}\sum_{n=1}^{\infty}\frac{(-1)^n}{n} exp\left(-\frac{an^2\pi^2 t}{R^2}\right) sin\left(\frac{n\pi r}{R}\right) \quad [3.27]$$

The temperature in the center is given by the limit of the relation [3.27] when r tends towards zero and is written as:

$$T(0,T) = T_0 + 2(T_0 - T_i)\sum_{n=1}^{\infty}(-1)^n exp\left(-\frac{an^2\pi^2 t}{R^2}\right)$$

As before, the problem can be treated by a Laplace transformation. The heat equation is written as:

$$\frac{1}{r}\frac{\partial^2(r\bar{T})}{\partial r^2} = \frac{1}{a}\frac{\partial \bar{T}}{\partial t} \qquad \text{(a)}$$

where: $\bar{T} = T - T_i$.

With the boundary conditions:

$$\bar{T}(r,0) = 0 \qquad \text{(b)}$$

$$\bar{T}(R,t) = T_0 - T_i \qquad \text{(c)}$$

The Laplace transform of equation (a) is written as:

$$\frac{d^2\theta}{dr^2} + \frac{2}{r}\frac{d\theta}{dr} = \frac{p}{a}\theta \qquad \text{(d)}$$

We perform a new change of variable: $\psi = r\theta$.

Equation (d) is then written as: $\frac{d^2\psi}{dr^2} - \frac{p}{a}\psi = 0$.

The general solution of this equation is written as: $\theta = \frac{\psi}{r} = A\frac{sinh(qr)}{r} + B\frac{cosh(qr)}{r}$ with $q = \sqrt{\frac{p}{a}}$.

$$\frac{cosh(qr)}{r} \to \infty \text{ when } r \to 0$$

hence $B = 0$.

We have the boundary condition: $\bar{T}(R,t) = T_0 - T_i$, hence: $A\frac{sinh(qR)}{R} = \frac{T_0 - T_i}{p}$.

We deduce: $A = \frac{T_0 - T_i}{p}\frac{R}{sinh(qR)}$, and finally:

$$\theta(r,p) = \frac{T_0 - T_i}{p}\frac{Rsinh(qr)}{rsinh(qR)} \qquad [3.28]$$

The temperature $T(r,t)$ can be deduced easily by applying a numerical method (Stehfest or other) to find the inverse Laplace transform of $\theta(r,p)$.

3.1.3.3.2. Second case: imposed heat flux

We consider a sphere of radius R at uniform initial temperature T_i to which a surface heat flux ϕ_0 is suddenly imposed.

The heat equation is written as:

$$\frac{1}{r}\frac{\partial^2(r\bar{T})}{\partial r^2} = \frac{1}{a}\frac{\partial \bar{T}}{\partial t} \qquad (a)$$

where $\bar{T} = T - T_0$ and $T_0 = \frac{3\emptyset_0 t}{\rho c R} + \frac{\emptyset_0(5r^2 - 3R^2)}{10\lambda R}$

With the boundary conditions:

$$\bar{T}(r,0) = T_i - T_0 \qquad (b)$$

$$\lambda\left(\frac{\partial \bar{T}}{\partial r}\right)_{r=R} = 0 \qquad (c)$$

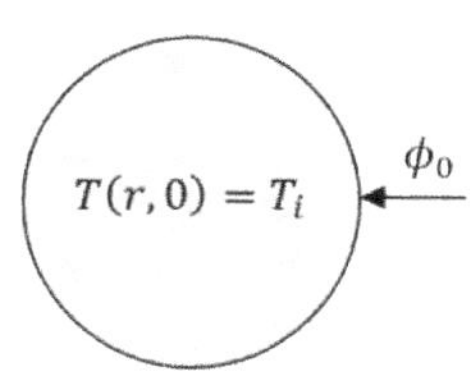

Figure 3.17. *Schematization of a sphere with imposed surface flux. For a color version of this figure, see www.iste.co.uk/jannot/heattransfer1.zip*

We perform the following change of variable: $U(r,t) = r\bar{T}(r,t)$, which, as before, allows us to return to the case of the infinite plate of thickness $2L$.

We finally get:

$$T(r,T) = \frac{3\phi_0 t}{\rho c R} + \frac{\phi_0(5r^2 - 3R^2)}{10\lambda R}$$

$$-\frac{2\phi_0 R}{\lambda r}\sum_{n=1}^{\infty} exp\left(-\frac{a\omega_n{}^2 t}{R^2}\right)\frac{sin\left(\frac{\omega_n r}{R}\right)}{\omega_n{}^2 sin(\omega_n)} \qquad [3.29]$$

where $\omega_n(n = 1.2.3 \ldots)$ are the eigenvalues, roots of the equation: $tan(\omega) = \omega$.

The problem can also be treated by a Laplace transform as in the case of an imposed surface temperature. It has been shown that the temperature is of the form: $\theta = \frac{\psi}{r} = A\frac{sinh(qr)}{r}$ with $q = \sqrt{\frac{p}{a}}$. The boundary condition at $r = R$ is written as: $\lambda\left(\frac{\partial \bar{T}}{\partial r}\right)_{r=R} = \phi_0$.

That is: $\lambda\left(\frac{d\theta}{dr}\right)_{r=R} = \frac{\phi_0}{p}$ or: $\lambda A\frac{qRcosh(qR) - sinh(qR)}{R^2} = \frac{\phi_0}{p}$ and:

$$A = \frac{\phi_0 R^2}{\lambda p}\frac{1}{qRcosh(qR) - sinh(qR)}$$

Hence:

$$\theta(r,p) = \frac{\phi_0 R^2}{\lambda p r}\frac{sinh(qr)}{qRcosh(qR) - sinh(qR)} \qquad [3.30]$$

The temperature $T(r,t)$ can be deduced easily by applying a numerical method (Stehfest or other) to find the inverse Laplace transform of $\theta(r,p)$.

3.1.3.3.3. Third case: transfer coefficient by imposed convection

We consider a sphere of radius R at the uniform initial temperature T_i on the surface of which a convective exchange (with a coefficient h) with the ambient medium at the temperature T_∞ is suddenly imposed.

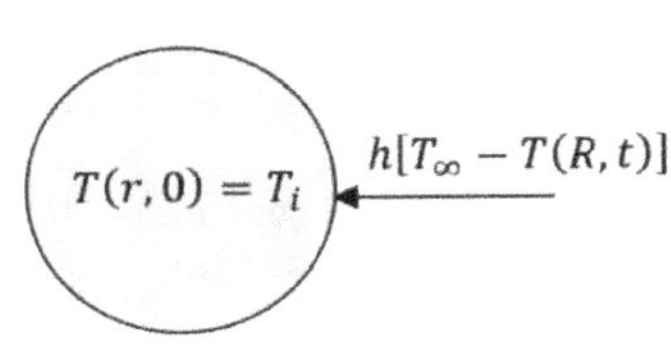

Figure 3.18. *Schematization of a sphere with imposed convective coefficient. For a color version of this figure, see www.iste.co.uk/jannot/heattransfer1.zip*

The heat equation is written as:

$$\frac{1}{r}\frac{\partial^2(r\bar{T})}{\partial r^2} = \frac{1}{a}\frac{\partial \bar{T}}{\partial t} \tag{a}$$

where: $\bar{T} = T - T_\infty$

With the boundary conditions:

$$\bar{T}(r,0) = T_i - T_\infty \tag{b}$$

$$-\lambda\left(\frac{\partial \bar{T}}{\partial r}\right)_{r=R} = h\bar{T}(R,t) \tag{c}$$

We perform the change of variable: $U(r,t) = r\bar{T}(r,t)$, which brings us back to the case of an infinite plate of thickness $2L$.

We finally get:

$$\frac{T(r,T) - T_\infty}{T_i - T_\infty} = \frac{2h}{\lambda R}\sum_{n=1}^{\infty} exp\left(-\frac{a\omega_n{}^2 t}{R^2}\right)\frac{R^2\omega_n{}^2 + \left(\frac{hR}{\lambda} - 1\right)}{\omega_n{}^2\left[R^2\omega_n{}^2 + \frac{hR}{\lambda}\left(\frac{hR}{\lambda} - 1\right)\right]} sin(\omega_n R)\, sin(\omega_n r) \tag{3.31}$$

where $\omega_n (n = 1.2.3\ ...)$ are the eigenvalues, roots of the equation: $\omega R cot(\omega R) + \frac{hR}{\lambda} - 1 = 0$.

We can also solve the problem by a Laplace transform as in the case of an imposed surface temperature.

It has been shown that the temperature is of the form: $\theta = \frac{\psi}{r} = A\frac{sinh(qr)}{r}$ with $q = \sqrt{\frac{p}{a}}$.

The boundary condition at $r = R$ is written as: $-\lambda\left(\frac{\partial T}{\partial r}\right)_{r=R} = h[T(R,t) - T_\infty]$.

therefore: $-\lambda\left(\frac{d\theta}{dr}\right)_{r=R} = h\theta(R) + \frac{h(T_i - T_\infty)}{p}$, where: $-\lambda A\frac{qRcosh(qR) - sinh(qR)}{R^2} = hA\frac{sinh(qR)}{R} + \frac{h(T_i - T_\infty)}{p}$ and: $A = \frac{T_i - T_\infty}{p}\frac{R}{\left(\frac{\lambda}{hR} - 1\right)sinh(qR) - \frac{\lambda q}{h}cosh(qR)}$.

Hence:

$$\theta(r,p) = \frac{T_i - T_\infty}{p}\frac{R}{r}\frac{sinh(qr)}{\left(\frac{\lambda}{hR} - 1\right)sinh(qR) - \frac{\lambda q}{h}cosh(qR)} \quad [3.32]$$

The temperature $T(r,t)$ can be deduced easily by applying a numerical method (Stehfest or other) to find the inverse Laplace transform of $\theta(r,p)$.

3.1.4. *Complex systems: quadrupole method*

In this section, we note:

– $\theta(x,p)$ the Laplace transform of the temperature $T(x,t)$ (see section A.6 of the Appendix);

– $\Phi(x,p)$ the Laplace transform of heat flow rate $\varphi(x,t)$.

A summary of the quadrupolar matrices associated with the systems most commonly encountered in practice will be found in section A.13 of the Appendix.

The quadrupole method is particularly well suited to solve 1D transient heat transfer problems in complex systems of the multilayer (flat or cylindrical) wall type. The resolution of this type of system by the methods described above can indeed quickly become very cumbersome, whereas the method of the quadrupoles leads very easily to the solution.

3.1.4.1. *Unidirectional flow in plane walls*

3.1.4.1.1. Single wall

We consider the case of unidirectional heat transfer in a wall of thickness e.

The temperature $T(x,t)$ within the wall verifies the equation:

$$\frac{\partial^2 T}{\partial x^2} = \frac{1}{a}\frac{\partial T}{\partial t} \quad (a)$$

By applying the Laplace transform to equation (a), we obtain: $\frac{d^2\theta}{dx^2} = \frac{p}{a}\theta$ (b)

if $T(x,0) = 0$.

Equation (b) admits a solution of the form:

$$\theta(x,p) = k_1 cosh(qx) + k_2 sinh(qx) \quad (c)$$

with $q = \sqrt{\frac{p}{a}}$.

The Laplace transform of the heat flow rate at any point in the wall is written as:

$$\Phi(x,p) = L\left[-\lambda S \frac{\partial T}{\partial x}(x,t)\right] = -\lambda S L\left[\frac{\partial T}{\partial x}(x,t)\right] = -\lambda S \frac{d\theta}{dx}(x,p) \quad \text{(d)}$$

This relation enables expressing $\Phi(x,p)$ as a function of k_1, k_2 and x:

$$\Phi(x,p) = -\lambda S q k_1 sinh(qx) - \lambda S q k_2 cosh(qx) \quad \text{(e)}$$

Relations (c) and (e) can be written at $x = 0$ and $x = e$, and we obtain:

$$\theta(0,p) = k_1;\ \Phi(0,p) = -\lambda S q k_2;\ \theta(e,p) = k_1 cosh(qe) + k_2 sinh(qe);\ \Phi(e,p) = -\lambda S q k_1 sinh(qe) - \lambda S q k_2 cosh(qe)$$

It is possible to eliminate k_1 and k_2 between these four equations, by expressing, for example, $[\theta(0,p), \Phi(0,p)]$ as a function of $[\theta(e,p), \Phi(e,p)]$, and we end up with:

$$\begin{bmatrix}\theta(0,p)\\ \Phi(0,p)\end{bmatrix} = \underbrace{\begin{bmatrix} cosh(qe) & \frac{1}{\lambda q S} sinh(qe) \\ \lambda q S\ sinh(qe) & cosh(qe)\end{bmatrix}}_{M\ =\ \text{quadrupolar matrix}} \begin{bmatrix}\theta(e,p)\\ \Phi(e,p)\end{bmatrix} \quad [3.33]$$

We have the property $det(M) = 1$, which allows us to establish the reciprocal relation:

$$\begin{bmatrix}\theta(e,p)\\ \Phi(e,p)\end{bmatrix} = \begin{bmatrix} cosh(qe) & -\frac{1}{\lambda q S} sinh(qe) \\ -\lambda q S sinh(qe) & cosh(qe)\end{bmatrix} \begin{bmatrix}\theta(0,p)\\ \Phi(0,p)\end{bmatrix}$$

We can also establish an analogy between the propagation of a current in sinusoidal regime and the unidirectional heat transfer in transient regime:

- intensity of the electric current $I \rightarrow$ heat flux in the Laplace space $\Phi(x,p)$;
- electrical potential $U \rightarrow$ temperature in Laplace space $\theta(x,p)$;
- electrical impedance $Z \rightarrow$ thermal impedance Z.

Ohm's law $U_1 - U_2 = ZI$ becomes: $\theta_1 - \theta_2 = Z\Phi$.

The node equation $\sum I = 0$ becomes: $\sum \Phi = 0$.

With these notations, the quadrupole relation [3.33] can be represented by the equivalent electric diagram of Figure 3.19.

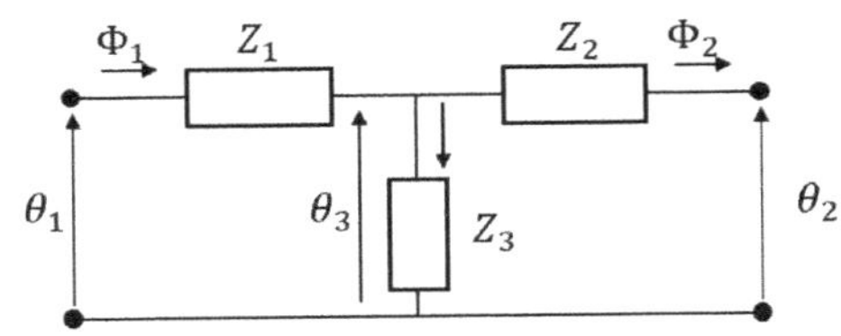

Figure 3.19. *Electrical diagram equivalent to a simple wall in variable regime*

In the case of the flat wall:

$$Z_1 = Z_2 = \frac{cosh(qe)-1}{\lambda qSsinh(qe)} \text{ and } Z_3 = \frac{1}{\lambda qSsinh(qe)}$$

3.1.4.1.2. Wall with convective exchange

We consider the case of a wall exchanging heat by convection with a fluid (see Figure 3.20).

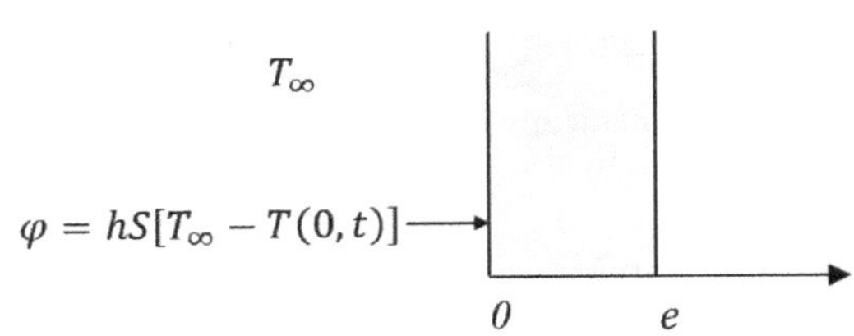

Figure 3.20. *Schematization of a simple wall with convective transfer. For a color version of this figure, see www.iste.co.uk/jannot/heattransfer1.zip*

The relation $\varphi = hS[T_\infty - T(0,t)]$ can also be written as: $T_\infty = \frac{\varphi}{hS} + T(0,t)$, whose Laplace transform is: $\theta_\infty = \frac{\Phi}{hS} + \theta(x=0)$.

We can therefore write in quadrupole matrix form:

$$\begin{bmatrix} \theta_\infty \\ \Phi_\infty \end{bmatrix} = \begin{bmatrix} 1 & \frac{1}{hS} \\ 0 & 1 \end{bmatrix} \begin{bmatrix} \theta(0,p) \\ \Phi(0,p) \end{bmatrix} \qquad [3.34]$$

Quadrupole relation [3.34] can be represented by the equivalent electric diagram of Figure 3.21.

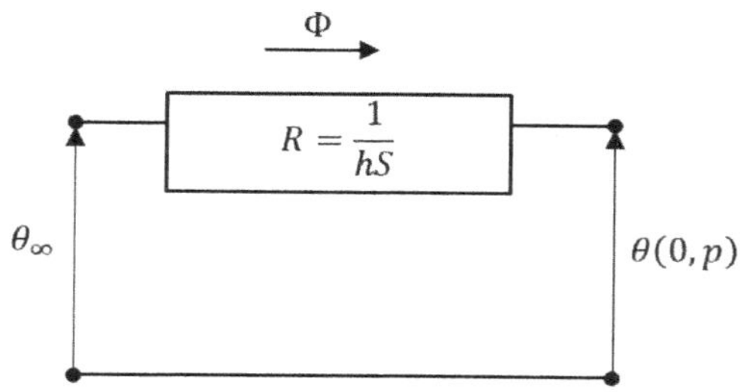

Figure 3.21. *Electrical diagram equivalent to a convective transfer in variable regime*

3.1.4.1.3. Contact resistance between two walls

Now consider the case of heat transfer through a thermal contact resistance R at the interface between two solid media as shown in Figure 3.22.

The equation of the heat flow rate $\varphi = \frac{T_{1(x=0)} - T_{2(x=0)}}{R}$ can also be written as: $T_{1(x=0)} = T_{2(x=0)} + R\varphi$, whose Laplace transform is: $\theta_{1(x=0)} = \theta_{2(x=0)} + R\Phi$.

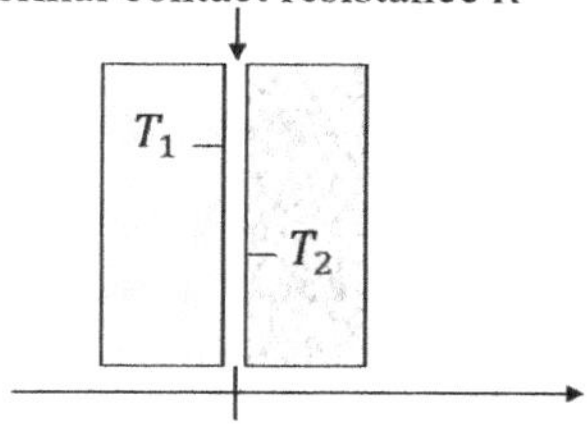

Figure 3.22. *Diagram of two walls with contact resistance. For a color version of this figure, see www.iste.co.uk/jannot/heattransfer1.zip*

We can therefore write in the quadrupole matrix form:

$$\begin{bmatrix} \theta_1 \\ \Phi_1 \end{bmatrix} = \begin{bmatrix} 1 & R \\ 0 & 1 \end{bmatrix} \begin{bmatrix} \theta_2 \\ \Phi_2 \end{bmatrix} \quad [3.35]$$

This expression is analogous to relation [3.34], and the equivalent electrical diagram is therefore of the same type as that presented in Figure 3.21.

3.1.4.1.4. Multilayer wall with convection and contact resistances

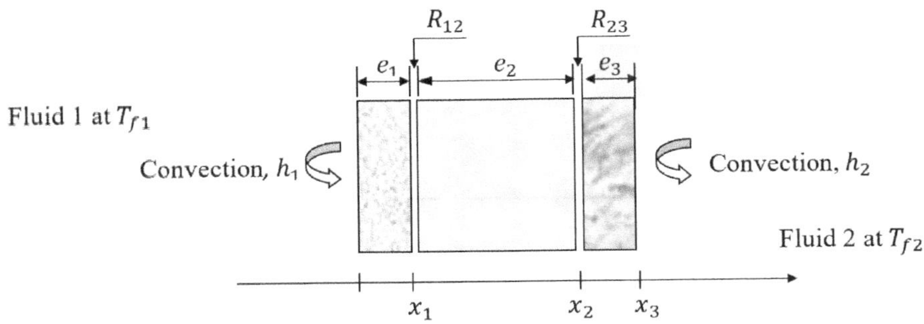

Figure 3.23. *Diagram of a multilayer wall with convection and contact resistances. For a color version of this figure, see www.iste.co.uk/jannot/heattransfer1.zip*

The quadrupole matrix equations previously established allow us to write:

$$\begin{bmatrix}\theta_{f1}\\ \Phi_1\end{bmatrix}=\begin{bmatrix}1 & \frac{1}{h_1S}\\ 0 & 1\end{bmatrix}\begin{bmatrix}A_1 & B_1\\ C_1 & D_1\end{bmatrix}\begin{bmatrix}1 & R_{12}\\ 0 & 1\end{bmatrix}\begin{bmatrix}A_2 & B_2\\ C_2 & D_2\end{bmatrix}\begin{bmatrix}1 & R_{23}\\ 0 & 1\end{bmatrix}\begin{bmatrix}A_3 & B_3\\ C_3 & D_3\end{bmatrix}\begin{bmatrix}1 & \frac{1}{h_2S}\\ 0 & 1\end{bmatrix}\begin{bmatrix}\theta_{f2}\\ \Phi_2\end{bmatrix}$$

with: $A_i = D_i = cosh(q_ie_i)$; $C_i = \lambda_iq_iSsinh(q_ie_i)$; $B_i = \frac{sinh(q_ie_i)}{\lambda_iq_iS}$ and $q_i = \sqrt{\frac{p}{a_i}}$.

The description of the problem in matrix form allows us to obtain a very simple formula, which shows the interest of the quadrupoles method.

3.1.4.1.5. Semi-infinite medium

It is shown in section 3.1.2.1 that the temperature in Laplace space of a semi-infinite medium is written as: $\theta(x,p) = Ae^{-qx}$, where: $q = \sqrt{\frac{p}{a}} = \sqrt{\frac{\rho cp}{\lambda}}$.

We deduce the value of the Laplace transform of the heat flow rate at a point of the semi-infinite medium:

$$\Phi(x,p) = -\lambda S\frac{d\theta}{dx}(x,p) = \lambda qSAe^{-qx} = \lambda qS\theta(x,p)$$

Φ can also be written as: $\Phi = \lambda q S \theta = \lambda \sqrt{\frac{\rho c p}{\lambda}} S\theta = \sqrt{\lambda \rho c} S \sqrt{p} \theta = ES\sqrt{p}\,\theta$.

where $E = \sqrt{\lambda \rho c}$ is the thermal effusivity.

We can therefore write at any point of a semi-infinite medium:

$$\begin{bmatrix} \theta \\ \Phi \end{bmatrix} = \begin{bmatrix} \theta \\ ES\sqrt{p}\theta \end{bmatrix} \qquad [3.36]$$

The quadrupole relation [3.36] can be represented by the equivalent electric diagram of Figure 3.24.

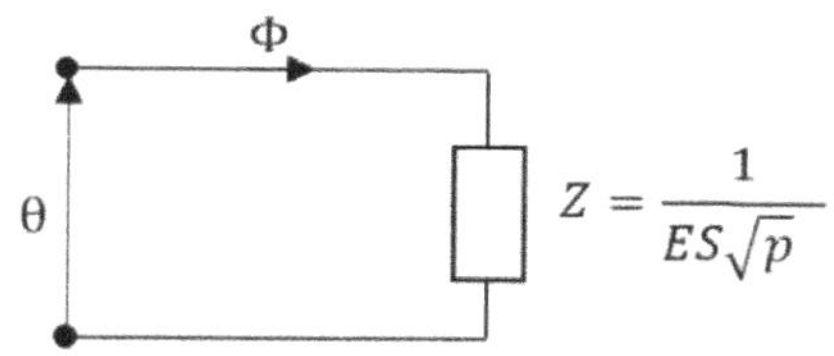

Figure 3.24. *Electrical diagram equivalent to a semi-infinite medium in variable regime*

3.1.4.1.6. Uniform temperature wall

In the case of a "thin system", a wall whose thickness and/or thermal conductivity allow its temperature to be considered as uniform ($Bi < 0.1$; see section 3.1.1), the difference between the incoming heat flow rate and the heat flow rate leaving the system is simply written as: $\varphi_1 - \varphi_2 = \rho c V \frac{dT}{dt}$; therefore, by applying the Laplace transform: $\Phi_1 - \Phi_2 = \rho c V p \theta$.

If $T(t = 0) = 0$, this relation can be expressed in the quadrupolar form by:

$$\begin{bmatrix} \theta_1 \\ \Phi_1 \end{bmatrix} = \begin{bmatrix} 1 & 0 \\ \rho c V p & 1 \end{bmatrix} \begin{bmatrix} \theta_2 \\ \Phi_2 \end{bmatrix} \qquad [3.37]$$

Quadrupole equation [3.37] can be represented by the equivalent electric diagram of Figure 3.25.

Modeling the hot plate method treated in Exercise 3.3.15 shows the interest of the quadrupole method in a complex system in Cartesian geometry.

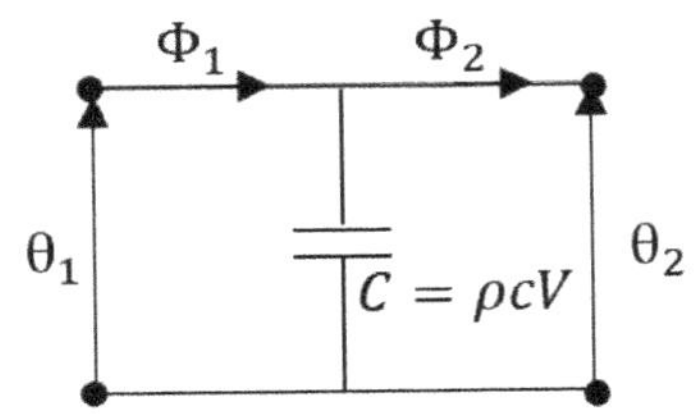

Figure 3.25. *Electrical diagram equivalent to a medium at uniform temperature in variable regime*

3.1.4.2. *Radial transfer*

3.1.4.2.1. Hollow cylinder

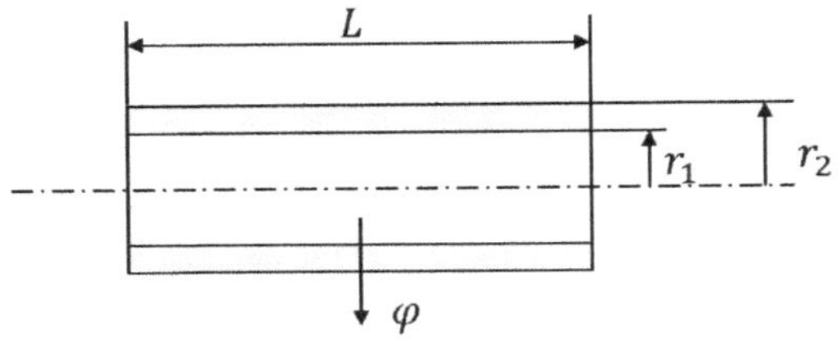

Figure 3.26. *Diagram of a hollow cylinder. For a color version of this figure, see www.iste.co.uk/jannot/heattransfer1.zip*

We show in the same way as in section 3.1.4.1 (Maillet et al. 2000) that temperatures and fluxes in Laplace space can be linked by a quadrupole equation:

$$\begin{bmatrix}\theta(r_1,p)\\ \Phi(r_1,p)\end{bmatrix} = \begin{bmatrix}A & B\\ C & D\end{bmatrix}\begin{bmatrix}\theta(r_2,p)\\ \Phi(r_2,p)\end{bmatrix}$$

$$A = qr_2[K_1(qr_2)I_0(qr_1) + K_0(qr_1)I_1(qr_2)]$$

$$B = \frac{1}{2\pi\lambda L}[K_0(qr_1)I_0(qr_2) - K_0(qr_2)I_0(qr_1)] \quad [3.38]$$

$$C = 2\pi L\rho c r_1 r_2 p[K_1(qr_1)I_1(qr_2) - K_1(qr_2)I_1(qr_1)]$$

$$D = qr_1[K_1(qr_1)I_0(qr_2) + K_0(qr_2)I_1(qr_1)]$$

I_0, I_1, K_0 and K_1 being Bessel functions (see section A.5 of the Appendix) and $q = \sqrt{\frac{p}{a}}$. The determinant of the quadrupole matrix is equal to 1.

3.1.4.2.2. Semi-infinite hollow cylinder

As in the case of a plane wall, we show that we can write at any point of a semi-infinite hollow cylinder ($r_2 \to \infty$) (Maillet et al. 2000):

$$\begin{bmatrix}\theta\\ \Phi\end{bmatrix} = \begin{bmatrix}\theta\\ 2\pi\lambda L\,\dfrac{qr_1K_1(qr_1)}{K_0(qr_1)}\theta\end{bmatrix} \qquad [3.39]$$

The quadrupole relationship [3.38] can be represented by the equivalent electrical diagram in Figure 3.27.

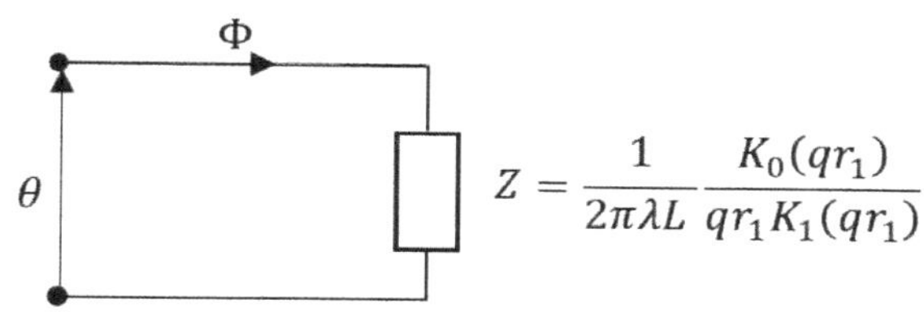

Figure 3.27. *Electrical diagram equivalent to a semi-infinite medium in variable regime*

Modeling the hot wire method treated in Exercise 3.3.18 shows the interest of the quadrupole method in a complex system in cylindrical geometry.

3.1.4.2.3. Hollow sphere

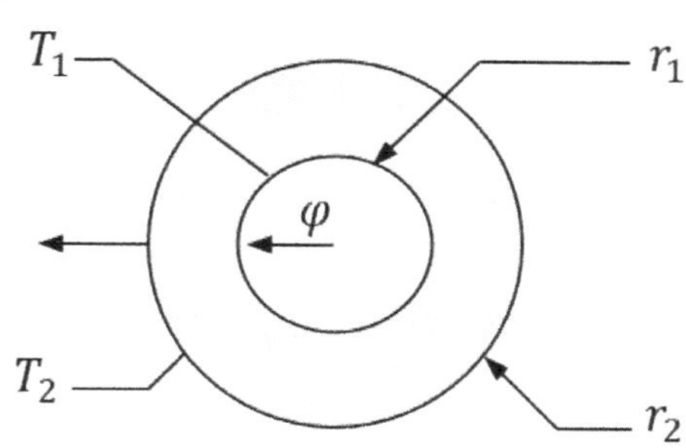

Figure 3.28. *Diagram of the hollow sphere. For a color version of this figure, see www.iste.co.uk/jannot/heattransfer1.zip*

We show in the same way as in section 3.1.4.1 (Maillet et al. 2000) that temperatures and heat flow rates in Laplace space can be linked by a quadrupole equation:

$$\begin{bmatrix}\theta(r_1,p)\\ \Phi(r_1,p)\end{bmatrix} = \begin{bmatrix}A & B\\ C & D\end{bmatrix}\begin{bmatrix}\theta(r_2,p)\\ \Phi(r_2,p)\end{bmatrix}$$

$$A = \frac{r_2}{r_1}cosh(q\Delta r) - \frac{sinh(q\Delta r)}{qr_1}; B = \frac{sinh(q\Delta r)}{4\pi\lambda q r_1 r_2}$$

$$C = 4\pi\lambda r_2\left[\left(1-\frac{r_1}{r_2}\right)cosh(q\Delta r) + \left(qr_1 - \frac{1}{qr_2}\right)sinh(q\Delta r)\right] \quad [3.40]$$

$$D = \frac{r_1}{r_2}cosh(q\Delta r) - \frac{sinh(q\Delta r)}{qr_2}$$

where: $\Delta r = r_2 - r_1$. The determinant of the quadrupole matrix is equal to 1.

3.1.4.2.4. Semi-infinite hollow sphere

As in the case of a plane wall, we show that we can write at any point of a semi-infinite hollow sphere ($r_2 \to \infty$):

$$\begin{bmatrix}\theta\\ \Phi\end{bmatrix} = \begin{bmatrix}\theta\\ 4\pi\lambda r_1(1+qr_1)\,\theta\end{bmatrix} \quad [3.41]$$

The equivalent electrical diagram is identical to that presented in Figure 3.24 with $Z = \frac{1}{4\pi\lambda r_1(1+qr_1)}$.

3.1.5. *Established periodic state*

In the case of a periodic excitation of pulsation ω around the (temporal) average temperature T_i, the temperature at any point of a system in the periodic steady state can be written in the following complex form:

$$T(x,t) = T_i + f(x)exp(i\omega t)$$

where f is a complex function of x: $f = X + iY$, which gives the amplitude $|f|$ and phase ψ of the response:

$$|f| = \sqrt{X^2+Y^2} \text{ and } \psi = arctang\left(\frac{Y}{X}\right)$$

By transferring this expression into the heat equation, we obtain:

$$exp(i\omega t)\frac{d^2 f}{dx^2} = \frac{i\omega}{a}exp(i\omega t)f$$

The complex function f is a solution of:

$$\frac{d^2f}{dx^2} - i\frac{\omega}{a}f = 0$$

The problem being linear, we consider either the real part or the imaginary part of the solution according to whether the temperature varies as $cos(\omega t)$ or $sin(\omega t)$.

This equation is analogous to the heat equation in Laplace's space by replacing p with $i\omega$. By analogy, by replacing p with $i\omega$, we can apply all the techniques developed to solve the heat equation in Laplace's space to its resolution in the established periodic state. In particular, the quadrupole method can be used by replacing p with $i\omega$ in all the quadrupoles.

Let us treat as an example the case (already solved in section 3.1.2.4) of a semi-infinite medium on the surface of which we impose the temperature $T_i + T_0 cos(\omega t)$.

We perform the change of variable $\bar{T} = T - T_i$ that we are looking for in the form:

$$\bar{T} = f(x)exp(i\omega t)$$

We write the quadrupole relation between the surface and a point located at depth x in the semi-infinite medium (see relations [3.33] and [3.36]):

$$\begin{bmatrix} f(0) \\ \mathrm{F}(0) \end{bmatrix} = \begin{bmatrix} cosh\left(\sqrt{\frac{i\omega}{a}}x\right) & \frac{1}{\lambda\sqrt{\frac{i\omega}{a}}S} sinh\left(\sqrt{\frac{i\omega}{a}}x\right) \\ \lambda\sqrt{\frac{i\omega}{a}} S sinh\left(\sqrt{\frac{i\omega}{a}}x\right) & cosh\left(\sqrt{\frac{i\omega}{a}}x\right) \end{bmatrix} \begin{bmatrix} f(x) \\ ES\sqrt{i\omega}f(x) \end{bmatrix}$$

with: $\bar{T}(0) = f(0)exp(i\omega t)$and $\varphi(0) = \mathrm{F}(0)exp(i\omega t)$.

We can deduce:

$$f(0) = \left[cosh\left(\sqrt{\frac{i\omega}{a}}x\right) + \frac{ES\sqrt{i\omega}}{\lambda\sqrt{\frac{i\omega}{a}}S} sinh\left(\sqrt{\frac{i\omega}{a}}x\right)\right] f(x)$$

with: $E = \frac{\lambda}{\sqrt{a}}$.

$$f(0) = \left[\frac{exp\left(\sqrt{\frac{i\omega}{a}}x\right) + exp\left(-\sqrt{\frac{i\omega}{a}}x\right)}{2} + \frac{exp\left(\sqrt{\frac{i\omega}{a}}x\right) - exp\left(-\sqrt{\frac{i\omega}{a}}x\right)}{2}\right] f(x)$$

$$= exp\left(\sqrt{\frac{i\omega}{a}}x\right) f(x)$$

$$f(x) = exp\left(-\sqrt{\frac{i\omega}{a}}x\right) f(0) = exp\left[-(1+i)\sqrt{\frac{\omega}{a}}x\right] f(0) \text{ because } \sqrt{i} = \frac{1+i}{\sqrt{2}}$$

$$\bar{T}(x) = T_0 exp(i\omega t) exp\left[-(1+i)\sqrt{\frac{\omega}{a}}x\right] = T_0 exp\left(-\sqrt{\frac{\omega}{a}}x\right) exp\left[i\left(\omega t - \sqrt{\frac{\omega}{2a}}x\right)\right]$$

The excitation being in cosine, we take the real part of the complex solution:

$$T(x) = T_i + T_1 exp\left(-\sqrt{\frac{\omega}{a}}x\right) cos\left(\omega t - \sqrt{\frac{\omega}{2a}}x\right)$$

We get equation [3.9].

The quadrupole method is of particular interest in the case of complex systems such as that of a multilayer wall subjected to periodic variations of the outside temperature. This case is treated in Exercise 3.3.21.

3.1.6. *Systems with temperature-dependent thermal properties*

All the methods described previously only apply if the equations have constant coefficients; i.e. if the thermal properties do not depend on the temperature. This assumption is generally acceptable over temperature intervals of the order of 10 degrees but can induce significant errors beyond that. In the latter case, the problem can be solved in a simple way by using a numerical method by finite differences with an explicit method, very easy to program. We will treat as an example a case of 1D transfer in Cartesian coordinates. The method can easily be transposed to other coordinate systems (cylindrical, spherical) and to a 2D transfer.

For a plate of thickness $2L$ with imposed temperature, the heat equation is written as:

$$\frac{\partial}{\partial x}\left(\lambda\frac{\partial T}{\partial x}\right) = \rho c\frac{\partial T}{\partial t} \qquad \text{(a)}$$

With the boundary conditions:

$$T(x,0) = T_i \qquad \text{(b)}$$

$$\frac{\partial T}{\partial x}(L,t) = 0 \qquad \text{(c)}$$

We can solve the problem in 1D by an explicit method based on the numerical scheme of Figure 3.29.

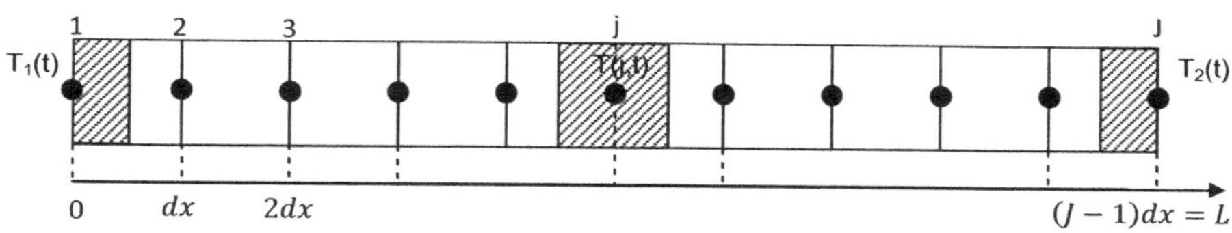

Figure 3.29. *Representation of the digital schematization*

The heat flow rate at any point is calculated by:

$$\varphi = -\lambda S\frac{\partial T}{\partial x}$$

We set: $T(j,t) = T1(j)$ and $T(j,t+dt) = T2(j)$.

Heat balances are written for each type of mesh:

– *Mesh 1*:

$$T1(1) = T2(1) = T_0$$

– *Mesh j*:

$$-\lambda\frac{T1(j) - T1(j-1)}{dx} = -\lambda\frac{T1(j+1) - T1(j)}{dx} + \rho c dx\frac{T2(j) - T1(j)}{dt}$$

Hence: $T2(j) = T1(j) + \frac{\lambda dt}{\rho c dx} \frac{T1(j+1)+T1(j-1)-2T1(j)}{dx}$ where the temperature-dependent properties λ and ρc are calculated for temperature $T1(j)$.

– *Mesh J* (middle of the plate therefore zero heat flow rate):

$$-\lambda \frac{T1(J) - T1(J-1)}{dx} dt = \rho c \frac{dx}{2} [T2(J) - T1(J)]$$

hence: $T2(J) = T1(J) - \frac{2\lambda dt}{\rho c dx} \frac{T1(J)-T1(J-1)}{dx}$

where the temperature-dependent properties λ and ρc are calculated for the mean temperature $T1(J)$ with the initial condition: $T1(j) = T_i, \forall j$ at $t = 0$.

This system can be solved very simply with a program of the type presented in section A.15 of the Appendix.

The time step dt is calculated as follows: $dt = \frac{dx^2}{ka}$. A value $k = 4$ is sufficient for the imposed temperature condition; it can be too low to ensure convergence in the case of an imposed flow. In this case, it is increased progressively until convergence is achieved.

REMARK.–

– For the two other types of boundaries conditions, the equation of mesh 1 is written as:

- imposed heat flux ϕ_0:

$$T2(1) = T1(1) + \frac{2dt}{\rho c\, dx} \left[\phi_0 + \lambda \frac{T1(2) - T1(1)}{dx} \right]$$

- imposed transfer coefficient h:

$$T2(1) = T1(1) + \frac{2dt}{\rho c\, dx} \left\{ h[T_0 - T1(1)] + \lambda \frac{T1(2) - T1(1)}{dx} \right\}$$

– It is very easy to take into account the thermodependence of the parameters by calculating them at each space and time step (see section A.15 of the Appendix).

As an example, Figure 3.30 represents the temperature at the center of a plate of thickness $2L$ initially at 20°C with an imposed surface temperature of 40°C by taking respectively:

1) $\lambda = \lambda_0 = 1\ \mathrm{W\,m^{-1}K^{-1}}$ and $\rho c = \rho c_0 = 10^6 \mathrm{J\,m^{-3}K^{-1}}$constants;

2) $\lambda = \lambda_0(1 + k_1 T)$ with $k_1 = 10^{-2}\mathrm{K^{-1}}$ and $\rho c = \rho c_0(1 + k_2 T)$ with $k_2 = 5 \times 10^{-3}\mathrm{K^{-1}}$.

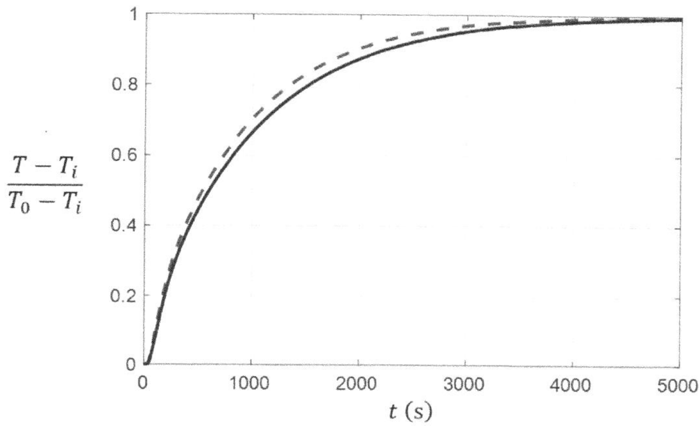

Figure 3.30. *Temperature at the center of a plate with constant (blue line) or thermodependent (red dotted line) properties. For a color version of this figure, see www.iste.co.uk/jannot/heattransfer1.zip*

3.2. Multidirectional conduction in transient regime

3.2.1. *Von Neuman's theorem*

Some two-dimensional or three-dimensional problems can be solved by combining two or three one-dimensional solutions.

Let us consider, for example, the case of an infinite rectangular bar (very great length compared to the sides $2L_1$and $2L_2$); it can be considered as the intersection of two infinite plates of respective thicknesses $2L_1$and $2L_2$. Von Neumann's theorem asserts that the dimensionless temperature of this bar can be expressed as the product of the adimensional temperatures of the two infinite plates of which it can be considered the intersection:

$$\left[\frac{T(x,y,t) - T_\infty}{T_i - T_\infty}\right]_{barre\ 2L_1 \times 2L_2} = \left[\frac{T(x,t) - T_\infty}{T_i - T_\infty}\right]_{plaque\ 2L_1} \times \left[\frac{T(y,t) - T_\infty}{T_i - T_\infty}\right]_{plaque\ 2L_2} \qquad [3.42]$$

Examples of its application can be found in exercises 3.3.8, 3.3.11 and 3.3.17.

REMARK.–

– It is necessary to verify that the initial and boundary conditions are satisfied in the adimensional form after decomposition of the geometry considered as the intersection of simple elements.

– More complex geometries can also be broken down into the intersection of simple elements, such as:

- semi-infinite cylinder = infinite cylinder ∩ semi−infinite medium;

- semi-infinite rectangular bar = semi-infinite medium ∩ infinite rectangular bar;

- cylinder height $2L$ =infinite cylinder ∩ plate thickness $2L$, etc.

3.2.2. *Integral transformations and separation of variables*

Multidirectional heat transfer problems can in some cases be treated as unidirectional by integral transformations and separation of variables. We will simply treat here as an example the heat transfer in a finite cylinder of thickness e and radius r_0, initially at uniform temperature, when one of its faces is subjected to a uniform heat flux $\phi_0(t)$. The cylinder exchanges heat by convection on all its faces with the surrounding environment (see Figure 3.31). If the heat flux is in the form of a short duration pulse, this system represents the Flash method used to measure thermal diffusivity.

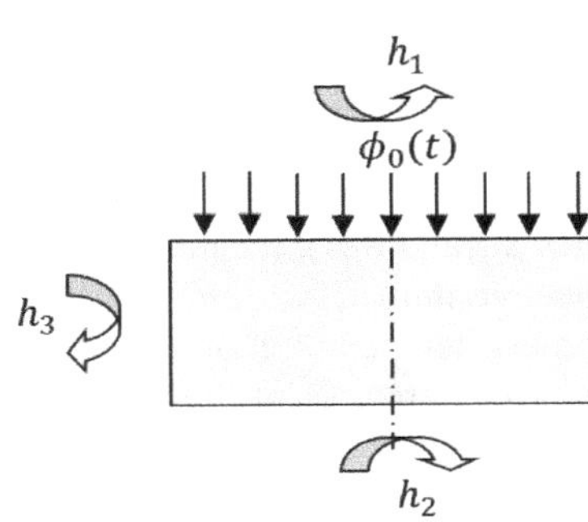

Figure 3.31. *Diagram of the modeled system. For a color version of this figure, see www.iste.co.uk/jannot/heattransfer1.zip*

The problem has cylindrical symmetry, so we use the heat equation in cylindrical coordinates:

$$\frac{\partial^2 T}{\partial r^2}+\frac{1}{r}\frac{\partial T}{\partial r}+\frac{\partial^2 T}{\partial z^2}=\frac{1}{a}\frac{\partial T}{\partial t} \quad \text{(a)}$$

The resolution method used is as follows:

– Laplace transformation;

– separation of variables.

We set: $\bar{T}(r,z,t)=T(r,z,t)-T_i$.

The boundary and initial conditions are as follows:

$$\lambda\left(\frac{\partial \bar{T}}{\partial z}\right)_{z=0}=h_1[\bar{T}(z=0)]-\phi_0(t) \quad \text{(b)}$$

$$-\lambda\left(\frac{\partial \bar{T}}{\partial z}\right)_{z=e}=h_2[\bar{T}(z=e)] \quad \text{(c)}$$

$$\left(\frac{\partial \bar{T}}{\partial r}\right)_{r=0}=0 \quad \text{(d)}$$

$$\lambda\left(\frac{\partial \bar{T}}{\partial r}\right)_{r=r_0}=h_3[\bar{T}(r=r_0)] \quad \text{(e)}$$

$$\bar{T}(r,z,0)=0 \quad \text{(f)}$$

It is noted that the boundary conditions (c), (d) and (e) are homogeneous, whereas the boundary condition (b) is not homogeneous.

We set: $\theta(r,z,p)=L[\bar{T}(r,z,t)]$.

The Laplace transform of (a) is written as: $\frac{\partial^2 \theta}{\partial r^2}+\frac{1}{r}\frac{\partial \theta}{\partial r}+\frac{\partial^2 \theta}{\partial z^2}=\frac{p}{a}\theta$.

The Laplace transform of the temperature can be written in the following form: $\theta(r,z,p)=R(r,p)Z(z,p)$. The previous equation leads to:

$$\frac{\partial^2 (RZ)}{\partial r^2}+\frac{1}{r}\frac{\partial (RZ)}{\partial r}+\frac{\partial (RZ)}{\partial z}=\frac{p}{a}(RZ)$$

or: $\frac{1}{R}\left(\frac{\partial^2 R}{\partial r^2}+\frac{1}{r}\frac{\partial R}{\partial r}\right)+\frac{1}{Z}\frac{\partial^2 Z}{\partial z^2}=\frac{p}{a}$.

The temperature at $z = 0$ increases suddenly, while the temperature at $z = e$ increases very slowly, so the curve $T(r, z, t)$ is concave with respect to the axis Oz and: $\frac{\partial^2 Z}{\partial z^2} > 0$.

We deduce: $\frac{1}{R}\left(\frac{\partial^2 R}{\partial r^2} + \frac{1}{r}\frac{\partial R}{\partial r}\right) = -\alpha^2$ and: $\frac{1}{Z}\frac{\partial^2 Z}{\partial z^2} - \gamma^2 = 0$ with: $\gamma^2 = \frac{p}{a} + \alpha^2$.

The solutions of the above equations are:

$$R(r, p) = AJ_0(\alpha r) + BY_0(\alpha r)$$
$$Z(r, p) = Csinh(\gamma z) + Dcosh(\gamma z)$$

Application of boundary conditions: $Y_0(\alpha r) \rightarrow \infty$ when $r \rightarrow 0$; or the temperature must therefore remain finite, therefore $B = 0$ and $R(r, p) = AJ_0(\alpha r)$.

At $r = r_0$: $-\lambda\left(\frac{\partial T}{\partial r}\right)_{r_0} = h_3[T(r = r_0) - T_i] \Rightarrow -\lambda\frac{\partial R}{\partial r}(r_0, p) = h_3 R(r_0, p)$.

$\frac{\partial R}{\partial z}(r, p) = -A\alpha J_1(\alpha r)$, therefore: $\lambda A\alpha J_1(\alpha r_0) = h_3 AJ_0(\alpha r_0)$

We set: $H_3 = \frac{h_3 r_0}{\lambda}$ and: $\omega = \alpha r_0$.

The eigenvalues ω_n are solutions of the transcendental equation: $\omega J_1(\omega) = H_3 J_0(\omega)$.

At $z = e$: $-\lambda\left(\frac{\partial T}{\partial z}\right)_{z=e} = h_2[T(z = e) - T_i] \Rightarrow -\lambda\frac{\partial Z}{\partial z}(e, p) = h_2 Z(e, p)$.

Hence: $-\lambda\gamma[Ccosh(\gamma e) + Dsinh(\gamma e)] = h_2[Csinh(\gamma e) + Dcosh(\gamma e)]$.

By setting: $\beta = \gamma e$ and $H_2 = \frac{h_2 e}{\lambda}$.

We get: $-\beta[Ccosh(\beta) + Dsinh(\beta)] = H_2[Csinh(\beta) + Dcosh(\beta)]$.

Hence: $\frac{D}{C} = -\frac{H_2 sinh(\beta) + \beta cosh(\beta)}{\beta sinh(\beta) + H_2 cosh(\beta)}$.

We set: $E = AC$.

$$\theta(r, z, p) = R(r, p)Z(z, p) = CAJ_0(\alpha r)\left[sinh(\gamma z) + \frac{D}{C}cosh(\gamma z)\right]$$

$$\theta(r, z, p) = EJ_0(\alpha r)\left[sinh(\gamma z) - \frac{H_2 sinh(\beta) + \beta cosh(\beta)}{\beta sinh(\beta) + H_2 cosh(\beta)}cosh(\gamma z)\right]$$

$$\theta(r,z,p) = EJ_0(\alpha r)\left[\frac{sinh(\gamma z)[\beta sinh(\beta) + H_2 cosh(\beta)] - [H_2 sinh(\beta) + \beta cosh(\beta)]cosh(\gamma z)}{\beta sinh(\beta) + H_2 cosh(\beta)}\right]$$

By setting: $F = \frac{-E}{\beta sinh(\beta)+H_2 cosh(\beta)}$, replacing β with γe and after development and factorization, we obtain:

$$\theta(r,z,p) = FJ_0(\alpha r)\{\beta cosh[\gamma(e-z)] + H_2 sinh[\gamma(e-z)]\}$$

By summing all the solutions, we get the general solution:

$$\theta(r,z,p) = \sum_{n=1}^{\infty} \theta_n(r,z,p) = \sum_{n=1}^{\infty} F_n J_0(\alpha_n r)\{\beta_n cosh[\gamma_n(e-z)] + H_2 sinh[\gamma_n(e-z)]\}$$

At $z = 0$: $\lambda\left(\frac{\partial T}{\partial z}\right)_{z=0} = h_1[T(z=0) - T_i] - \phi_0(t)$.

Either: $\lambda\left(\frac{\partial\theta}{\partial z}\right)(r,0,p) = h_1\theta(r,0,p) - \Phi_0(\mathrm{p})$.

$$\lambda\sum_{n=1}^{\infty} F_n J_0(\alpha_n r)\{-\beta_n\gamma_n sinh[\gamma_n(e-z)] - H_2\gamma_n cosh[\gamma_n(e-z)]\} = h_1$$

$$\sum_{n=1}^{\infty} F_n J_0(\alpha_n r)\{\beta_n cosh[\gamma_n(e-z)] + H_2 sinh[\gamma_n(e-z)]\} - \Phi_0(p)$$

At $z = 0$ and by setting $H_1 = \frac{h_1 e}{\lambda}$:

$$\sum_{n=1}^{\infty} F_n J_0(\alpha_n r)\frac{\lambda}{e}\left[\left({\beta_n}^2 + H_1 H_2\right)sinh(\beta_n) + \beta_n(H_1+H_2)cosh(\beta_n)\right] = \Phi_0(p)$$

If we set: $G_n = F_n\frac{\lambda}{e}\left[\left({\beta_n}^2 + H_1 H_2\right)sinh(\beta_n) + \beta_n(H_1+H_2)cosh(\beta_n)\right]$.

We get: $\sum_{n=1}^{\infty} G_n J_0(\alpha_n r) = \Phi_0(p)$.

The eigenfunctions (see section A.14 of the Appendix) are therefore orthogonal: $\int_0^{ro} J_0(\alpha_n r)J_0(\alpha_m r)rdr = 0$ if $\alpha_n \neq \alpha_m$.

We can write: $\int_0^{r_0} rJ_0(\alpha_m r)dr \sum_{n=1}^{\infty} G_n J_0(\alpha_n r) = \int_0^{r_0} rJ_0(\alpha_m r)dr\Phi_0(p)$.

Hence: $G_m \int_0^{r_0} r{J_0}^2(\alpha_m r)dr = \int_0^{r_0} \Phi_0(p)rJ_0(\alpha_m r)dr$.

$$G_m = \frac{\int_0^{r_0} \Phi_0(p)rJ_0(\alpha_m r)dr}{\int_0^{r_0} r{J_0}^2(\alpha_m r)dr} = \frac{\Phi_0(p)\frac{1}{\alpha_m}J_1(\alpha_n r_0)}{\frac{{r_0}^2}{2}\left[{J_0}^2(\alpha_m r_0) + {J_0}^2(\alpha_m r_0)\right]}$$

$$= \frac{2\Phi_0(p)}{\omega_m\left(1+\frac{{\omega_m}^2}{{H_3}^2}\right)J_1(\alpha_m r_0)}$$

We finally deduce:

$$\theta(r,z,p) = \sum_{n=1}^{\infty} F_n J_0(\alpha_n r)\{\beta_n cosh[\gamma_n(e-z)] + H_2 sinh[\gamma_n(e-z)]\} \qquad [3.43]$$

where: $F_n = \frac{2\Phi_0(p)\frac{e}{\lambda}}{\omega_n\left(1+\frac{{\omega_n}^2}{{H_3}^2}\right)J_1(\omega_n)\left[({\beta_n}^2+H_1H_2)sinh(\beta_n)+\beta_n(H_1+H_2)cosh(\beta_n)\right]}$

with ω_n being the solutions of the transcendent equation $J_0(\omega_n) = \frac{\omega_n J_1(\omega_n)}{H_3}$ solved numerically. A hundred terms are enough to calculate $\theta(r,z,p)$. We then calculate $T(r,z,t)$ by inverse Laplace transform carried out numerically.

REMARK.– If we consider that $\phi_0(t)$ is a Dirac, we are in the case of the flash method in 3D. In this case, $\Phi_0(p) = \phi_0$.

3.3. Corrected exercises

3.3.1. *Age of the Earth: "Kelvin ambiguity" (1864)*

Originally, the Earth mass was at a uniform temperature T_i. Currently, the thermal gradient at the surface of the Earth is of the order of 0.03 K m^{-1}.

1) Deduce the age of the Earth by considering Kelvin's hypotheses: $T_i = 3{,}000°C$ and $a = 10^{-6} m^2 s^{-1}$.

2) The age of the Earth is estimated at 5 billion years. Where does the difference between the two values come from?

3) Assuming that a constant energy generation $\dot{q}$ takes place, calculate this value assuming that the age of the Earth is 5 billion years. Compare it to the currently estimated value: $\dot{q} = 3.32 \times 10^{-8}$ W m^{-3}.

Data: $\lambda = 1$ W m^{-1}K^{-1}.

Corrected

1) It will be shown that the radiation balance of the Earth leads to find its average temperature of the order of 5°C (see exercise on "Solar radiation" in the chapter on radiation in Volume 2). We will therefore make the simplifying assumption here that the Earth has cooled as a semi-infinite medium with an imposed surface temperature $T_0 = 0$°C. The temperature therefore verifies: $\bar{T}(x,t) = \frac{T(x,t)-T_0}{T_i-T_0} = erf\left(\frac{x}{2\sqrt{at}}\right)$.

The gradient is obtained by derivation of this expression, with the relation: $erf(u) = \frac{2}{\sqrt{\pi}}\int_0^u exp(-v^2)dv$ hence: $\frac{d[erf(u)]}{dx} = \frac{du}{dx}\frac{2exp(-u^2)}{\sqrt{\pi}}$.

The thermal gradient at the surface of the Earth (at $x = 0$) is therefore given by: $\frac{dT}{dx} = \frac{T_i-T_0}{\sqrt{\pi a t}}$.

We deduce that: $t = \frac{(T_i-T_0)^2}{\pi a\left(\frac{dT}{dx}\right)^2} = \frac{3000^2}{\pi \times 10^{-6} \times 0.03^2} = 3.18 \times 10^{15}$s, i.e. about 101 million years.

2) The age of the Earth is currently estimated at 5 billion years. Kelvin had neglected in his calculation the generation of radioactive energy within the Earth. Furthermore, phase change phenomena (solidification) were not taken into account.

3) Assuming constant and uniform heating, the problem is written as:

$$\rho c\frac{\partial T}{\partial t} = \lambda\frac{\partial^2 T}{\partial x^2} + \dot{q} \quad 0 < x < +\infty$$

$$T = T_0 \quad x = 0$$

$$T = T_i \quad t = 0$$

Setting $\theta(p) = L[T(t) - T_i]$, where p is the Laplace variable, we obtain:

$$\frac{d^2\theta}{dx} = \frac{p}{a}\theta + \frac{\dot{q}}{\lambda p} \quad (0 < x < +\infty)$$

$\theta = \frac{T_0 - T_i}{p}$ $for\ x = 0$

θ finite for $x \to \infty$

The solution is written as: $\theta(p) = \frac{T_0 - T_i}{p} exp\left(-\sqrt{\frac{p}{a}}x\right) + \frac{\dot{q}}{\rho c p^2}\left[1 - exp\left(-\sqrt{\frac{p}{a}}x\right)\right]$.

Hence: $\left(\frac{\partial \theta}{\partial x}\right)_{x=0} = \frac{T_0 - T_i}{\sqrt{ap}} + \frac{\dot{q}}{\sqrt{\lambda \rho c}p^{3/2}}$.

And by performing the inverse Laplace transform, the geothermal gradient at the surface of the Earth G is:

$$G = \frac{T_i - T_0}{\sqrt{\pi a t}} + \frac{2}{\sqrt{\pi}}\frac{\dot{q}}{\lambda}\sqrt{at}$$

with $t = 5 \times 10^9$years and $G = 0.03\ \text{K m}^{-1}$, we find: $\dot{q} = 5.74 \times 10^{-8}\ \text{W m}^{-3} = 1.81\ \text{J m}^{-3}\text{yr}^{-1}$.

The current estimated value is $\dot{q} = 3.32 \times 10^{-8}\text{W m}^{-3} = 1.05\ \text{J m}^{-3}\text{yr}^{-1}$.

3.3.2. *Periodic variation of temperature in the ground*

1) It is assumed that the ground surface temperature varies sinusoidally. At what depth is there an inversion of the temperatures *maxima* and *minima?* What is then the damping of the oscillations? We will solve the problem in the cases of an annual period and a daily period.

2) How deep should you bury your cellar if you want to limit temperature variations to 2°C in a climate in which the temperature varies annually by 20°C? What is then the value of the phase shift?

Data: $a = 3.5 \times 10^{-7}\ \text{m}^2\text{s}^{-1}$.

Corrected

1) The temperature is calculated by: $T(x,t) = T_0 exp\left(-\sqrt{\frac{\omega}{2a}}x\right) cos\left(\omega t - \sqrt{\frac{\omega}{2a}}x\right)$.

The damping A is equal to $exp\left(-\sqrt{\frac{\omega}{2a}}x\right)$, and the phase shift is equal to $\sqrt{\frac{\omega}{2a}}x$. The inversion of the *maxima* corresponds to a phase shift equal to π obtained for $x = \pi\sqrt{\frac{2a}{\omega}}$.

Daily period:

$$\omega = \frac{2\pi}{24\times 3600} = 7.27\times 10^{-5}\text{s}^{-1}, x = \pi\sqrt{\frac{2\times 3.5.10^{-7}}{7.27\times 10^{-5}}} = 0.31\text{ m}$$

and $A = exp\left(-\sqrt{\frac{7.27\times 10^{-5}}{2\times 3.5\times 10^{-7}}}\times 0.31\right) = 0.042.$

Annual period: $\omega = \frac{2\pi}{24\times 3600\times 365} = 1.99\times 10^{-7}\text{s}^{-1}$, $x = \pi\sqrt{\frac{2\times 3.5\times 10^{-7}}{1.99\times 10^{-7}}} =$ 5.89 m and A unchanged.

2) We want to obtain a damping $A = \frac{2°\,C}{20°\text{C}} = 0.1$, from which: $exp\left(-\sqrt{\frac{\omega}{2a}}x\right) = 0.1$, we deduce: $x = \sqrt{\frac{2a}{\omega}}ln(A) = \sqrt{\frac{2\times 3.5\times 10^{-7}}{1.99\times 10^{-7}}}ln(0.1) = 4.31$ m.

At this depth, the phase shift is: $\sqrt{\frac{\omega}{2a}}x = \sqrt{\frac{1.99\times 10^{-7}}{2\times 3.5\times 10^{-7}}}\times 4.31 = 2.29$ rd corresponding to a phase delay of: $12\times\frac{2.29}{2\pi} = 4.4$ months.

3.3.3. *Measurement of thermal diffusivity by sinusoidal excitation*

The temperature T_p of the end $x = 0$ of a very long circular cylindrical bar of small diameter is varied sinusoidally, while maintaining uniform convection exchange conditions along the bar, the ambient temperature being T_∞.

1) Under what condition (which we will assume a priori verified) is the temperature field of the bar only a function of x and of t?

2) Then write the equation that is verified by $T(x,t)$ by generalizing the equation of the fin to the transient regime.

3) Give the shape of the solution for an established periodic state if the temperature of the end varies sinusoidally in the form: $T(0,t) - T_\infty = \Delta T exp(i\omega t)$.

4) Specify the boundary conditions and give the expression for $T(x,t) - T_\infty$.

5) Show that knowledge of the attenuation (ratio of amplitudes) and of the phase shift between the temperatures of two points distant from $\ell = x_2 - x_1$ enables the thermal diffusivity of the medium to be measured.

6) Calculate the thermal diffusivity value with the following data: period: 24 minutes; $\ell = 5$ cm; amplitude ratio: $\frac{A_2}{A_1} = \frac{1}{3}$; phase shift: $\frac{\pi}{4}$.

What is the value of the exchange coefficient if the bar of diameter $d = 1$ cm is made of steel? ($\rho = 7{,}800$ kg m^{-3} and $c = 460$ J kg^{-1}K^{-1}).

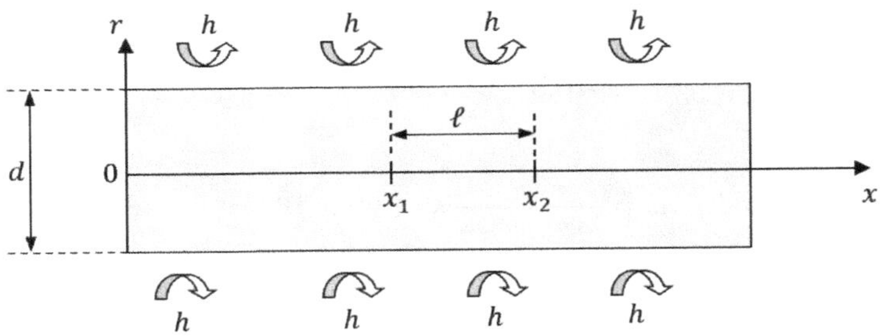

Corrected

1) $Bi = \frac{hd}{\lambda} < 0.1$.

2) If $S = \frac{\pi d^2}{4}$ denotes the cross-section of the bar and $p_e = \pi d$ its perimeter, the equation of the bar in transient regime is written as:

$$\lambda S \frac{\partial^2 T}{\partial x^2} - hp_e(T - T_\infty) = \rho c S \frac{\partial T}{\partial t}$$

3) By setting: $T(x,t) - T_\infty = \Delta T exp(i\omega t) f(x)$, it comes:

$$f''(x) - \left(\frac{hp_e}{\lambda S} + \frac{i\omega}{a}\right) f(x) = 0$$

4) Boundary conditions: $f(0) = 1$ and $f(+\infty) = 0$.

5) The solution is: $f(x) = exp(-rx)$ with $r^2 = \frac{hp_e}{\lambda S} + \frac{i\omega}{a}$ and $r = r_1 + ir_2$ choosing the root with a positive real part ($r_1 > 0$) with:

$$r_1{}^2 - r_2{}^2 = \frac{hp_e}{\lambda S}$$

$$2r_1r_2 = \frac{\omega}{a}$$

We then have:

$$T(x_1) = \Delta T A_1 exp[i(\omega t + \phi_1)] = \Delta T A_1 exp(i\omega t) exp[-(r_1 + ir_2)x_1]$$

$$T(x_2) = \Delta T A_2 exp[i(\omega t + \phi_2)] = \Delta T A_2 exp(i\omega t) exp[-(r_1 + ir_2)x_2]$$

hence: $\frac{T(x_2)}{T(x_1)} = \frac{A_2}{A_1} exp[i(\phi_2 - \phi_1)] = exp[-r_1(x_2 - x_1)] exp[-ir_2(x_2 - x_1)]$.

Therefore: $ln\left(\frac{A_2}{A_1}\right) = -r_1\ell$.

$$\phi_2 - \phi_1 = -r_2\ell$$

and: $a = \frac{\omega\ell^2}{2ln\left(\frac{A_2}{A_1}\right)(\phi_2 - \phi_1)}$.

6) Numerical application

$$a = \frac{\frac{2\pi}{24 \times 60} \times 0.05^2}{2ln(3) \times \frac{\pi}{4}} = 6.32 \times 10^{-6} \text{m}^2\text{s}^{-1}$$

Calculation of the exchange coefficient:

$$\frac{hp_e}{\lambda S} = r_1{}^2 - r_2{}^2 = \frac{\left[ln\left(\frac{A_2}{A_1}\right)\right]^2 - (\phi_2 - \phi_1)^2}{\ell^2} = \frac{[ln(3)]^2 - \left(\frac{\pi}{4}\right)^2}{0.05^2} = 236.04 \text{ m}^2$$

$$\frac{p_e}{S} = \frac{4}{d} = \frac{4}{0.01} = 400 \text{ m}^{-1}$$

$$\lambda = \rho c a = 7\,800 \times 460 \times 6.32 \times 10^{-6} = 22.68 \text{ W m}^{-1}\text{K}^{-1}$$

$$h = \left(\frac{hp_e}{\lambda S}\right)\frac{\lambda S}{p_e} = \frac{236.04 \times 22.68}{400} = 13.38 \text{ W m}^{-2}\text{K}^{-1}$$

3.3.4. *Freezing a lake*

We consider a lake where the liquid water is permanently at freezing temperature $T_0 = 273$ K. The air above the lake is at constant temperature $T_a = 263$ K. Free of ice at the initial time $t = 0$, the lake is gradually covered with a layer of ice of thickness $e(t)$.

The following values are given for ice: $\rho = 920$ kg m^{-3}; $\lambda = 1.88$ W m^{-1}K^{-1}; $c = 2\,040$ J K^{-1}kg^{-1}; latent heat of fusion: $L_f = 334$ kJ kg^{-1}. The air–ice transfer coefficient has the value $h = 20$ W m^{-2}K^{-1}.

1) Show that, in a first approximation, the heat capacity of ice can be neglected in the study of the freezing phenomenon.

2) Show that with this hypothesis the temperature $T(x,t)$ in the ice varies linearly with x and give its expression as a function of $T(0,t)$, T_a, $e(t)$, h and λ.

3) By a heat balance between two instants t and $t + dt$ at the air–ice interface, establish a relationship between T_0, T_a, $e(t)$, ρ, L_f, h and λ.

4) Deduce an expression of $e(t)$. Give a simplified expression for long times.

5) It was said that to skate in Paris you needed 30°C of cold, i.e. 5 days at –6°C or 3 days at –10°C. How to interpret this saying?

Corrected

1) We write that the heat flux produced by the solidification of water is distributed as follows:

– a part $\lambda \frac{T_a - T(0,t)}{e(t)}$ is transmitted to the outside air by conduction through the layer of ice;

– one part $\rho c e \frac{dT}{dt}$ is stored in the layer of ice.

For low values of the ice cover (e small), the first term is preponderant in front of the second because the thermal conductivity of the ice is quite high ($\lambda = 1.88$ W m^{-1}K^{-1}). We can therefore, as a first approximation, neglect the capacitive term.

2) Within this assumption, we write that the heat flux transmitted by conduction through the ice layer is then entirely transmitted by convection to the outside air:

$$\phi = h[T(0,t) - T_a] = \lambda \frac{T_0 - T(0,t)}{e(t)} \qquad [3.44]$$

3) A heat balance is then carried out between the times t and $t + dt$ when the layer of ice has increased in thickness from $e(t)$ to $e(t) + de(t)$:

$$\phi = h[T(0,t) - T_a] = \rho \frac{de(t)}{dt} L_f \qquad [3.45]$$

4) From relation [3.44], we compute the flux ϕ as a function of $(T_0 - T_a)$. Using this expression in relation [3.45] and integrating lead to:

$$e^2 + \frac{2h}{\lambda}e - \frac{2\lambda}{\rho L_f}(T_0 - T_a)t = 0$$

Hence: $e\left(e + \frac{2h}{\lambda}\right) - \frac{2\lambda}{\rho L_f}(T_0 - T_a)t = 0.$

We have: $\frac{2h}{\lambda} = \frac{2\times 20}{1.88} = 21.2$ m.

We can therefore make the approximation: $e + \frac{2h}{\lambda} \approx \frac{2h}{\lambda}$.

We then deduce: $e(t) = \frac{h(T_0 - T_a)t}{\rho L_f}$.

5) After 5 days at –6°C (or 3 days at –10°C), this relationship leads to the value $e = 16.9$ cm, which seems to be a sufficient ice thickness for skating.

3.3.5. *Freezing water pipes in dry ground*

In the installation of underground water pipes, it is important to determine the depth at which a temperature variation on the surface of the ground is felt during a period of 12 hours. Knowing that the initial temperature of the ground is 4°C and that the temperature drops suddenly to -4°C, determine the depth to which the temperature of the freezing point penetrates.

We will consider a dry soil: $\rho = 1{,}700\ \text{kg m}^{-3}$; $c = 840\ \text{J K}^{-1}\text{kg}^{-1}$; $\lambda = 0.4\ \text{W m}^{-1}\text{K}^{-1}$.

Corrected

We come back to the problem of a semi-infinite medium initially at a uniform temperature of 4°C on which a surface temperature of –4°C is suddenly imposed. We seek the critical depth x such that $T(x, 12h) = 0$°C.

The solution to this problem is: $\overline{T}(x,t) = \frac{T(x,t) - T_0}{T_i - T_0} = erf\left(\frac{x}{2\sqrt{at}}\right)$ with $T_0 = -4$°C; $T_i = 4$°C and $a = \frac{\lambda}{\rho c} = 2.74 \times 10^{-7}\text{m}^2\text{s}^{-1}$.

We can deduce:

$$x = 2\sqrt{at}\, erf^{-1}\left(\frac{T(x.12h) - T_0}{T_i - T_0}\right) = 2\sqrt{2.74 \times 10^{-7} \times 12 \times 3{,}600}\, erf^{-1}\left(\frac{0+4}{4+4}\right)$$

hence: $x = 10.4$ cm.

The calculation is only valid for one day of frost, and the necessary depth is greater if the days of frost are repeated.

3.3.6. *Freezing water pipes in wet ground*

Repeat Exercise 3.3.5 if the soil is wet ($\rho = 1{,}800$ kg m^{-3}; $c = 1{,}000$ J K^{-1}kg^{-1}; $\lambda = 0.6$ W m^{-1}K^{-1}) and contains $w = 10\%$ water by mass. The heat of fusion of ice is $L_f = 333$ kJ kg^{-1}.

1) Write the problem to be solved in the frozen ($0 < x < X(t)$) and unfrozen ($x > X(t)$) zones, the boundary conditions at $x = 0$ and $x \to \infty$ and the conditions to be checked at the interface $X(t)$ between the two zones.

2) Show that the solution for the temperature field can be found in the form:

$T_1(x,t) = Aerf\left(\frac{x}{2\sqrt{at}}\right) + T_0$ in the frozen area

$T_2(x,t) = T_i - B\left[1 - erf\left(\frac{x}{2\sqrt{at}}\right)\right]$ in the unfrozen zone

where x is the depth calculated from the ground surface, $T_i = 4$°C is the initial ground temperature, $T_f = 0$°C is the freezing temperature, a is the ground thermal diffusivity, and A and B are constants to be determined.

3) By considering the conditions at the interface $X(t)$, we show that necessarily $X(t) = 2\delta\sqrt{at}$, where δ is a constant satisfying the transcendental equation:

$$\sqrt{\pi}Sn\delta = \frac{1}{exp(\delta^2)erf(\delta)} - \frac{\theta}{exp(\delta^2)[1 - erf(\delta)]}$$

where $\theta = \frac{T_i - T_f}{T_f - T_0}$ and $Sn = \frac{wL_f}{c(T_f - T_0)}$ is the Stefan–Neumann number.

4) Solve the previous transcendental equation assuming a priori δ is small.

5) Determine the depth to which the freezing point temperature penetrates.

Corrected

1) The heat equation is written in phases 1 and 2:

$$\frac{\partial^2 T_1}{\partial x^2} = \frac{1}{a}\frac{\partial T_1}{\partial t} \quad \text{(a)}$$

in phase 1 [for $x < X(t)$].

$$\frac{\partial^2 T_2}{\partial x^2} = \frac{1}{a}\frac{\partial T_2}{\partial t} \quad \text{(b)}$$

in phase 2 [for $x > X(t)$].

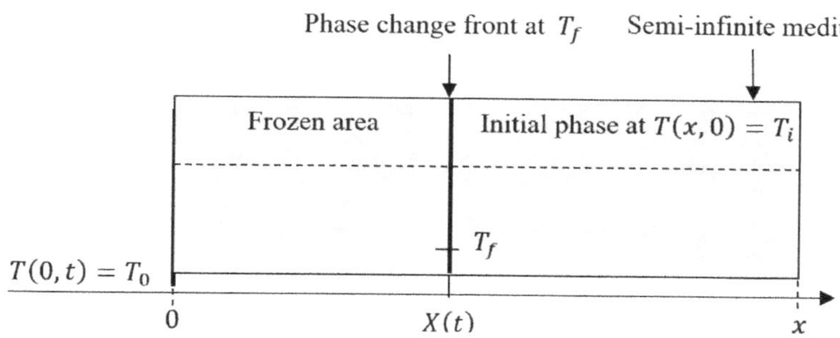

The boundary conditions are written as:

$$T_1(0,t) = T_0 \quad \text{(c)}$$

$$T_2(\infty,t) = T_i \quad \text{(d)}$$

$$T_1(X,t) = T_1(X,t) = T_f \quad \text{(e)}$$

$$\lambda\left(\frac{\partial T_1}{\partial x}\right)_X - \lambda\left(\frac{\partial T_2}{\partial x}\right)_X = wL_f\rho\frac{dX}{dt} \quad \text{(f)}$$

The relation (f) results from a thermal balance carried out at the level of the phase change front (at $x = X$) between the instants t and $t + dt$: the difference of the heat fluxes exchanged by conduction is equal to the flux necessary for the phase change of a layer with a thickness dX.

2) From the relationships established in section 3.1.2, the function $erf\left(\frac{x}{2\sqrt{at}}\right)$ satisfies the heat equation $T_1(x,t) = Aerf\left(\frac{x}{2\sqrt{at}}\right) + T_0$, where A is an arbitrary

constant that satisfies equations (a) and (c). Similarly, the function $T_2(x,t) = T_i - B\left[1 - erf\left(\frac{x}{2\sqrt{at}}\right)\right]$, where B is an arbitrary constant, satisfies (b) and (d).

Equation (e) then leads to: $Aerf\left(\frac{X}{2\sqrt{at}}\right) + T_0 = T_i - B\left[1 - erf\left(\frac{X}{2\sqrt{at}}\right)\right] = T_f$.

3) This relation must be verified for all values of t, we deduce that: $X(t) = 2\delta\sqrt{at}$, where δ is a constant.

Taking this form of $X(t)$, equation (f) allows us to write:

$$\frac{\lambda A exp(-\delta^2)}{\sqrt{\pi a}} - \frac{\lambda B exp(-\delta^2)}{\sqrt{\pi a}} = wL_f\rho\sqrt{a} \text{ with: } A = \frac{T_f - T_0}{erf(\delta)} \text{and:} B = \frac{T_i - T_f}{1 - erf(\delta)}$$

So, δ is the solution of the equation:

$$\sqrt{\pi} Sn\delta = \frac{1}{exp(\delta^2)erf(\delta)} - \frac{\theta}{exp(\delta^2)[1 - erf(\delta)]}$$

where: $\theta = \frac{T_i - T_f}{T_f - T_0}$ and $Sn = \frac{wL_f}{c(T_f - T_0)}$.

The position $X(t)$ of the phase change front is finally calculated by: $X(t) = 2\delta\sqrt{at}$.

4) If δ is small: $exp(\delta^2) = 1 + \delta^2 + O(\delta^2)$ and:

$$erf(\delta) = \frac{2}{\sqrt{\pi}}\int_0^\delta exp(-u^2)du = \frac{2}{\sqrt{\pi}}\left(\delta - \frac{\delta^3}{3}\right) + O(\delta^4)$$

$$exp(\delta^2)erf(\delta) = \frac{2\delta}{\sqrt{\pi}}\left[1 + \frac{2\delta^2}{3}\right] + O(\delta^4)$$

$$exp(\delta^2)[1 - erf(\delta)] = 1 - \frac{2\delta}{\sqrt{\pi}} + O(\delta^4)$$

$$\sqrt{\pi}\, Sn\, \delta = \frac{\sqrt{\pi}}{2\delta} - \frac{1}{3}\sqrt{\pi}\delta - \theta\left[1 + \frac{2\delta}{\sqrt{\pi}}\right] + O(\delta^2)$$

$$\frac{\sqrt{\pi}}{2\delta} - \delta\sqrt{\pi}\left(Sn + \frac{1}{3} + \frac{2\theta}{3\pi}\right) - \theta = 0$$

By setting $C = Sn + \frac{1}{3} + \frac{2\theta}{3\pi}$, it comes: $C\sqrt{\pi}\delta^2 + \theta\delta - \frac{\sqrt{\pi}}{2} = 0$ whose solution is: $\delta = \frac{-\theta+\sqrt{\theta^2+2\pi C}}{2C\sqrt{\pi}}$.

Numerical application:

$$a = 3.33 \times 10^{-7}\text{m}^2\text{s}^{-1};\ \theta = 0.5;\ Sn = 8.325;\ \delta = 0.223$$

We deduce: $X(t) = 2\delta\sqrt{at}$ with $t = 12$ h.

Hence: $X(12\text{h}) = 2 \times 0.223\sqrt{12 \times 3\,600 \times 3.33 \times 10^{-7}} = 5.3$ cm.

We find a lower value than that calculated in dry soil (10.4 cm) because the release of heat at the level of the freezing front slows down the propagation of the disturbance.

3.3.7. *Firewall*

A firewall must meet the following requirements: from the ambient temperature, if the temperature on the hot side suddenly increases by 700 K, the temperature on the cold side must not rise by more than 140 K. What must be the thickness L of the wall (thermal diffusivity $a = 5 \times 10^{-7}\text{m}^2\text{s}^{-1}$) to meet this condition after 3, 6 or 12 hours respectively?

Corrected

The boundary condition on the cold side is not specified, we place ourselves in the most unfavorable case where it would be isolated, i.e. a condition of null flux. The problem is then reduced to finding the temperature at the center of a $2L$ thick wall whose faces would be maintained at a constant temperature. The solution is written as:

$$\frac{T(x,t) - T_0}{T_i - T_0} = \frac{4}{\pi}\sum_{n=0}^{\infty}\frac{1}{2n+1} exp\left[-(2n+1)^2\frac{\pi^2}{4}\frac{at}{L^2}\right] sin\left[\frac{(2n+1)\pi x}{2L}\right]$$

where x is the distance to the heated face. This solution converges quickly at long times, and we will only use the first term to solve this problem. We obtain:

$$\frac{T(L,t)-T_0}{T_i-T_0} = \frac{4}{\pi} exp\left[-\frac{\pi^2}{4}\frac{at}{L^2}\right] \text{ for } n = 0$$

According to the data: $\frac{T(L,t)-T_0}{T_i-T_0} = \frac{140-700}{0-700} = 0.8$, therefore:

$$L = \frac{\pi}{2}\sqrt{\frac{-at}{ln\left[\frac{\pi}{4}\frac{T(L,t)-T_0}{T_i-T_0}\right]}} = \frac{\pi}{2}\sqrt{\frac{-5\times10^{-7}\times3\times3600}{ln\left(\frac{\pi}{4}\times0.8\right)}} = 16.9 \text{ cm for } t = 3 \text{ h}.$$

We find: for $= 6$ h: $L = 23.9$ cm and for $t = 12$ h: $L = 33.9$ cm.

To validate the calculation based on the first term of the series, we can calculate the second term of the (alternating) series and we find that it is 100 times smaller than the first. A resolution by inverse Laplace transformation using the Stehfest method (see section A.7 of the Appendix) of the solution: $\theta(x,p) = (T_0 - T_i)\frac{cosh(qx)}{pcosh(qL)}$ with $q = \sqrt{\frac{p}{a}}$ leads to the same values.

3.3.8. *Fire from a wooden beam*

To fight fires, it is necessary to know how wooden beams can support the fire before igniting. The beams are long with a straight section of 100 × 100 mm^2 and initially at a uniform temperature of 16°C. At the moment the fire breaks out, the beams are exposed to a temperature of 650°C and the convective transfer coefficient is equal to $h = 12$ W m^{-2}K^{-1}. Assess the time before the wood reaches its ignition temperature of 427°C. The physical properties of wood are as follows: $\rho = 800$ kg m^{-3}; $c = 2{,}400$ J K^{-1}kg^{-1}; $\lambda = 0.3$ W m^{-1}K^{-1}.

Corrected

According to Neumann's theorem, we have:

$$\left[\frac{T(x,y,t) - T_\infty}{T_i - T_\infty}\right]_{beam} = \left[\frac{T(x,t) - T_\infty}{T_i - T_\infty}\right]_{plate1}\left[\frac{T(y,t) - T_\infty}{T_i - T_\infty}\right]_{plate2}$$

The maximum temperature is reached on the edges for $x = y = L$.

We deduce from the data: $\left[\frac{T(L,L,t)-T_\infty}{T_i-T_\infty}\right]_{beam} = \frac{427-650}{16-650} = 0.352$ and:

$$\left[\frac{T(L,t) - T_\infty}{T_i - T_\infty}\right]_{plate} = \sqrt{0.352} = 0.593$$

We will therefore use the abacuses (section A.12 of the Appendix), which give $\frac{T(x,t)-T_\infty}{T_i-T_\infty} = f\left(\frac{1}{Bi}, \frac{x}{L}, Fo\right)$ for a plate.

In this case: $Bi = \frac{hL}{\lambda} = 2$; $\frac{T(x,t)-T_\infty}{T_i-T_\infty} = 0.593$; $\frac{x}{L} = 1$; we deduce from this by reading the second chart of section A.12 of the Appendix: $Fo \approx 0.075$.

And finally: $t = Fo\frac{L^2}{a}$ with $a = \frac{\lambda}{\rho c} = \frac{0.3}{800 \times 2{,}400} = 1.56 \times 10^{-7} \mathrm{m^2 s^{-1}}$.

Hence: $t = 0.075 \times \frac{0.075^2}{1.56 \times 10^{-7}} = 1{,}202$ s about 20 minutes.

A resolution by inverse Laplace transform using the Stehfest method (see section A.7 of the Appendix) of the solution:

$\theta(x,p) = \frac{T_0 - T_i}{p} \frac{cosh(qx)}{cosh(qL) + \frac{\lambda q}{h} sinh(ql)}$ leads to $t = 1{,}190$ s

3.3.9. *Flash method*

To determine the thermal diffusivity a of a sample of thickness e, its upper surface is exposed for a very short time to a high-power incident energy flux (Flash), the opaque material absorbing energy Q per unit area.

The sample is assumed to be isolated on all its faces. We assimilate the pulse source to an initial condition $T(x, t = 0) = T_0$ for $x < \varepsilon$ and $T(x, t = 0) = 0$ for $x > \varepsilon$, by choosing the origin of the times after the flash.

1) Using the method of separation of variables, establish the expression of the temperature field $T(x,t)$ in the sample. We will tend ε towards 0 to determine the integration constants.

2) Calculate the temperature field after an infinite time and show that the thermal diffusivity a can be estimated from $t_{1/2}$ (Parker's method) such that: $T(e, t_{1/2}) = \frac{1}{2} max[T(e,t)]$.

3) How are the previous results qualitatively modified if the faces are not perfectly isolated?

4) Interpret the two thermograms proposed using the Parker method and the Degiovanni partial time method, which takes into account the convective losses:

$$a = \frac{1}{3}(a_1 + a_2 + a_3) \text{ with } a_1 = \frac{e^2}{t_{5/6}}\left(1.131 - 1.222\frac{t_{2/3}}{t_{5/6}}\right)$$

$$a_2 = \frac{e^2}{t_{5/6}}\left[0.954 - 1.581\frac{t_{1/2}}{t_{5/6}} + 0.558\left(\frac{t_{1/2}}{t_{5/6}}\right)^2\right]$$

and: $a_3 = \frac{e^2}{t_{5/6}}\left[0.818 - 1.708\frac{t_{1/3}}{t_{5/6}} + 0.885\left(\frac{t_{1/3}}{t_{5/6}}\right)^2\right]$.

The real values are $a_{PVC} = 1.2 \times 10^{-7}\mathrm{m^2s^{-1}}$and $a_{Steel} = 5 \times 10^{-6}\mathrm{m^2s^{-1}}$. Comment on the results.

5) Consider the case where the duration of the pulse is no longer infinitely short. It will be possible in particular to deal with the case of a pulse in the form of a slot.

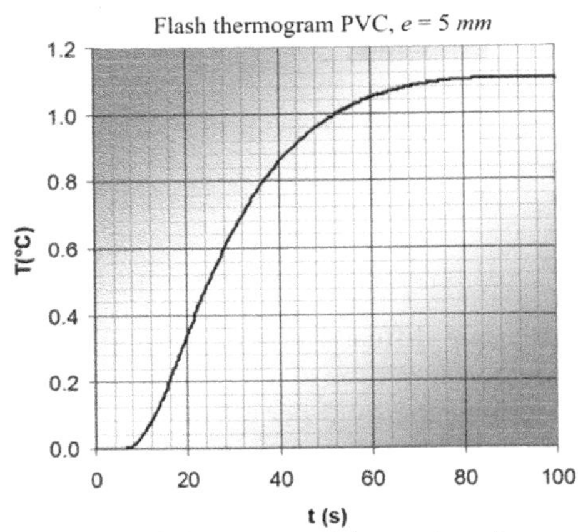

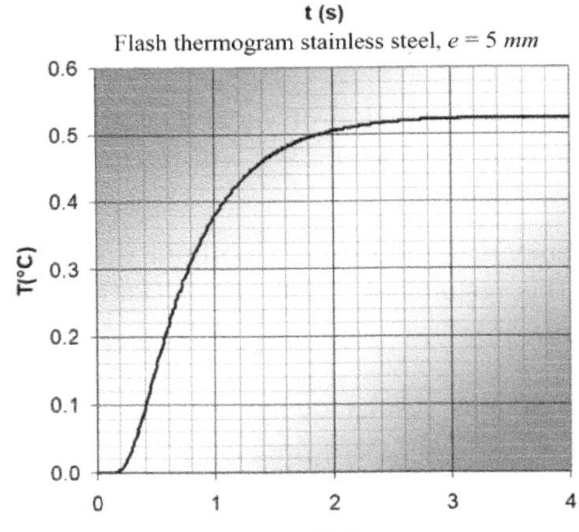

Corrected

1) The heat equation is written as:

$$\frac{\partial^2 T}{\partial x^2} = \frac{1}{a}\frac{\partial T}{\partial t} \quad \text{(a)}$$

With the boundary conditions:

$$T(x,0) = 0 \text{ if } x > \varepsilon \quad \text{(b)}$$

$$T(x,0) = \frac{Q}{\rho c \varepsilon} \text{ if } x < \varepsilon \quad \text{(c)}$$

$$\frac{\partial T}{\partial x}(0,t) = \frac{\partial T}{\partial x}(e,t) = 0 \quad \text{(d)}$$

The temperature is decomposed into a product of functions in the form: $T(x,t) = X(x) \times Y(t)$. The heat equation leads to the following equation:

$$X"Y = \frac{1}{a} XY' \text{ or } \frac{X"}{X} = \frac{1}{a}\frac{Y'}{Y} = -\alpha^2$$

where α is a constant because the two functions X and Y are independent. We deduce:

$$X" + \alpha^2 X = 0$$
$$Y' + a\alpha^2 Y = 0$$

The solutions are of the form: $T(x,t) = exp(-a\alpha^2 t)[Acos(\alpha x) + Bsin(\alpha x)]$.

The boundary condition $\frac{\partial T}{\partial x}(0,t) = 0$ imposes $B = 0$.

The boundary condition $\frac{\partial T}{\partial x}(e,t) = 0$ is written as: $sin(\alpha e) = 0$, and the eigenvalues of the system are therefore: $\alpha_n = \frac{n\pi}{e}$.

The solution superposition theorem allows us to write the general solution in the form:

$$T(x,t) = \sum_{n=0}^{\infty} A_n exp(-a{\alpha_n}^2 t)cos(\alpha_n x) = A_0 + \sum_{n=1}^{\infty} A_n exp(-a{\alpha_n}^2 t)cos(\alpha_n x)$$

It is necessary to treat separately the case $n = 0$ where the eigenfunction is a constant (equal to 1) and not a cosine.

Therefore, calculate $\int_0^e T(x,0)dx = eA_0$; or $\int_0^e T(x,0)dx = \int_0^\varepsilon T(x,0)dx + \int_\varepsilon^e T(x,0)dx = \frac{Q}{\rho c}$.

We deduce: $A_0 = \frac{Q}{\rho c e}$.

Calculation of A_m:

$$\int_0^e T(x,0)cos(\alpha_m x)dx = \int_0^e \sum_{n=0}^{\infty} A_n cos(\alpha_n x)cos(\alpha_m x)dx$$

By using the orthogonality of the eigenfunctions, we obtain:

$$\int_0^e T(x,0)cos(\alpha_m x)dx = A_m \int_0^e cos^2(\alpha_m x)dx = A_m \frac{e}{2}$$

Furthermore:

$$\int_0^e T(x,0)cos(\alpha_m x)dx = \int_0^{\varepsilon} \frac{Q}{\rho c\varepsilon} cos(\alpha_m x)dx = \frac{Q}{\rho c\varepsilon}\int_0^{\varepsilon} cos(\alpha_m x)dx$$
$$= \frac{Q}{\rho c\varepsilon}\left[\frac{e}{m\pi} sin\left(\frac{m\pi\varepsilon}{e}\right)\right]$$

We deduce: $A_m = \frac{2Q}{\rho c\varepsilon}\frac{e}{m\pi} sin\left(\frac{m\pi\varepsilon}{e}\right)$ and if $\varepsilon \to 0$: $A_m = \frac{2Q}{\rho ce}$ and finally:

$$T(x,t) = \frac{Q}{\rho ce}\left[1 + 2\sum_{n=1}^{\infty} cos\left(\frac{n\pi x}{e}\right) exp\left(-\frac{an^2\pi^2}{e^2}t\right)\right]$$

2) The sample is assumed to be perfectly insulated, and the energy is distributed uniformly in the sample after an infinite time, and we have: $T_{max} = T_\infty = \frac{Q}{\rho ce}$.

The temperature on the lower face ($x = e$) is written as:

$$T(e,t) = T_{max}\left[1 + 2\sum_{n=1}^{\infty} cos(\pi x)exp\left(-\frac{an^2\pi^2}{e^2}t\right)\right]$$
$$= T_{max}\left[1 + 2\sum_{n=1}^{\infty}(-1)^n exp(-n^2\pi^2 Fo)\right]$$

where: $Fo = \frac{at}{e^2}$ is the Fourier number.

The plot of the function $T(e,t)$ (which tends to T_{max} when t tends to infinity) shows that $\frac{\bar{T}(e,t)}{\bar{T}_{max}} = \frac{1}{2}$ for $Fo = 0.139$, hence: $a = 0.139\ \frac{e^2}{t_{1/2}}$.

3) Due to heat losses, the temperature of the lower face passes through a maximum before tending towards its initial value after an infinite time. The curves $T = f(t)$ are processed in the same way by considering the ratio $\frac{\bar{T}(e,t)}{\bar{T}_{max}}$. The losses slightly modify the results of (2) depending on the values of the thermal diffusivity a of the sample and of the convective exchange coefficient. The model established in (1) represents the temperature accurately at short times but deviates from it at long times.

4) The supplied thermograms give: for PVC: $T_{max} = 1.1°\text{C}$ and $t_{1/3} = 21$ s, $t_{1/2} = 25$ s, $t_{2/3} = 34$ s, $t_{5/6} = 44$ s.

We deduce:

$$a_{Parker} = 1.39 \times 10^{-7} \text{m}^2\,\text{s}^{-1}$$

$a_1 = 1.06 \times 10^{-7}\text{m}^2\text{s}^{-1}$, $a_2 = 1.40 \times 10^{-7}\text{m}^2\text{s}^{-1}$, $a_3 = 1.16 \times 10^{-7}\text{m}^2\text{s}^{-1}$

hence: $a = 1.21 \times 10^{-7}\text{m}^2\text{s}^{-1}$.

For steel: $T_{max} = 0.525°\text{C}$ and $t_{1/3} = 0.55$ s, $t_{1/2} = 0.7$ s, $t_{2/3} = 0.9$ s, $t_{5/6} = 1.25$ s.

We deduce:

$$a_{Parker} = 4.96 \times 10^{-7}\text{m}^2\text{s}^{-1}$$

$a_1 = 5.02 \times 10^{-7}\text{m}^2\text{s}^{-1}$, $a_2 = 5.06 \times 10^{-7}\text{m}^2\text{s}^{-1}$, $a_3 = 4.76 \times 10^{-7}\text{m}^2\text{s}^{-1}$

hence: $a = 4.95 \times 10^{-7}\text{m}^2\text{s}^{-1}$.

The partial times method allows an estimate with less than 1% error in these two cases. Parker's method is all the more precise as the diffusivity (and therefore in general the conductivity) is high. A more conductive material leads to a lower Biot number, the internal resistance is lower compared to the external convection resistance and the relative importance of thermal losses is lower.

5) Let: $q(\tau) = \frac{dQ}{dt}(\tau)$ and: $F(x,t) = \frac{Q}{\rho c e}\left[1 + 2\sum_{n=1}^{\infty} cos\left(\frac{n\pi x}{e}\right) exp\left(-\frac{an^2\pi^2}{e^2}t\right)\right]$ the response to an infinitely short unit pulse.

The energy supplied to the front face between the times τ and $\tau + d\tau$ is equal to: $q(\tau)d\tau$.

The field of temperature resulting from this elementary source is written as: $dT_\tau(x,t) = F(x, t-\tau)q(\tau)d\tau$.

The temperature field resulting from a slot between $t = 0$ and $t = \tau_c$ is the sum of the fields $dT_\tau(x,t)$, and we deduce:

$$T(x,t) = \int_0^t F(x, t-\tau)q(\tau)d\tau \text{ if } t < \tau_c$$

and: $T(x,t) = \int_0^{\tau_c} F(x, t-\tau)q(\tau)d\tau \text{ if } t > \tau_c$.

This is the convolution product of the excitation by the response to a Dirac.

3.3.10. *Heat treatment of landing gear*

The heat treatment of landing gear consists of rapidly cooling the piece to obtain a metal structure and satisfactory mechanical characteristics. It is also known that the structure obtained is satisfactory for any point having reached or exceeded a critical cooling rate of 800°C s^{-1} in the vicinity of 250°C. What is the thickness of the heat-treated layer if it is assumed that the cooling is carried out by imposing a constant temperature of 20°C on the surface of the piece initially brought to a uniform temperature of 470°C? The piece is made of aluminum and will be likened to a cylinder 60 cm in diameter and 3 m high. The calculations will be approached assuming that the thickness of the treated layer is very small compared to the radius of the cylinder (this assumption will be verified a posteriori).

Corrected

It is assumed that the treated layer is thin and that the medium is therefore semi-infinite for the duration of the treatment. We are brought back to the problem of a semi-infinite medium with imposed surface temperature whose solution is: $\bar{T}(x,t) = \frac{T(x,t)-T_0}{T_i-T_0} = erf\left(\frac{x}{2\sqrt{at}}\right)$.

The time derivative is written as: $\frac{\partial T(x,t)}{\partial t} = (T_i - T_0)\frac{x}{2\sqrt{a}}\left(\frac{-1}{2t^{3/2}}\right)\frac{2}{\sqrt{\pi}}exp\left(-\frac{x^2}{4at}\right)$.

Let us calculate the value of $x(t)$ which at t fixed leads to $T(x,t) = 250$ C:

$$\bar{T}(x,t) = \frac{250-20}{470-20} = erf\left(\frac{x}{2\sqrt{at}}\right)$$

we deduce: $\frac{x}{2\sqrt{at}} = 0.489$.

The time corresponding to a speed $V_0 = 800$ K s^{-1} in the vicinity of 250°C is given by the relationship:

$$V_0 = (T_i - T_0)\frac{x}{2\sqrt{a}}\left(\frac{-1}{2t^{3/2}}\right)\frac{2}{\sqrt{\pi}}exp\left(-\frac{x^2}{4at}\right)$$

in which $\frac{x}{2\sqrt{at}} = 0.489$.

Therefore: $V_0 = (T_i - T_0) \times 0.489\sqrt{t}\left(\frac{-1}{2t^{3/2}}\right)\frac{2}{\sqrt{\pi}}exp(-0.489^2)$.

Or again: $t = \frac{0.489\times 450}{800\sqrt{\pi}}exp(-0.489^2) = 0.197$ s.

Corresponding to:

$$x = 2 \times 0.489\sqrt{at} = 2 \times 0.489\sqrt{5\times 10^{-5} \times 0.197} = 3.06 \text{ mm}.$$

Let us verify that the temperature at $x = 0.3$ m(at the center) has not varied after $t = 0.197$ s: $erf\left(\frac{0.3}{2\sqrt{5\times10^{-5}\times0.197}}\right) = 1$, and the condition of the semi-infinite medium is well satisfied.

3.3.11. *Heat treatment of a carbon block*

Graphite is a material that has many qualities for use in process equipment: resistance to corrosion, resistance to high temperatures and very good thermal conductivity. However, it is porous, so it must be made watertight through an impregnation treatment to be used in exchangers. The graphite blocks are impregnated with liquid resin and then introduced into an oven whose temperature is kept constant at 200°C. The purpose of the heat treatment is to polymerize the resin. The treatment will be considered finished if all the points of the block have reached the temperature of 180°C. We will focus here on the kinetics of this treatment.

We consider cubic blocks with a side of 0.35 m, an emissivity $\varepsilon = 0.9$ and average thermal properties: $\lambda = 78\ \mathrm{W\,m^{-1}K^{-1}}$; $\rho c = 1.24 \times 10^6\ \mathrm{J\,m^{-3}K^{-1}}$. The convective heat transfer coefficient is $h = 7\ \mathrm{W\,m^{-2}K^{-1}}$.

1) Calculate the minimum and maximum values of the radiation coefficient of the bricks if their emissivity is equal to 0.9.

2) Calculate the temperature at the center of the brick after one hour of cooking. Specify the assumptions and approximations used to carry out the calculation.

3) Calculate the maximum temperature of the brick after one hour of cooking.

4) After how long will the heat treatment be finished?

5) Show that the block can be considered as a body at uniform temperature and recalculate the heat treatment time by a simplified method.

6) Show that we can take into account a linear variation of the global transfer coefficient with this h simplified model.

7) Calculate the time for the block to reach the temperature of 100°C.

Corrected

1) $h_r = 4\varepsilon\sigma T^3$.

For $T = 20$°C, we get $h_r = 5.1\ \mathrm{W\,m^{-2}\,K^{-1}}$.

For $T = 200$°C, we get $h_r = 21.6\ \mathrm{W\,m^{-2}K^{-1}}$.

2) The calculations will be made with an average radiation transfer coefficient:

$$\bar{h}_r = \frac{5.1 + 21.6}{2} = 13.4\ \mathrm{W\,m^{-2}\,K^{-1}}$$

and an overall transfer coefficient: $h = h_c + \bar{h}_r = 20.4\ \mathrm{W\,m^{-2}\,K^{-1}}$.

The dimensionless temperature is the product of three dimensionless temperatures:

$$\frac{T(x,y,z,t) - T_\infty}{T_i - T_\infty} = \left[\frac{T(x,t) - T_\infty}{T_i - T_\infty}\right]_{Plate\ 2L} \times \left[\frac{T(y,t) - T_\infty}{T_i - T_\infty}\right]_{Plate\ 2L} \times \left[\frac{T(z,t) - T_\infty}{T_i - T_\infty}\right]_{Plate\ 2L}$$

with $2L = 0.35$ m.

The temperature at the center of the plate depends on two dimensionless numbers:

$$Bi = \frac{hL}{\lambda} = \frac{20.4 \times 0.175}{78} = 0.045$$

$$Fo = \frac{at}{L^2} = \frac{78 \times 3600}{1.24 \times 10^6 \times 0.175^2} = 7.39$$

We read on the abacus: $\left[\frac{T(0,3600)-T_\infty}{T_i-T_\infty}\right]_{Plate\ 2L} \approx 0.73.$

Hence: $\frac{T(0.0.0.3600)-T_\infty}{T_i-T_\infty} = 0.73^3 = 0.39$ and: $T(0,0,0,3600) = 130°C.$

3) The maximum temperature is reached on the corners:

$$\frac{T(L,L,L.3600) - T_\infty}{T_i - T_\infty} = \left[\frac{T(L,t) - T_\infty}{T_i - T_\infty}\right]^3_{Plate\ 2L}$$

We read on the abacus: $\left[\frac{T(L.3600)-T_\infty}{T_i-T_\infty}\right]_{Plate\ 2L} \approx 0.71.$

Hence: $\frac{T(L,L,L.3600)-T_\infty}{T_i-T_\infty} = 0.71^3 = 0.358$ and: $T(L,L,L,3600) = 135.6°C.$

4) After how long will the heat treatment be finished?

The heat treatment will be finished when the temperature in the center is 180°C, i.e.:

$$\frac{T(0,0,0,t) - T_\infty}{T_i - T_\infty} = \frac{180 - 200}{20 - 200} = 0.11$$

From this, we deduce: $\left[\frac{T(0,t)-T_\infty}{T_i-T_\infty}\right]_{Plate\ 2L} = 0.11^{1/3} = 0.48.$

On the abacus, we read the approximate value: $Fo \approx 17.$

From this, we deduce: $t = \frac{FoL^2}{a} \approx \frac{78 \times 17 \times 0.175^2}{1.24 \times 10^6} \approx 8{,}276$ s. An exact 3D numerical calculation of the temperature evolution at the center leads to the value $t = 8{,}000$ s, the difference is due to the imprecision of the reading on the abacus.

5) We have: $Bi = 0.045$; therefore, we can make the hypothesis of the "small body" and write:

$$\rho V c \frac{dT}{dt} = -hS(T - T_\infty)$$

$$\frac{T - T_\infty}{T_i - T_\infty} = exp\left(-\frac{hS}{\rho c V}t\right)$$

with: $S = 6 \times 4L^2$and: $V = 8L^3$.

From this, we deduce: $t = -\frac{\rho c V}{hS} ln\left(\frac{T-T_\infty}{T_i-T_\infty}\right) = -\frac{1.24\times10^6\times0.175}{3\times20.4} ln\left(\frac{180-200}{20-200}\right) =$ 7,790 s.

An exact 3D numerical calculation of the temperature evolution at the center leads to the value $t = 8{,}000$ s, i.e. a difference of less than 3%.

6) We have: $\rho V c \frac{dT}{dt} = -h_0[1 + \alpha(T - T_\infty)]S(T - T_\infty)$.

With: $h_0 = 21.6 + 7 = 28.6$ W m^{-2} K^{-1} and: $\alpha = \frac{\frac{12.1}{28.6}-1}{-180} = 0.0032$.

If: $\bar{T} = T - T_\infty$ then: $\frac{d\bar{T}}{(1+\alpha\bar{T})\bar{T}} = -\frac{h_0 S}{\rho V c}dt$

where: $\frac{d\bar{T}}{(1+\alpha\bar{T})\bar{T}} = \frac{d\bar{T}}{\bar{T}} - \alpha\frac{d\bar{T}}{1+\alpha\bar{T}}$.

$$ln(\bar{T}) - ln(1 + \alpha\bar{T}) = -\frac{h_0 S}{\rho c V}t$$

$$\left[\frac{\bar{T}}{1 + \alpha\bar{T}}\right]_{T_i-T_\infty}^{T-T_\infty} = exp\left(-\frac{h_0 S}{\rho c V}t\right)$$

7) $-\frac{h_0 S}{\rho c V}t = \left[ln\left(\frac{\bar{T}}{1+\alpha\bar{T}}\right)\right]_{T_i-T_\infty}^{T-T_\infty}$.

with: $T - T_\infty = 180 - 200 = -20$°C and: $T_i - T_\infty = 20 - 200 = -180$°C.

$$t = -\frac{1.24 \times 10^9 \times 0.175}{3 \times 28.6} \times ln\left[\frac{-20 \times (1 - 0.0032 \times 180)}{-80 \times (1 - 0.0032 \times 20)}\right] = 2{,}681 \text{ s}$$

3.3.12. *Heat treatment of a thin layer*

A thick layer e of material 1 is deposited on material 2 sufficiently thick to be considered semi-infinite. A heat treatment must be applied to the layer, which consists of increasing its temperature to a predefined value T_t. For this, a uniform and constant heat flux $\phi_0 = 5{,}000\ \mathrm{W\ m^{-2}}$ is imposed on its surface. It will be assumed that before heating, the two media are in equilibrium with the external medium at the temperature T_e and that the layer is sufficiently thin so that its temperature can be considered as uniform at all times. The transverse dimensions are large compared to the thickness, which allows unidirectional heat transfer to be considered. It will be assumed that the heat flow is provided by a very thin heating element bonded to the layer and perfectly insulated on its other side.

1) Write the system of equations verified by the temperature $T(x, t)$.

2) Establish the expression $\theta(p)$ of the Laplace transform of the temperature difference $[T(0, t) - T_e]$ of the layer.

3) Using a limited expansion of $\theta(p)$, establish an approximate expression for the temperature $T(0, t)$ at a long time ($p \to 0$).

4) Deduce the time after which the temperature T_t will be reached.

Data:

– material 1: $\rho c_1 = 2.0 \times 10^6 \mathrm{J\ m^{-3} K^{-1}}$; $e_1 = 0.5$ mm;

– material 2: $\rho c_2 = 1.4 \times 10^6 \mathrm{J\ m^{-3} K^{-1}}$; $\lambda_2 = 0.18\ \mathrm{W\ m^{-1} K^{-1}}$;

– $T_e = 20°\mathrm{C}$; $T_t = 120°\mathrm{C}$; $\phi_0 = 5{,}000\ \mathrm{W\ m^{-2}}$.

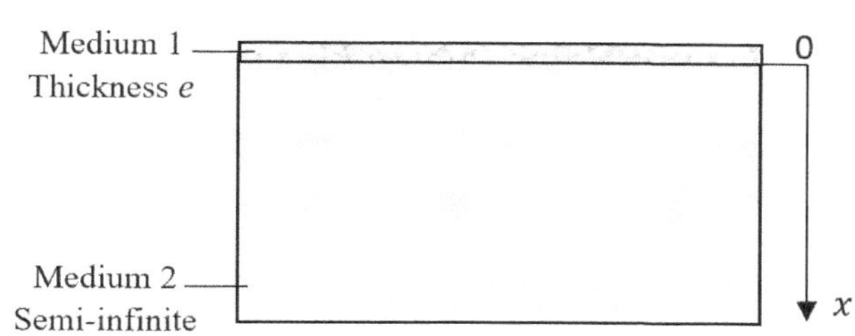

Corrected

1) $\frac{\partial^2 T}{\partial x^2} = \frac{1}{a_2}\frac{\partial T}{\partial t}$ and $\phi_0 = (\rho c e)_1 \frac{\partial T(0,t)}{\partial t} - \lambda_2 \frac{\partial T(0,t)}{\partial x}$.

2) The solution in the Laplace space for a semi-infinite medium is written in the form:

$$\theta(x,p) = Aexp(-qx) \text{ with } q = \sqrt{\frac{p}{a_2}}$$

The Laplace transform of the boundary condition is written as: $\frac{\phi_0}{p} = (\rho ce)_1 p\theta(0,p) - \lambda_2 \frac{d\theta(0,p)}{\partial x}$.

Hence:

$$\frac{\phi_0}{p} = (\rho ce)_1 pA + \lambda_2 q_2 A \text{ thus: } A = \frac{\phi_0}{p} \frac{1}{(\rho ce)_1 p + \frac{\lambda_2}{\sqrt{a_2}}\sqrt{p}}$$

$$\theta(0,p) = A = \frac{\phi_0}{p} \frac{1}{(\rho ce)_1 p + E_2\sqrt{p}} = \frac{\phi_0}{E_2 p^{3/2}} \frac{1}{1 + \frac{(\rho ce)_1}{E_2}\sqrt{p}}$$

3) $\theta(0,p) \neq \frac{\phi_0}{E_2 p^{3/2}}\left[1 - \frac{(\rho ce)_1}{E_2}\sqrt{p}\right] \neq \frac{\phi_0}{E_2 p^{3/2}} - \frac{\phi_0(\rho ce)_1}{p(E_2)^2}$.

By inverse Laplace transform, we finally obtain: $T(0,t) = \frac{2\phi_0\sqrt{t}}{E_2\sqrt{\pi}} - \frac{\phi_0(\rho ce)_1}{(E_2)^2} + T_e$.

4) $t = \left[\frac{E_2\sqrt{\pi}}{2\phi_0}\left(T(0,t) - T_e + \frac{\phi_0(\rho ce)_1}{(E_2)^2}\right)\right]^2$.

Numerical application: t =113 s. The exact calculation would give t =109 s.

3.3.13. *Quenching of a ball*

A spherical ball of radius R, homogeneous, isotropic, of thermal diffusivity a, of thermal conductivity λ, is initially in equilibrium with the external medium at temperature T_0. At the initial time, the ball is immersed in a homogeneous oil bath at temperature T_1. The exchange between the sphere and the oil is modeled by a global exchange coefficient h.

1) Using the method of resolution by separation of variables, express the temperature at the center of the ball in the general case.

2) Express the temperature at the center of the ball in the case $Bi \gg 1$.

3) Express the temperature of the ball in the case $Bi \ll 1$.

4) What case would we approach in the case of the oil quenching of a steel ball with a diameter of 1 cm?

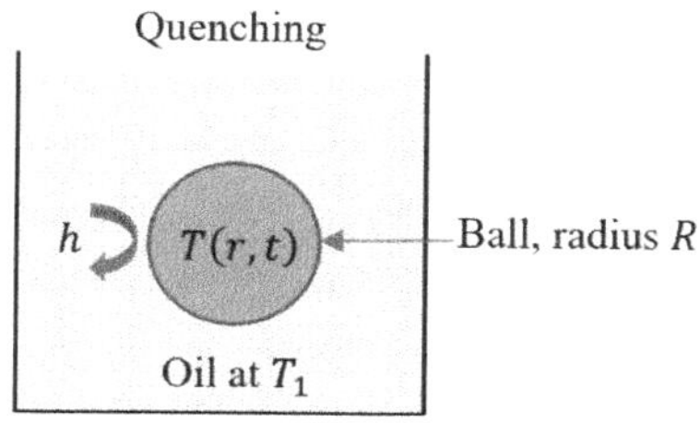

Corrected

1) The spherical ball is assumed to be homogeneous and isotropic, with no internal source.

The heat equation in spherical coordinates is written as [2.7]:

$$\frac{1}{r}\frac{\partial^2}{\partial r^2}(rT) + \frac{1}{r^2 \, sin(\theta)}\frac{\partial}{\partial \theta}\left[sin(\theta)\frac{\partial T}{\partial \theta}\right] + \frac{1}{r^2 sin^2(\theta)}\frac{\partial^2 T}{\partial \varphi^2} + \frac{\dot{q}}{\lambda} = \frac{1}{a}\frac{\partial T}{\partial t}$$

$$\frac{1}{r}\frac{\partial^2}{\partial r^2}(rT) = \frac{1}{r}\frac{\partial}{\partial r}\left[\frac{\partial (rT)}{\partial r}\right] = \frac{1}{r}\frac{\partial}{\partial r}\left(r\frac{\partial T}{\partial r} + T\right) = \frac{1}{r}\left(\frac{\partial T}{\partial r} + r\frac{\partial^2 T}{\partial r^2} + \frac{\partial T}{\partial r}\right)$$
$$= \frac{1}{r}\left(2\frac{\partial T}{\partial r} + r\frac{\partial^2 T}{\partial r^2}\right)$$

Taking into account the spherical symmetry of the problem, the heat equation is therefore written as:

$$\frac{\partial^2 T}{\partial r^2} + \frac{2}{r}\frac{\partial T}{\partial r} = \frac{1}{a}\frac{\partial T}{\partial t} \qquad \text{(a)}$$

with the following initial and boundary conditions:

$$T(r, t = 0) = T_0 \qquad \text{(b)}$$

$$-\lambda\frac{\partial T}{\partial t} = h(T - T_1) \text{ in } r = R \qquad \text{(c)}$$

$$T \text{ finite in } r = 0 \qquad \text{(d)}$$

The resolution method by separation of variables only applies if the boundary conditions are homogeneous ($\lambda\frac{\partial T}{\partial n} + h\,T = 0$), which is not the case here. We are therefore going to make a change of variable to make the relation (c) homogeneous. The system to be solved becomes:

$$\frac{\partial^2 \theta}{\partial r^2} + \frac{2}{r}\frac{\partial \theta}{\partial r} = \frac{1}{a}\frac{\partial \theta}{\partial t} \qquad \text{(e)}$$

$$\theta(r, t = 0) = T_0 - T_1 \qquad \text{(f)}$$

$$-\lambda \frac{\partial \theta}{\partial r} = h\theta \text{ in } r = R \qquad \text{(g)}$$

The method consists of first determining the set of solutions of the homogeneous system, i.e. equations (e), (d) and (g).

We set: $\theta(r,t) = \mathrm{X}(r)Y(t) = XY$.

Equation (e) becomes:

$$\frac{1}{X}\left(\frac{\partial^2 X}{\partial r^2} + \frac{2}{r}\frac{\partial X}{\partial r}\right) = \frac{1}{Y}\frac{1}{a}\frac{\partial Y}{\partial t} \qquad \text{(h)}$$

The two parameters $r$and t being independent, we necessarily have:

$$\frac{1}{X}\left(\frac{\partial^2 X}{\partial r^2} + \frac{2}{r}\frac{\partial X}{\partial r}\right) = -\alpha^2 \qquad \text{(i)}$$

$$\frac{1}{Y}\frac{1}{a}\frac{\partial Y}{\partial t} = -\alpha^2 \qquad \text{(j)}$$

The general solution of equation (j) is written as:

$$Y(t) = Aexp(-a\alpha^2 t)$$

This justifies a posteriori the choice of a negative constant $(-\alpha^2)$ because the temperature remains finite in the ball.

The general solution of equation (i) is written as:

$$X(r) = B\frac{sin(\alpha r)}{r} + C\frac{cos(\alpha r)}{r}$$

The temperature must remain finite in the ball, in particular in $r = 0$ so $C = 0$.

Hence: $X(r) = B\frac{sin(\alpha r)}{r}$.

The general solution of the homogeneous system is therefore of the form:

$$\theta(r,t) = Kexp(-a\alpha^2 t)\frac{sin(\alpha r)}{r}$$

The boundary condition (g) in $r = R$ is written as: $-\lambda \frac{\partial(XY)}{\partial t} = hXY$, from where: $-\lambda \frac{\partial Y}{\partial t} = hY$ and:

$$-\lambda \frac{\partial \left[B \frac{sin(\alpha R)}{R} \right]}{\partial R} = hB \frac{sin(\alpha R)}{R}$$

$$-\lambda \left[-\frac{sin(\alpha R)}{R^2} + \alpha \frac{cos(\alpha R)}{R} \right] = h \frac{sin(\alpha R)}{R}$$

By setting: $\omega = \alpha R$ and $Bi = \frac{hR}{\lambda}$, we obtain the following transcendental equation:

$$\omega cos(\omega) = (1 - Bi) sin(\omega)$$

which admits an infinity of solutions ω_n (numerically computable), so the homogeneous system also admits an infinity of solutions of the form:

$$\theta_{h,n}(r,t) = K_n exp(-a\alpha_n{}^2 t) \frac{sin(\alpha_n r)}{r}$$

$\alpha_n = \frac{\omega_n}{R}$ are the eigenvalues and $\psi_n = \frac{sin(\alpha_n r)}{r}$ are the eigenfunctions of the system.

The solution of the complete system ((e), (d), (g)) is a linear combination of the solutions of the homogeneous system, i.e.:

$$\theta_n(r,t) = \sum_{n=1}^{\infty} K_n exp(-a\alpha_n{}^2 t) \frac{sin(\alpha_n r)}{r}$$

We will now use the inhomogeneous condition to determine the expression for the coefficients K_n. This initial condition at $t = 0$ is written as:

$$\sum_{n=1}^{\infty} K_n \frac{sin(\alpha_n r)}{r} = T_0 - T_1$$

The method consists of multiplying this condition by a specific function $\frac{sin(\alpha_m r)}{r}$ and integrating this product over its domain of validity, i.e. in this case over the volume of the ball. We obtain:

$$\sum_{n=1}^{\infty}\int_{0}^{R} K_n \frac{sin(\alpha_n r)}{r}\frac{sin(\alpha_m r)}{r} dV = \int_0^R \frac{sin(\alpha_m r)}{r}(T_0 - T_1)dV$$

with $dV = \pi r^2 dr$.

The property of orthogonality of eigenfunctions (see section A.14 of the Appendix) allows us to simplify the expression of the first member:

$$K_m \int_0^R sin^2(\alpha_m r)dr = (T_0 - T_1)\int_0^R sin(\alpha_m r)\, rdr$$

hence: $K_m = \frac{(T_0 - T_1)\int_0^R sin(\alpha_m r)rdr}{\int_0^R sin^2(\alpha_m r)dr}$.

After integration by parts, the calculation of the two integrals leads to:

$$K_m = 2\,\frac{T_0 - T_1}{\alpha_m}\left[\frac{-\alpha_m Rcos(\alpha_m R) + sin(\alpha_m R)}{\alpha_m R - sin(\alpha_m R)cos(\alpha_m R)}\right]$$

The expression for the temperature at any point is:

$$T(r,t) = T_1 + 2\,(T_0 - T_1)\sum_{n=1}^{\infty}\left[\frac{-\alpha_n Rcos(\alpha_n R) + sin(\alpha_n R)}{\alpha_n R - sin(\alpha_n R)cos(\alpha_n R)}\right] exp(-a{\alpha_n}^2 t)\frac{sin(\alpha_n r)}{\alpha_n r}$$

And we get at the center of the ball:

$$T(0,t) = T_1 + 2\,(T_0 - T_1)\sum_{n=1}^{\infty}\left[\frac{-\alpha_n Rcos(\alpha_n R) + sin(\alpha_n R)}{\alpha_n R - sin(\alpha_n R)cos(\alpha_n R)}\right] exp(-a{\alpha_n}^2 t)$$

2) If the Biot number defined by $Bi = \frac{hR}{\lambda}$ is large, this means that h is very large in front of λ. The external resistance is negligible compared to the internal resistance and therefore, and we are in the case where $T(R) = T_1$.

The system to be solved is the same as before by simply replacing equation (g) with the equation:

$$T(R) = T_1 \qquad \text{(k)}$$

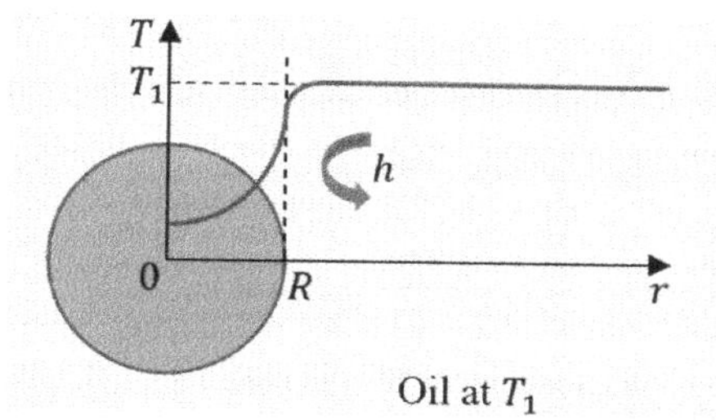

The general form of the solution of the homogeneous system is identical with the same eigenfunctions, but the use of the relation (k) instead of (g) leads to the following eigenvalues:

$$\alpha_n = \frac{n\pi}{R}$$

By calculating as previously the integral over the volume of the initial condition multiplied by a eigenfunction $\frac{sin(\alpha_m r)}{r}$, we end up with:

$$T(0,t) = T_1 + 2\,(T_0 - T_1)\sum_{n=1}^{\infty}(-1)^{n+1}exp(-a\alpha_n{}^2 t)$$

Another method consists of noting that the transcendent equation becomes: $tan(\omega) = 0$, hence $\omega_n = n\pi$ and $\alpha_n = \frac{n\pi}{R}$. By replacing α_n with $\frac{n\pi}{R}$ in the solution obtained in 1, we find the expression established above.

3) If the defined Biot number is small, λ is very large in front of h. The internal resistance is negligible compared to the external resistance and therefore the temperature of the ball is uniform and depends only on time.

We can write the heat balance of this ball between two times t and $t + dt$:

$$-hS(T - T_0) = \rho c V\frac{dT}{dt}$$

thus: $\frac{dT}{T-T_0} = -\frac{hS}{\rho c V}dt.$

hence: $\frac{T-T_0}{T_i-T_0} = exp\left(-\frac{3h}{\rho c r}t\right).$

4) The thermal conductivity of carbon steel is of the order of 50 W m^{-1} K^{-1} and the order of magnitude of the transfer coefficient with a liquid is 1,000 W m^{-2} K^{-1}. With a radius of 1 cm, the order of magnitude of the Biot number would be:

$$Bi = \frac{hR}{\lambda} = \frac{1000 \times 0.01}{50} = 0.2$$

According to the accommodation criterion, as $Bi > 0.1$ we cannot consider that the temperature in the ball is uniform and we will have to use the solution established in 1.

3.3.14. *Brake pad heating*

We propose to study the heating of brake pads and disks during braking. It is assumed that a vehicle with a mass of 1,200 kg travels at 108 km/h. The braking time to bring the vehicle to a complete stop is 5 s. Braking is obtained by friction (on the four wheels) of two pads on a 1 cm thick steel disk. Given the short duration of braking, the pads and the disk will be assimilated to semi-infinite media with a uniform initial temperature of 20°C. One directional heat transfer will be assumed. We will suppose that the surface of the disk is equal to that of the pads.

1) Calculate the total energy dissipated during braking (passing from 108 to 0 km/h). Deduce the value of the heat flux (assumed to be uniform) on each pad/disk interface during the braking period.

2) Write the equations of the system corresponding to a sudden contact between two semi-infinite media with imposed flux.

3) Braking is carried out on the four wheels by two pads made of composite material that a jaw applies to a disk. The surface of a wafer is 30 cm^2. Give the expression of the contact temperature between the disk and the pad at the end of braking. Calculate its value at the end of braking.

Data: effusivity $E_{disk} = 10{,}000 \text{ W m}^{-2}\text{K}^{-1}\text{s}^{1/2}$; $E_{pad} = 1{,}000 \text{ W m}^{-2}\text{K}^{-1}\text{s}^{1/2}$. After reading the information and photo below, do you think this temperature is higher or lower than the actual temperature?

Principle of operation of pads.

Brake pads are fitted to all vehicles fitted with disk brakes. It is a wear part whose friction against the disk provides braking. The plates are composed of two elements:

– *A sheet steel base.*

– *A friction material (lining)* composed of several organic or metallic elements. The composition implies differences in performance, comfort or noise during braking.

Diagram and photo of a brake disk and pads

Corrected

1) Quantity of heat dissipated: $Q = \frac{1}{2}mv^2 = \frac{1}{2} \times 1{,}200 \times 30^2 = 540$ kJ.

Heat flow rate dissipated by plate: $\varphi = \frac{1}{8}\frac{Q}{t} = \frac{1}{8}\frac{540{,}000}{5} = 13.5$ kW.

2) Write the equations of the system corresponding to a sudden contact between two semi-infinite media with an imposed flow rate.

$$\frac{\partial^2 T_1}{\partial x^2} = \frac{1}{a_1}\frac{\partial T_1}{\partial t} \quad \text{(a)}$$

and:

$$\frac{\partial^2 T_2}{\partial x^2} = \frac{1}{a_2}\frac{\partial T_2}{\partial t} \quad \text{(b)}$$

The origin of the abscissa is taken at the point of contact between the two media. The boundary conditions are written as:

$$T_1(x,0) = T_i \text{ for } x < 0 \quad \text{(c)}$$

$$T_2(x,0) = T_i \text{ for } x > 0 \quad \text{(d)}$$

$$\phi = \lambda_1 \frac{\partial T_1}{\partial x}(0,t) - \lambda_2 \frac{\partial T_2}{\partial x}(0,t) \quad \text{(e)}$$

$$T_1(0,t) = T_2(0,t) \quad \text{(f)}$$

3) We perform the following changes of variable:

$\bar{T}_1 = T_1 - T_i$ and $\bar{T}_2 = T_2 - T_i$

Equations (a) and (b) can then be written as: $\frac{\partial^2 \bar{T}_1}{\partial x^2} = \frac{1}{a_1}\frac{\partial \bar{T}_1}{\partial t}$ and $\frac{\partial^2 \bar{T}_2}{\partial x^2} = \frac{1}{a_2}\frac{\partial \bar{T}_2}{\partial t}$.

And the boundary conditions become:

$$\bar{T}_1(x,0) = 0 \text{ for } x < 0 \quad \text{(c)}$$

$$\bar{T}_2(x,0) = 0 \text{ for } x > 0 \quad \text{(d)}$$

$$\phi = \lambda_1 \frac{\partial \bar{T}_1}{\partial x}(0,t) - \lambda_2 \frac{\partial \bar{T}_2}{\partial x}(0,t) \quad \text{(e)}$$

$$\bar{T}_1(0,t) = \bar{T}_2(0,t) \quad \text{(f)}$$

The Laplace transforms of equations (a) and (b) lead to:

$$\theta_1(x,p) = A_1 e^{-q_1 x} + B_1 e^{q_1 x} \text{and } \theta_2(x,p) = A_2 e^{-q_2 x} + B_2 e^{q_2 x}$$

The temperature remains finite when x tends towards $\pm\infty$, therefore $A_1 = 0$ and $B_2 = 0$, and we deduce that: $\theta_1(x,p) = B_1 e^{q_1 x}$ and $\theta_2(x,p) = A_2 e^{-q_2 x}$.

Condition (f) imposes: $B_1 = A_2 = A$.

The Laplace transform of equation (e) is then written as: $\frac{\phi}{p} = A(\lambda_1 q_1 + \lambda_2 q_2)$.

Hence: $\theta_1(x,p) = \frac{\phi}{p}\frac{e^{q_1 x}}{(E_1+E_2)\sqrt{p}}$ and: $\theta_2(x,p) = \frac{\phi}{p}\frac{e^{-q_2 x}}{(E_1+E_2)\sqrt{p}}$.

At the point of contact ($x = 0$), these relations become:

$$\theta_1(0,p) = \theta_2(0,p) = \frac{\phi}{p}\frac{1}{(E_1 + E_2)\sqrt{p}}$$

The use of the inverse Laplace transform tables presented in section A.7 of the Appendix leads to the following result: $T(0,t) = \frac{2\phi}{(E_1+E_2)\sqrt{\pi}}\sqrt{t} = 1{,}032°\text{C}$.

4) The temperature is overvalued because:

– the layer having an effusivity of 1,000 W m^{-2} K^{-1} s$^{1/2}$ is thin and rests on a steel support which has a much greater effusivity;

– the disk rotates during braking; therefore, the surface of the disk is much larger than the surface of the pads, contrary to the assumption made.

3.3.15. *Hot plate method*

The purpose of the hot plate method is to measure the thermal effusivity E of a material. The method consists of inserting a flat heating element, the temperature evolution of which is measured, between two very thick samples of the material to be characterized (initially at the uniform temperature T_i). A voltage step is sent to it, which causes the release of a heat flow rate φ_0. The heat produced diffuses into the samples and the evolution of the temperature $T(t)$ of the heating element over time is recorded.

1) Write the expression for the Laplace transform of the temperature of the heating element using the quadrupole method. It will be assumed that the temperature is uniform in the heating element of mass $2m$ and heat capacity c; we note R_c the thermal contact resistance between the heating element and the medium.

2) Study, using series expansions, the asymptotic behavior of this temperature. Deduce a simple method for determining effusivity.

3) Estimate the values of E, mc and R_c from the thermogram below (represented in several systems of axes) obtained on PVC of thickness $e = 1$ cm.

4) Show that after 60 s, the condition of the semi-infinite medium is still valid. It will be assumed that the unheated face of the sample is perfectly insulated.

Data: heating element area $S = 20.3\ \text{cm}^2$, electrical resistance: $R = 40.5\ \Omega$, electrical voltage $U = 7$ V. Thermal diffusivity of PVC measured by the flash method: $a = 1.2 \times 10^{-7} \text{m}^2\text{s}^{-1}$.

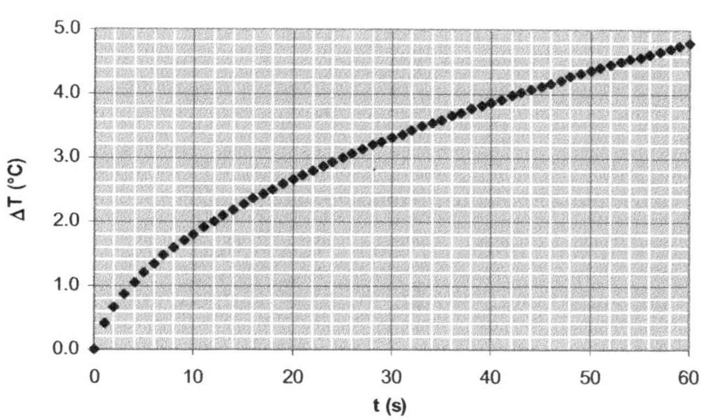

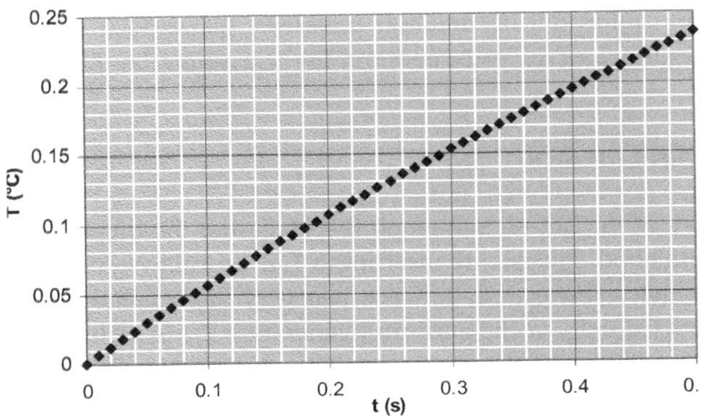

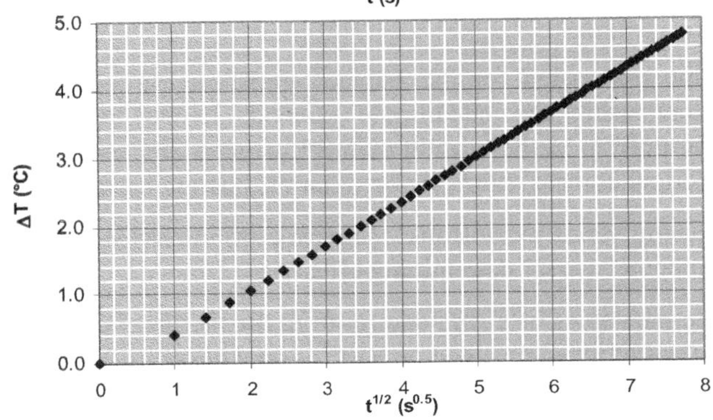

Corrected

1) Modeling the system using the formalism of the quadrupoles allows us to write:

$$\begin{bmatrix} \theta_s \\ \dfrac{\varphi_0}{2p} \end{bmatrix} = \begin{bmatrix} 1 & 0 \\ mcp & 1 \end{bmatrix} \begin{bmatrix} 1 & R_c \\ 0 & 1 \end{bmatrix} \begin{bmatrix} \theta \\ ES\sqrt{p}\theta \end{bmatrix}$$

with:

– θ_s, Laplace transform of the difference $T_s(t) - T_i$;

– θ, Laplace transform of the difference $T(x = 0, t) - T_i$;

– R_c, contact resistance at the heating element/material interface to be characterized;

– $2m$, thermocouple + heating resistor mass;

– c, thermocouple + heating resistor specific heat;

– E, thermal effusivity of the material to be characterized;

– a, thermal diffusivity of the material to be characterized;

– p, Laplace variable;

– S, area of the heating resistor;

– φ_0, power dissipated in the heating resistor.

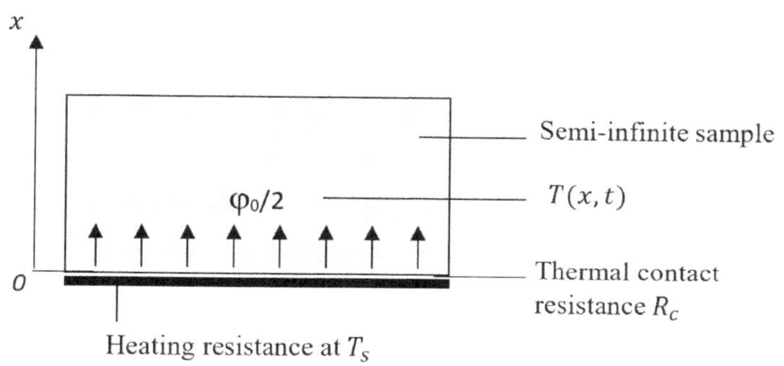

Schematization of a hot plane and notations

We deduce:

$$\theta_s = \theta + R_c ES\sqrt{p}\theta$$

$$\frac{\varphi_0}{2p} = mcp\theta + (1 + R_c mcp)ES\sqrt{p}\theta$$

hence: $\theta_s(p) = \frac{\varphi_0}{2p}\frac{1+R_c ES\sqrt{p}}{mcp+(1+R_c mcp)ES\sqrt{p}}$.

The unknown parameters to be determined experimentally are:

– the effusivity E of the sample;

– the thermal contact resistance R_c between the heating element and the sample;

– the thermal capacitance mc of the probe.

1) The temperature of the probe is written in Laplace space:

$$\theta_s(p) = \frac{\varphi_0}{2p}\frac{1 + R_c ES\sqrt{p}}{mcp + (1 + R_c mcp)ES\sqrt{p}}$$

At long times ($p \to 0$), this expression becomes:

$$\theta_s(p) = \frac{\varphi_0}{2p} \frac{1 + R_c ES\sqrt{p}}{mcp + (1 + R_c mcp)ES\sqrt{p}} = \frac{\varphi_0}{2ESp^{3/2}} \frac{1 + R_c ES\sqrt{p}}{1 + \frac{mc}{ES}\sqrt{p}}$$

where: $\theta_s(p) = \frac{\varphi_0}{2ESp^{3/2}}\left(1 + R_c ES\sqrt{p}\right)\left(1 - \frac{mc}{ES}\sqrt{p}\right)$.

Thus: $\theta_s(p) = \frac{\varphi_0}{2ESp^{3/2}}\left[1 + \left(R_c ES - \frac{mc}{ES}\right)\sqrt{p}\right] = \frac{\varphi_0}{2ESp^{3/2}} + \frac{\varphi_0}{2p}\left[R_c - \frac{mc}{(ES)^2}\right]$.

Hence: $T_s(t) - T_s(0) = \frac{\varphi_0}{2}\left[R_c - \frac{mc}{(ES)^2}\right] + \frac{\varphi_0}{ES\sqrt{\pi}}\sqrt{t}$.

The plot of $T_s(t) - T_s(0)$ as a function of $\sqrt{t}$ is therefore a straight line with a slope $\frac{\varphi_0}{ES\sqrt{\pi}}$ whose determination makes it possible to calculate the thermal effusivity E. The inertia of the probe and the contact resistance do not influence the temperature at long times. To apply this estimation method, we must be sure that the assumption of the semi-infinite medium remains valid over the estimation interval chosen.

Moreover, at short times ($p \to \infty$), the expression for $\theta_s(p)$ becomes:

$$\theta_s(p) = \frac{\varphi_0}{2p} \frac{1 + R_c ES\sqrt{p}}{mcp + (1 + R_c mcp)ES\sqrt{p}} = \frac{\varphi_0}{2p} \frac{1}{mcp}$$

hence: $T_s(t) - T_s(0) = \frac{\varphi_0}{2mc} t$.

2) We determine the slope α of the experimental curve $T = f(\sqrt{t})$ between 5 and 60 s, we find $\alpha = 0.65$.

We can deduce:

$$E = \frac{\varphi_0}{\alpha S\sqrt{\pi}} = \frac{U^2}{R\alpha S\sqrt{\pi}} = \frac{7^2}{40.5 \times 0.65 \times 2.03 \times 10^{-3}\sqrt{\pi}} = 517.3 \text{ W m}^{-2}\text{K}^{-1}\text{s}^{-1/2}$$

We determine the slope β of the experimental curve $T = f(t)$ between 0 and 0.5 s, and we find $\beta = 0.55$.

From this, we deduce: $mc = \frac{\varphi_0}{2\beta} = \frac{7^2}{240.5 \times 0.55} = 1.10 \text{ J K}^{-1}$.

We determine the ordinate at the origin δ of the experimental curve $T = f(\sqrt{t})$ between 5 and 60 s, and we find $\delta = -0.3$, where: $\delta = \frac{\varphi_0}{2}\left[R_c - \frac{mc}{(ES)^2}\right]$.

We deduce: $R_c = \frac{2\delta}{\varphi_0} + \frac{mc}{(ES)^2} = \frac{-0.6\times 40.5}{7^2} + \frac{1.10}{(517.3\times 2.03\times 10^{-3})^2} = 0.502\ \mathrm{K\,W^{-1}}$.

Or, for a unit area: $R_c = 0.004\ \mathrm{K\,m^2\,W^{-1}}$.

3) We are only looking for an order of magnitude and not a precise value of $T(e, 60\ s)$, and we will therefore place ourselves in the simple case where the mass of the probe and the thermal contact resistance are negligible. For a sample of thickness e perfectly thermally insulated on its unheated face, we then have the quadrupole relation:

$$\begin{bmatrix}\theta_s \\ \frac{\varphi_0}{2p}\end{bmatrix} = \begin{bmatrix}A & B \\ C & D\end{bmatrix}\begin{bmatrix}\theta(e,p) \\ 0\end{bmatrix} \text{ where: } \begin{bmatrix}A & B \\ C & D\end{bmatrix} = \begin{bmatrix}\cosh(qe) & \frac{1}{\lambda qS}\sinh(qe) \\ \lambda qS\sinh(qe) & \cosh(qe)\end{bmatrix}$$

with: $q = \sqrt{\frac{p}{a}}$.

From this, we deduce: $\theta(e,p) = \frac{\varphi_0}{2p}\frac{1}{\lambda qS\sinh(qe)}$.

At short times ($p \to \infty$), this expression becomes: $\theta(e,p) = \frac{\varphi_0}{p}\frac{1}{\lambda qS\exp(qe)}$.

Or: $\theta(e,p) = \frac{\varphi_0}{\lambda S}\frac{\exp(-qe)}{pq}$.

Hence, using section A.7 of the Appendix:

$$T(e,t) - T_s(0) = \frac{\varphi_0}{\lambda S}\left[2\sqrt{\frac{at}{\pi}}\exp\left(-\frac{e^2}{4at}\right) - erfc\left(\frac{e}{2\sqrt{at}}\right)\right].$$

$$T(e,t) - T_s(0) = \frac{7^2}{0.18\times 40.5\times 2.03\times 10^{-3}}\left[2\sqrt{\frac{1.2.10^{-7}\times 60}{\pi}}\exp\left(-\frac{0.01^2}{4\times 1.2\times 10^{-7}\times 60}\right) - 0.01\times erfc\left(\frac{0.01}{2\sqrt{1.2\times 10^{-7}\times 60}}\right)\right]$$

We obtain: $T(e,t) - T_s(0) = 0.017°\mathrm{C}$, which validates the hypothesis of the semi-infinite medium.

3.3.16. *Measurement of the thermal diffusivity of a thin plate*

The setup described in the figure below is carried out to measure the thermal diffusivity of a thin plate. A heat flux of any shape is applied to one end of the plate for a time t_c and the evolutions of the temperatures $T_0(t)$ and $T_1(t)$ at two points on the plate separated by d are recorded.

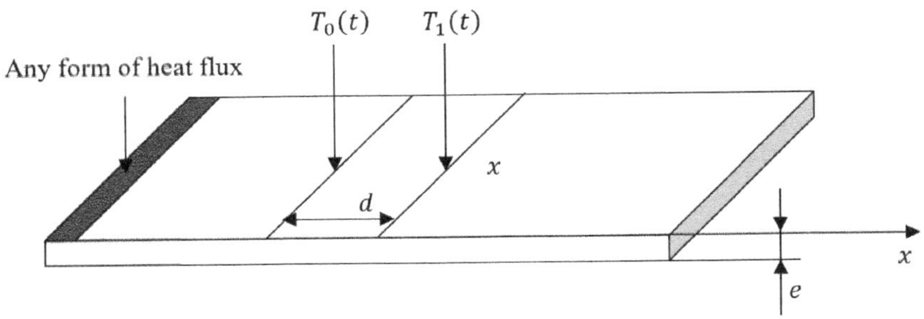

The following assumptions are made:

– the temperature distribution is unidirectional along x;

– the material is homogeneous and isotropic;

– the thermophysical properties are constant;

– at the initial moment, the material is at the same temperature T_a as the ambient air;

– the sample is long enough to be considered semi-infinite along x.

1) Write the differential equation verified by the temperature $T(x,t)$ in the transient regime.

Give the expression for the ratio of the Laplace transforms $\theta_0(p)$ and $\theta_1(p)$ of the temperatures $T_0(t) - T_a$ and $T_1(t) - T_a$.

2) Show that this ratio can be used to estimate the value of the thermal diffusivity of the plate. Find a function of this ratio that is linear with respect to p to simplify the estimation.

Corrected

1) A thermal balance is carried out on the piece of sample between x and $x + dx$:

$$-\lambda e\ell\left(\frac{\partial T}{\partial x}\right)_x = -\lambda e\ell\left(\frac{\partial T}{\partial x}\right)_{x+dx} + hp_e(T - T_a) + \rho c e\ell\, dx\left(\frac{\partial T}{\partial t}\right)$$

with: $P = 2\,(e + \ell) \approx 2\ell$, where ℓ is the width of the plate.

Hence: $\frac{\partial^2 T}{\partial x^2} - \frac{2h}{\lambda e}(T - T_a) = \frac{1}{a}\frac{\partial T}{\partial t}$.

2) After Laplace transform: $\frac{d^2\theta}{dx^2} - q^2\theta = 0$ with $q^2 = \frac{p}{a} + \frac{2h}{\lambda e}$.

From this, we deduce: $\theta(x,p) = Aexp(-qx)$.

Hence: $\frac{\theta(x_1,p)}{\theta(x_0,p)} = \frac{exp(-qx_1)}{exp(-qx_0)} = exp[-q(x_1 - x_0)] = exp(-qd)$.

And: $\left\{ln\left[\frac{\theta(x_1,p)}{\theta(x_0,p)}\right]\right\}^2 = -\frac{d^2}{a}p - \frac{2hd^2}{\lambda e}$.

It is therefore necessary to calculate the Laplace transforms of the measured temperatures $T_0(t) - T_a$ and $T_1(t) - T_a$ then to draw the graph of $\left\{ln\left[\frac{\theta(x_1,p)}{\theta(x_0,p)}\right]\right\}^2$ as a function of p, which is a straight line whose slope allows us to directly calculate the thermal diffusivity a.

3.3.17. *Regular regime method*

The regular regime method is a method for measuring the thermal diffusivity of a solid material. The principle of the method consists of making a parallelepipedal sample, inside which a thermocouple is inserted to measure the temperature $T(t)$ at a fixed point in the sample. At the initial time, the sample is at a uniform temperature T_i and it is suddenly plunged into a water bath that is stirred and thermostated at a temperature T_0. We note the evolution of the temperature $T(t)$ and use it "for long times" to estimate the thermal diffusivity of the sample.

1) Explain why we can assume that the surface temperature of the sample is imposed and equal to T_0. For what type of sample will this assumption be the most accurate?

2) Give an exact literal expression of the temperature $T(x,y,z,t)$ in the case of an anisotropic medium.

3) Give *without calculation* the shape of the curve $T(x,y,z,t)$ for a point close to the center of the sample.

4) Give a simplified approximate expression for long times and show that the curve $ln[T(x,y,z,t) - T_0] = f(t)$ is then a straight line whose slope does not depend on the geometric position of the measurement point but only on the thermal diffusivity a.

5) Propose a method for estimating the thermal diffusivity a from a recording of $T(t)$.

6) Apply this method to determine the thermal diffusivity a from the following experimental points obtained by immersing a 6 × 6 × 2 cm sample in a bath at 20°C. The temperature was measured at the coordinate point (1 cm, 1 cm, 0 cm) taking the center of the sample as the origin of the axes.

t	s	50	100	150	200	250	300
$T(t)$	°C	6.4	13.2	16.7	18.5	19.3	19.7

Corrected

1) The convective transfer coefficient is high because the fluid is water in forced convection. The resistance to external transfer is therefore weak compared to the resistance to internal transfer. This will be even more true if the material is insulating and if the temperature is measured far from the surface.

2) We can apply Von Neumann's theorem:

$$\left[\frac{T(x,y,z,t)-T_0}{T_i-T_0}\right]_{L_1,L_2,L_3} = \left[\frac{T(x,t)-T_0}{T_i-T_0}\right]_{PlateL_1} \times \left[\frac{T(y,t)-T_0}{T_i-T_0}\right]_{PlateL_2} \times \left[\frac{T(y,t)-T_0}{T_i-T_0}\right]_{PlateL_3}$$

For a thick plate $2L$, we obtained:

$$\bar{T}(x,t) = \frac{T(x,t)-T_0}{T_i-T_0} = \frac{4}{\pi}\sum_{n=0}^{\infty}\frac{1}{2n+1} sin\left[(2n+1)\frac{\pi}{2}\frac{x}{L}\right] exp\left[-(2n+1)^2\frac{\pi^2}{4}\frac{at}{L^2}\right]$$

and given that at long times, the first term is sufficient to obtain a good estimate, i.e.:

$$\bar{T}(x,t) = \frac{T(x,t)-T_0}{T_i-T_0} = \frac{4}{\pi} sin\left(\frac{\pi}{2}\frac{x}{L}\right) exp\left[-\frac{\pi^2}{4}\frac{at}{L^2}\right]$$

We deduce:

$$\left[\frac{T(x,y,z,t)-T_0}{T_i-T_0}\right]_{L_1,L_2,L_3} = \left(\frac{4}{\pi}\right)^3 sin\left(\frac{\pi}{2}\frac{x}{L_1}\right) sin\left(\frac{\pi}{2}\frac{y}{L_2}\right) sin\left(\frac{\pi}{2}\frac{z}{L_3}\right) exp\left(-\frac{\pi^2}{4}\frac{a_x t}{L_1^2}\right) exp\left(-\frac{\pi^2}{4}\frac{a_y t}{L_2^2}\right)$$

$$exp\left(-\frac{\pi^2}{4}\frac{a_z t}{L_3^2}\right) \left[\frac{T(x,y,z,t)-T_0}{T_i-T_0}\right]_{L_1,L_2,L_3} = \left(\frac{4}{\pi}\right)^3 sin\left(\frac{\pi}{2}\frac{x}{L_1}\right) sin\left(\frac{\pi}{2}\frac{y}{L_2}\right) sin\left(\frac{\pi}{2}\frac{z}{L_3}\right)$$

$$exp\left[-\frac{\pi^2 t}{4}\left(\frac{a_x}{L_1^2}+\frac{a_y}{L_2^2}+\frac{a_z}{L_3^2}\right)\right]$$

3) Shape of the curve $T(x,y,z,t)$ for a point close to the center of the sample:

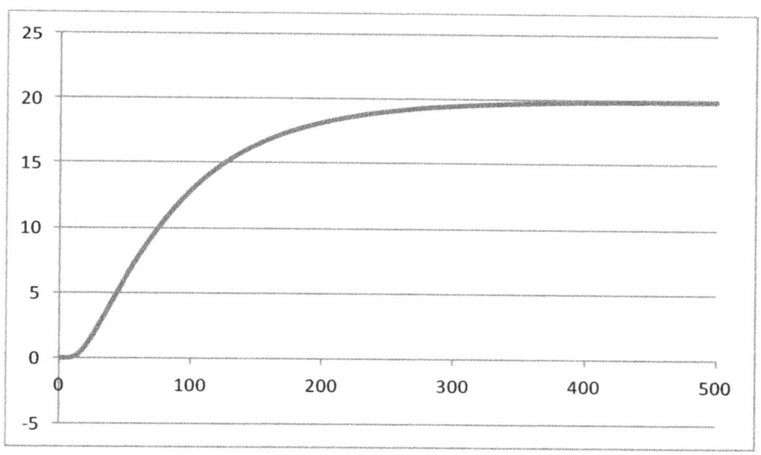

4) If the medium is isotropic: $a_x = a_y = a_z = a$, the solution is of the form:

$$\left[\frac{T(x,y,z,t)-T_0}{T_i-T_0}\right]_{L_1,L_2,L_3} = F(x,y,z) \times exp\left[-\frac{\pi^2 at}{4}\left(\frac{1}{L_1^2}+\frac{1}{L_2^2}+\frac{1}{L_3^2}\right)\right]$$

Or by taking the logarithm: $ln[\overline{T}(x,y,z,t)] = ln[F(x,y,z)] - \frac{\pi^2 at}{4}\left(\frac{1}{L_1^2}+\frac{1}{L_2^2}+\frac{1}{L_3^2}\right)$.

It is a straight line whose slope $-\frac{\pi^2 a}{4}\left(\frac{1}{L_1^2}+\frac{1}{L_2^2}+\frac{1}{L_3^2}\right)$ only depends on the thermal diffusivity a and does not depend on the position of the thermocouple in the sample.

5) Proposed method.

Plot $ln[T(x,y,z,t) - T_0] = f(t)$ and perform a linear regression on the linear portion of the curve to estimate its slope. Deduce from the slope α the value of a using: $a = \frac{\frac{4\alpha}{\pi^2}}{\frac{1}{L_1^2}+\frac{1}{L_2^2}+\frac{1}{L_3^2}}$.

6) Estimation of thermal diffusivity a.

A linear regression between 50 s and 300 s allows us to estimate a slope $\alpha = -0.0152$. A value of the thermal diffusivity is deduced therefrom $a = 5.04 \times 10^{-7} \mathrm{m^2s^{-1}}$.

3.3.18. *Hot wire modeling*

The purpose of the hot wire method is to measure the thermal conductivity λ of a material. The method consists of inserting a heating wire whose temperature evolution is measured between two very thick samples of the material to be characterized. A voltage step is sent to it, which causes the release of a heat flow rate φ_0. The heat produced diffuses radially in the samples, and the evolution of the temperature $T_s(t)$ of the heating wire over time is recorded.

Modeling this heat transfer allows us to calculate the evolution of the temperature $T_s(t)$ of the wire. A parameter estimation method is applied to calculate the values of:

- the thermal conductivity λ;
- the thermal capacitance mc of the heating wire;
- the contact resistance R_c at the wire/sample interface,

which minimize the difference between the theoretical and experimental curves $T_s(t)$.

1) Write the expression of the Laplace transform of this temperature using the quadrupole method. It will be considered that the temperature is uniform in the heating element.

2) Study, using limited expansions, the asymptotic behavior of this temperature. Deduce a simple method for determining thermal conductivity.

Corrected

1) The vicinity of the heating wire can be schematized as follows.

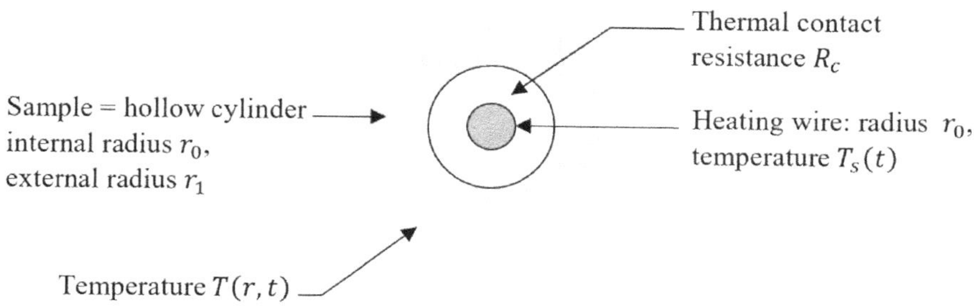

The heat equation is written in the sample:

$$\frac{\partial^2 T}{\partial r^2}+\frac{1}{r}\frac{\partial T}{\partial r}=\frac{1}{a}\frac{\partial T}{\partial t} \quad \text{(a)}$$

With the boundary conditions:

$$T(x.0)=T_i \quad \text{(b)}$$

$$T(\infty,t)=T_i \quad \text{(c)}$$

$$\frac{T_S(t)-T(0,t)}{R_C}=-\lambda\frac{\partial T}{\partial r}(0,t) \quad \text{(d)}$$

$$\varphi_0=mc\frac{dT_S}{dt}+\frac{T_S(t)-T(0,t)}{R_C} \quad \text{(e)}$$

Modeling the system using the formalism of the quadrupoles allows us to write in the Laplace space:

$$\begin{bmatrix}\theta_s\\ \dfrac{\varphi_0}{p}\end{bmatrix}=\begin{bmatrix}1 & 0\\ mcp & 1\end{bmatrix}\begin{bmatrix}1 & R_c\\ 0 & 1\end{bmatrix}\begin{bmatrix}\theta_0\\ 2\pi\lambda L\dfrac{qr_0K_1(qr_0)}{K_0(qr_0)}\theta_0\end{bmatrix}$$

$$\theta_s=\frac{\varphi_0}{p}\frac{A_0+B_0/Z}{C_0+D_0/Z}, \text{ where: } Z=\frac{K_0(qr_0)}{2\pi\lambda Lqr_0K_1(qr_0)}$$

where $A_0=1$; $B_0=R_c$; $C_0=mcp$; $D_0=R_cmcp+1$ and $mcp=\rho c\pi {r_0}^2L$

with:

– θ_s, the Laplace transform of the difference $T_s(t)-T_i$;

– θ_0, the Laplace transform of the difference $T(r_0,t)-T_i$;

– R_c, the contact resistance at the hotwire/sample interface;

– ρc, the volumetric heat capacity of the thermocouple + resistance;

– λ, the thermal conductivity of the sample;

– a, the thermal diffusivity of the sample;

– p, the Laplace variable;

– r_0, the radius of the heating wire;

– L, the length of the heating wire;

– φ_0, the power dissipated in the heating wire;

– K_0, K_1, Bessel functions.

2) If we consider a thin wire (r_0 small), and if we consider long times ($p \to 0$), we can use the limited expansions of Bessel's functions:

$$K_0(x) \approx -ln(x) + ln(2) - \gamma;\ K_1(x) \approx \frac{1}{x}$$

which leads to: $Z = -\frac{ln\left(\frac{qr_0}{2}\right)}{2\pi\lambda L} - \frac{\gamma}{2\pi\lambda L}$.

We deduce:

$$\theta_s(p) = \frac{\varphi_0}{p} \frac{Z + R_c}{mcp(Z + R_c) + 1} \approx \frac{\varphi_0}{p}\left[-\frac{ln(p)}{4\pi\lambda L} - \frac{ln\left(\frac{r_0}{2\sqrt{a}}\right)}{2\pi\lambda L} - \frac{\gamma}{2\pi\lambda L} + R_c\right]$$

The use of tables of the inverse Laplace transform (see section A.7 of the Appendix) enables calculating the temperature $T_s(t)$ at long times:

$$T_s(t) - T_i \approx \frac{\varphi_0}{4\pi\lambda L} ln(t) + R_c\varphi_0 - \frac{\gamma\varphi_0}{4\pi\lambda L} - \frac{\varphi_0 ln\left(\frac{r_0}{2\sqrt{a}}\right)}{2\pi\lambda L}$$

where $\gamma = 0.57721$ is the Euler constant.

The plot of $(T_s(t) - T_i)$ as a function of $ln(t)$ is therefore a straight line with slope $\frac{\varphi_0}{4\pi\lambda L}$, whose determination allows us to calculate the thermal conductivity λ. At long times, the inertia of the probe and the contact resistance have the only effect of shifting the temperature by a constant value with respect to the perfect probe model (probe with zero mass and no contact resistance). To apply this estimation method, the assumption of the semi-infinite medium has to remain valid over the chosen estimation interval. The measurement noises on the temperature values at the different measurement times being constant and uncorrelated, the linear least squares method is used to estimate the slope.

3.3.19. *Intermittent heating of a chalet*

A mountain chalet is not permanently occupied, and the heating is turned off during non-occupancy periods. The front walls of the chalet are made of stone with a thickness $e_2 = 20$ cm. The owners want to insulate them with a thickness $e_1 = 10$ cm of polystyrene and wonder if it is better to put the insulation outside (1) or inside (2) of the walls. The global internal and external transfer coefficients are h_i =10 W m^{-2} K^{-1} and h_e =15 W m^{-2} K^{-1}. The ceiling is assumed to be perfectly insulated and without thermal inertia. The floor is made up of a thick ($e_3 = 6$ cm) concrete screed assumed to be isothermal and perfectly insulated on its underside (with a layer of 10 cm of polyurethane). The height of the rooms is 2.5 m, and the floor area of the chalet is $S_3 = 8 \times 4\ \text{m}^2$.

Data: polystyrene: $\rho_1 c_1 = 6 \times 10^4\ \text{J m}^{-3}\text{K}^{-1}$; $\lambda_1 = 0.035\ \text{W m}^{-1}\text{K}^{-1}$;

stone: $\rho_2 c_2 = 2 \times 10^6\ \text{J m}^{-3}\text{K}^{-1}$; $\lambda_2 = 2.5\ \text{W m}^{-1}\text{K}^{-1}$;

concrete: $\rho_3 c_3 = 2 \times 10^6\ \text{J m}^{-3}\text{K}^{-1}$.

Steady-state study.

The temperature inside the chalet is assumed to be constant and equal to $T_i = 20$°C, and the outside temperature is assumed to be constant $T_e = 0$°C.

The following questions will be answered for *each* of the two configurations (1) and (2).

1) Calculate the four thermal resistances involved in the system.

2) Give the equivalent electrical diagram of the system.

3) Calculate the heating power required to maintain $T_i = 20$°C.

4) Calculate and represent the temperature profile obtained.

5) How much does insulation save?

Study of the transient regime.

It is assumed that on arrival in the chalet, the temperature of the chalet is uniform and equal to the assumed constant outside temperature $T_e = 0$°C, the constant power heating system $P_0 = 2{,}000$ W is switched on.

6) Show that we can in first approximation consider the thermal insulation as a pure resistance ($\rho c_i = 0$) and the stone wall as a pure thermal capacity (λ infinity).

Questions 7 to 9 will be answered for *each* of the two configurations (1) and (2).

7) Using the electric analogy (quadrupoles), establish in Laplace space the expression of the interior temperature of the chalet and deduce $T_i(t)$.

8) Calculate the time taken to reach $T_i(t) = 20°\text{C}$ and deduce the amount of energy expended.

9) Calculate the temperature T_{pi} of the inner wall (surface of the stone for case (1) and surface of the insulation for case (2)).

10) Conclude on the type of insulation (interior or exterior) to be chosen in this specific case of occupancy of the chalet. Would we choose the same configuration by considering the thermal comfort of an occupant as a criterion? Justify your answer.

Data: inverse Laplace transform (see section A.7 of the Appendix).

$$L^{-1}\left[\frac{1}{(p+\alpha)(p+\beta)}\right] = \frac{exp(-\beta t) - exp(-\alpha t)}{\alpha - \beta}$$

$$L^{-1}\left[\frac{1}{(p+\alpha)(p+\beta)p}\right] = \frac{1}{\alpha\beta}\left[1 + \frac{\beta exp(-\alpha t) - \alpha exp(-\beta t)}{\alpha - \beta}\right]$$

Corrected

1) The total surface of the walls is: $S = 2 \times (8 + 4) \times 2.5 = 60 \text{ m}^2$.

$$R_{ci} = \frac{1}{h_i S} = \frac{1}{10 \times 60} = 1.7 \times 10^{-3} \text{ K W}^{-1}$$

$$R_1 = \frac{e_1}{\lambda_1 S} = \frac{0.1}{0.035 \times 60} = 4.76 \times 10^{-2} \text{ K W}^{-1}$$

$$R_2 = \frac{e_2}{\lambda_2 S} = \frac{0.2}{2.5 \times 60} = 1.33 \times 10^{-3} \text{ K W}^{-1}$$

$$R_{ce} = \frac{1}{h_{es}} = \frac{1}{15 \times 60} = 1.11 \times 10^{-3} \text{ K W}^{-1}$$

2) Case (1):

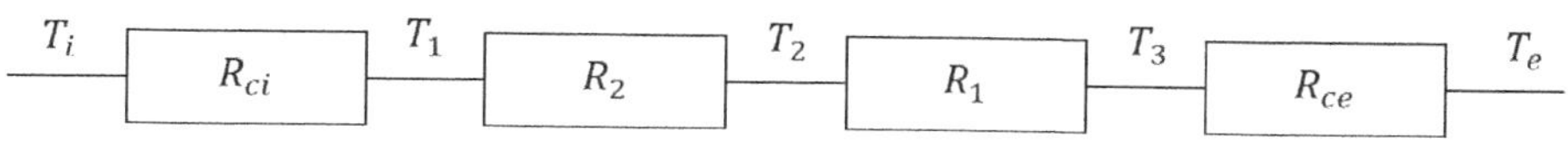

Case (2):

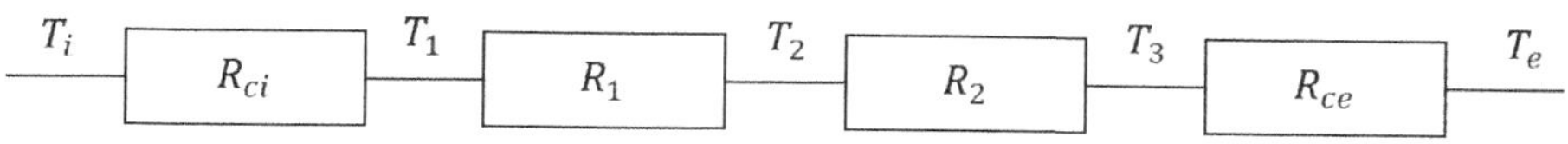

3) The flow rate which crosses the system is written in the two cases:

$$\varphi = \frac{T_i - T_e}{R_{ci} + R_1 + R_2 + R_{ce}}$$
$$= \frac{20}{1.67 \times 10^{-3} + 4.76 \times 10^{-2} + 1.33 \times 10^{-3} + 1.11 \times 10^{-3}} = 386.6 \text{ W}$$

4)

– Case (1):

$$T_1 = T_i - \varphi R_{ci} = 20 - 386.6 \times 1.67 \times 10^{-3} = 19.36°\text{C}$$

$$T_2 = T_i - \varphi(R_{ci} + R_2) = 20 - 386.6 \times (1.67 \times 10^{-3} + 1.33 \times 10^{-3}) = 18.84°\text{C}$$

$$T_3 = T_e + \varphi R_{ce} = 0 + 386.6 \times 1.11 \times 10^{-3} = 0.43°\text{C}$$

– Case (2):

$$T_1 = T_i - \varphi R_{ci} = 20 - 20 - 386.6 \times 1.67 \times 10^{-3} = 19.36°\text{C}$$

$$T_2 = T_i - \varphi(R_{ci} + R_1) = 20 - 386.6 \times (1.67 \times 10^{-3} + 4.76 \times 10^{-2}) = 0.95°\text{C}$$

$$T_3 = T_e + \varphi R_{ce} = 0 + 386.6 \times 1.11 \times 10^{-3} = 0.43°\text{C}$$

5) In the absence of insulation:

$$\varphi = \frac{T_i - T_e}{R_{ci} + R_2 + R_{ce}} = \frac{20}{1.67 \times 10^{-3} + 1.33 \times 10^{-3} + 1.11 \times 10^{-3}} = 4{,}865 \text{ W}$$

The insulation allows us to divide the losses and therefore the energy consumption by 12.6 when the steady state is reached regardless of its position in the wall.

6) $\frac{R_1}{R_2} = \frac{2.86}{0.08} = 35.8 \gg 1$, so we can consider as a first approximation $R_2 = 0$.

$\frac{\rho_2 c_2 e_2}{\rho_1 c_1 e_1} = \frac{2.10^6 \times 0.2}{6.10^4 \times 0.1} = 66.7 \gg 1$, so we can consider as a first approximation $\rho_1 c_1 e_1 = 0$.

7) The electrical analogy, with the previous hypotheses, gives:

– case (1):

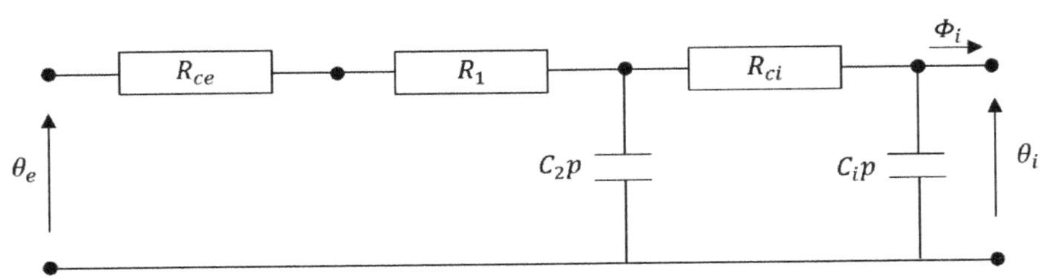

– case (2):

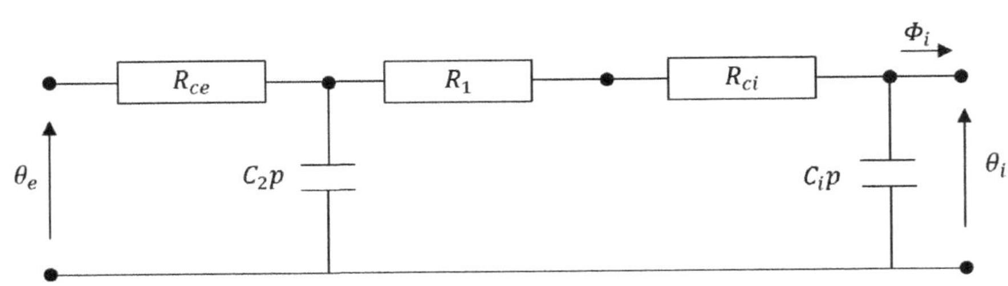

with: $\Phi_i = L(P_0)$, $\theta_i = L[T_i - T(t=0)]$, $\theta_e = L[T_e - T(t=0)]$, $C_2 = \rho_2 c_2 e_2 S$,$C_i = \rho_3 c_3 e_3 S_3$.

Now: $T(t=0) = T_e$, the schematization becomes:

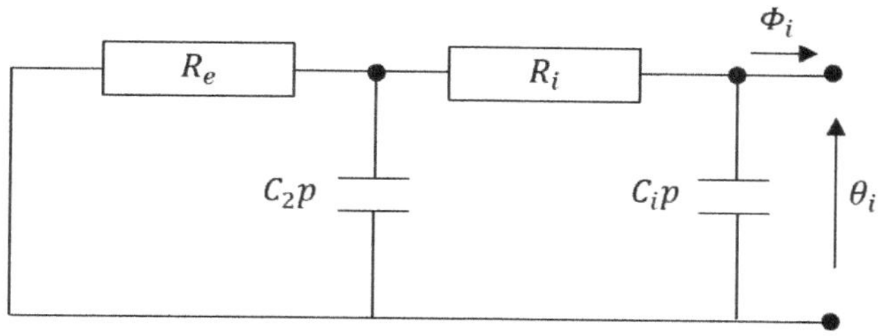

with:

– for case (1): $R_e = R_{ce} + R_1$ and $R_i = R_{ci}$;

– for case (2): $R_e = R_{ce}$ and $R_i = R_{ci} + R_1$.

We can write the following quadrupole equation:

$$\begin{bmatrix} 0 \\ \Phi_e \end{bmatrix} = \begin{bmatrix} 1 & R_e \\ 0 & 1 \end{bmatrix} \begin{bmatrix} 1 & 0 \\ C_2p & 1 \end{bmatrix} \begin{bmatrix} 1 & R_i \\ 0 & 1 \end{bmatrix} \begin{bmatrix} 1 & 0 \\ C_ip & 1 \end{bmatrix} \begin{bmatrix} \theta_i \\ \Phi_i \end{bmatrix}$$
$$= \begin{bmatrix} 1 + R_eC_2p & R_e \\ C_2p & 1 \end{bmatrix} \begin{bmatrix} 1 + R_iC_ip & R_i \\ C_ip & 1 \end{bmatrix} \begin{bmatrix} \theta_i \\ \Phi_i \end{bmatrix}$$

$$\begin{bmatrix} 0 \\ \Phi_e \end{bmatrix} = \begin{bmatrix} A & B \\ C & D \end{bmatrix} \begin{bmatrix} \theta_i \\ \Phi_i \end{bmatrix}$$

with:

$$A = (1 + R_eC_2p)(1 + R_iC_ip) + R_eC_ip; B = R_i(1 + R_eC_2p) + R_e$$

and: $\Phi_i = -\frac{P_0}{p}$.

We deduce: $\theta_i = -\frac{B}{A}\Phi_i = \frac{R_i(1+R_eC_2p)+R_e}{(1+R_eC_2p)(1+R_iC_ip)+R_eC_ip}\frac{P_0}{p}$.

We can also find this result using the impedance corresponding to the equivalent circuit:

$$Z = \cfrac{1}{C_ip + \cfrac{1}{R_i + \cfrac{1}{C_2p + \cfrac{1}{R_e}}}}$$

$$\theta_i = \frac{R_i + R_e}{1 + (R_iC_i + R_eC_2 + R_eC_i)p + R_iR_eC_2C_ip^2}\frac{P_0}{p} + \frac{R_iR_eC_2p}{1 + (R_iC_i + R_eC_2 + R_eC_i)p + R_iR_eC_2C_ip^2}\frac{P_0}{p}$$

$$\theta_i = \frac{(R_i + R_e)P_0}{R_iR_eC_2C_i}\frac{1}{p^3 + p^2\frac{R_iC_i + R_eC_2 + R_eC_i}{R_iR_eC_2C_i} + p\frac{1}{R_iR_eC_2C_i}} + \frac{R_iR_eC_2P_0}{R_iR_eC_2C_i}\frac{p}{p^3 + p^2\frac{R_iC_i + R_eC_2 + R_eC_i}{R_iR_eC_2C_i} + p\frac{1}{R_iR_eC_2C_i}}$$

of the form:

$$\theta_i = \left[K_1\frac{1}{(p+\alpha)(p+\beta)(p+\gamma)} + K_2\frac{p}{(p+\alpha)(p+\beta)(p+\gamma)}\right]P_0$$

with:

– $K_1 = \frac{(R_i+R_e)}{R_iR_eC_2C_i}$;

– $K_2 = \frac{1}{C_i}$

where: $(p+\alpha)(p+\beta)(p+\gamma) = p^3 + (\alpha+\beta+\gamma)p^2 + (\alpha\beta+\alpha\gamma+\beta\gamma)p + \alpha\beta\gamma$.

By identification:

$$\alpha + \beta = \frac{R_iC_i + R_eC_2 + R_eC_i}{R_iR_eC_2C_i} = K_3$$

$$\alpha\beta = \frac{1}{R_iR_eC_2C_i} = K_4$$

We deduce:

$$\alpha = \frac{K_3 + \sqrt{K_3 - 4K_4}}{2}$$

$$\beta = \alpha K_4$$

$$\gamma = 0$$

hence:

$$\theta_i = \left[K_1 \frac{1}{p(p+\alpha)(p+\beta)} + K_2 \frac{1}{(p+\alpha)(p+\beta)}\right] P_0$$

Knowing that:

$$L^{-1}\left[\frac{1}{(p+\alpha)(p+\beta)}\right] = \frac{exp(-\beta t) - exp(-\alpha t)}{\alpha - \beta}$$

$$L^{-1}\left[\frac{1}{(p+\alpha)(p+\beta)p}\right] = \frac{1}{\alpha\beta}\left[1 + \frac{\beta exp(-\alpha t) - \alpha exp(-\beta t)}{\alpha - \beta}\right]$$

We finally get:

$$T_i - T_e = \left[K_1 \frac{\alpha - \beta + \beta exp(-\alpha t) - \alpha exp(-\beta t)}{\alpha\beta(\alpha - \beta)} + K_2 \frac{exp(-\beta t) - exp(-\alpha t)}{\alpha - \beta}\right] P_0$$

8) $C_2 = \rho_2 c_2 e_2 S = 2 \times 10^6 \times 0.2 \times 60 = 2.4 \times 10^7$ J K^{-1}.

$C_i = \rho_3 c_3 e_3 S_3 = 2 \times 10^6 \times 0.06 \times 8 \times 4 = 3.84 \times 10^6$ J K^{-1}

– For case (1):

$R_e = R_{ce} + R_1 = 1.11 \times 10^{-3} + 4.76 \times 10^{-2} = 4.87 \times 10^{-2}$ K W^{-1}

$R_i = R_{ci} = 1.67 \times 10^{-3}$ K W^{-1}

– For case (2):

$R_e = R_{ce} = 1.11 \times 10^{-3}$ K W^{-1}

$R_i = R_{ci} + R_1 = 1.67 \times 10^{-3} + 4.76 \times 10^{-2} = 4.93 \times 10^{-2}$ K W^{-1}

The curves in Figure (4) show the evolution of $T_i(t)$ over time; we obtain 20°C for:

– $t = 73.9$ h for case (1);

– $t = 11.9$ h for case (2),

which results in an energy expenditure of:

– $Q = 73.9 \times 2 = 147.8$ kWh for case (1);

– $Q = 11.9 \times 2 = 23.8$ kWh for case (2).

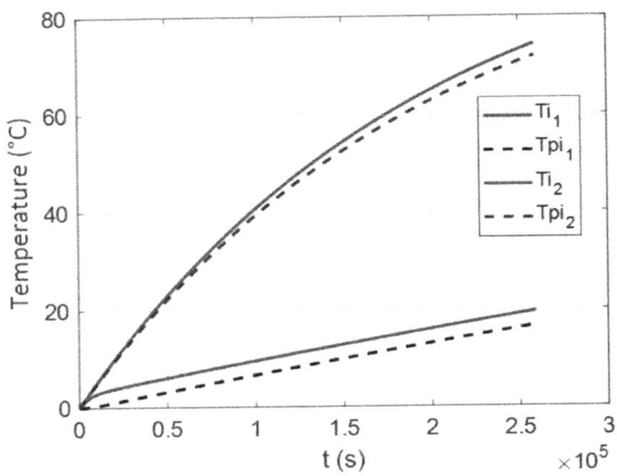

An even more approximate calculation can be obtained: if we consider R_1 infinite (no losses towards the outside) and neglect R_{ci}.

– Case (1): $\varphi = (C_i + C_2)\frac{dT_i}{dt}$ hence: $T_i = \frac{\varphi}{C_i + C_2}t$.

Time taken to reach 20°C: $t = \frac{C_i + C_2}{\varphi}T_i = \frac{2.4\times10^7 + 3.84\times10^6}{2000} \times 20 = 77.3$ h.

– Case (2): $\varphi = C_i\frac{dT_i}{dt}$ hence: $T_i = \frac{\varphi}{C_i}t$.

Time taken to reach 20°C: $t = \frac{C_i}{\varphi}T_i = \frac{3.84\times10^6}{2000} \times 20 = 10.7$ h.

This gives an order of magnitude to within 10%.

In the Laplace domain, we can write the quadrupole relationship between the interior face of the wall and the air inside the room:

$$\begin{bmatrix}\theta_{pi}\\ \Phi_{pi}\end{bmatrix} = \begin{bmatrix}1 & R_{ci}\\ 0 & 1\end{bmatrix}\begin{bmatrix}1 & 0\\ C_i p & 1\end{bmatrix}\begin{bmatrix}\theta_i\\ \Phi_i\end{bmatrix}$$

hence: $\theta_{pi} - \theta_i = R_{ci}(\Phi_i + C_i p\theta_i)$.

thus: $\theta_{pi} = -R_{ci}\frac{P_0}{p} + \theta_i(1 + R_{ci}C_i p)$.

$$T_{pi}(t) = -R_{ci}P_0 + T_i(t) + R_{ci}C_i L^{-1}(p\theta_i)$$

However, according to the properties of the Laplace transform (see section A.6 of the Appendix):

$$L\left[\frac{dT}{dt}(t)\right] = p\theta(p) - T(t=0)$$

Here: $T(t=0) = 0$ (otherwise, we would have made a change of variable by subtracting its initial value T_i from the temperature); therefore:

$$T_{pi}(t) = -R_{ci}P_0 + T_i(t) + R_{ci}C_i\frac{dT_i}{dt}$$

with: $T_i - T_e = \left[K_1\frac{\alpha-\beta+\beta exp(-\alpha t)-\alpha exp(-\beta t)}{\alpha\beta(\alpha-\beta)} + K_2\frac{exp(-\beta t)-exp(-\alpha t)}{\alpha-\beta}\right]P_0$.

and: $\frac{dT_i}{dt} = \left[K_1\frac{\alpha\beta exp(-\beta t)-\alpha\beta exp(-\alpha t)}{\alpha\beta(\alpha-\beta)} + K_2\frac{\alpha exp(-\alpha t)-\beta exp(-\beta t)}{\alpha-\beta}\right]P_0$.

thus: $\frac{dT_i}{dt} = \left[K_1\frac{exp(-\beta t)-exp(-\alpha t)}{\alpha-\beta} + K_2\frac{\alpha exp(-\alpha t)-\beta exp(-\beta t)}{\alpha-\beta}\right]P_0$.

The results are shown in the previous figure.

In case (1), the difference between the air temperature of the chalet and the wall is around 3°C (for an air temperature of 20°C); in case (2), the difference is only 0.7°C.

10. Interior insulation is much more economical in this type of situation. In addition, the wall temperature with external insulation is significantly lower than T_i for a long time. Heat exchanges by radiation between an occupant and the wall will therefore produce a feeling of discomfort.

3.3.20. *Heat loss through the floor of a house*

The floor of a house is made of a 20 cm thick concrete slab placed directly on the ground. After a long period of vacancy, the heating is restored to power $\varphi_0 = 1{,}500$ W in a room with a surface $S = 30$ m^2. It is assumed that all the other walls (walls and ceiling) are perfectly insulated, and that when the heating is turned on, the temperature in the room is 12°C. The ground temperature is assumed to be uniform and equal to $T_i = 12$°C. It is assumed that air and soil exchange with a transfer coefficient $h = 8\ \mathrm{W\,m^{-2}K^{-1}}$. The concrete and the ground are assumed to have similar thermal properties taken equal to: $\lambda = 1.8\ \mathrm{W\,m^{-1}K^{-1}}$ and $a = 10^{-6}\mathrm{m^2s^{-1}}$.

We first assume that the heat transfer from the slab to the ground is purely 1D.

1) Calculate the time t_c after which the air temperature T_a in the room will be equal to 22°C. Then, calculate the value of the temperature T_s at the surface of the slab and the heating energy expended at this instant.

2) What would be the heat flow rate lost by the slab 48 hours after the start of heating? We can consider for this calculation that the air temperature has been kept constant and equal to 22°C since the start of heating (the preheating time t_c is neglected).

3) Repeat the same calculation after a week. What can be said about losses through the ground?

In fact, the previous calculations are only acceptable for short times because in reality:

– the heat transfer in the ground is 3D (flow φ_2 on the diagram below);

– the ground surface outside is at a lower temperature than the inside temperature (flow φ_1 on the diagram below).

4) Calculate the total flux lost by the surface of the slab after 48 h and in the steady state. For this, the constriction impedances given in section A.13 of the Appendix will be used, and for this calculation, it will be considered that the temperature of the interior air has been kept constant and equal to 22°C since the beginning of the heating.

Data: inverse Laplace transform (see section A.7 of the Appendix).

$$L^{-1}\left[\frac{\sqrt{p}+c}{p\left(\sqrt{p}+d\right)}\right] = L^{-1}\left[\frac{1}{\sqrt{p}\left(\sqrt{p}+d\right)}\right] + L^{-1}\left[\frac{c}{p\left(\sqrt{p}+d\right)}\right]$$
$$= \frac{c}{d} + \frac{d-c}{d}\, erfc\left(d\sqrt{t}\right) exp\left(d^2 t\right)$$

5) Calculate the losses towards the outside in the steady state if the wall delimiting the room is insulated with 10 cm of polystyrene. We will use the following formula (established in cylindrical coordinates) to calculate the thermal resistance between the surface of the interior ground at the temperature T_s and the surface of the exterior ground at the temperature T_i (Laraqi and Monier-Vinard 2013):

$$R_c = \frac{\sqrt{k}}{2\,\lambda r_1 ln\left(\frac{1+\sqrt{k}}{1-\sqrt{k}}\right)} \text{ with } k = \frac{r_0}{r_1}$$

corresponding to the following configuration shown in the figure below.

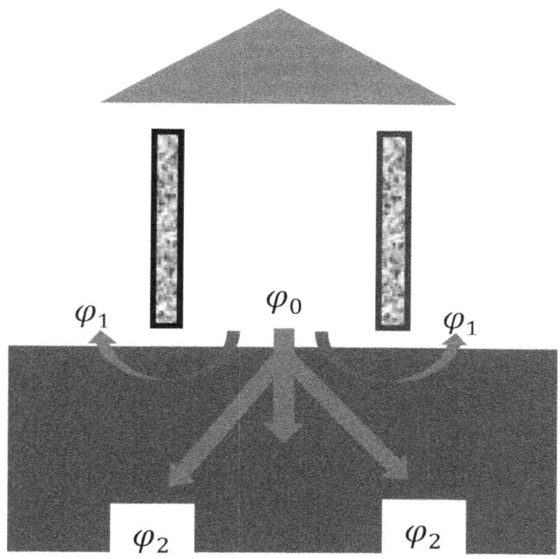

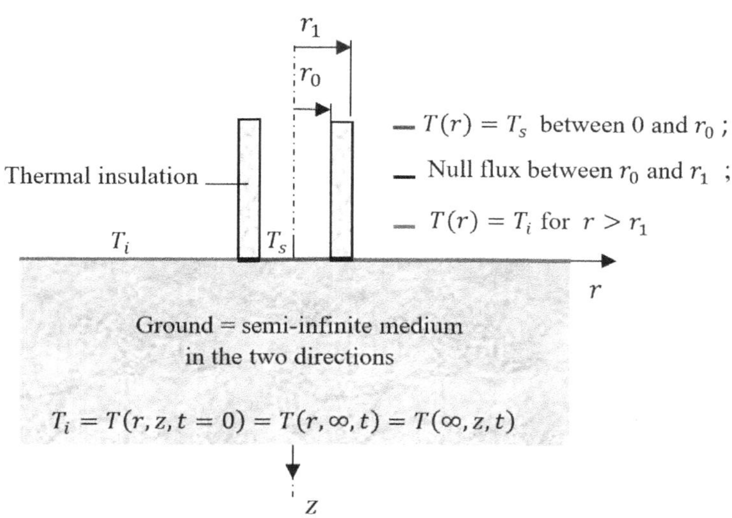

Corrected

1) The heat flux produced is entirely transmitted to the semi-infinite medium slab + floor through the convection resistance $R = \frac{1}{h}$. Using the quadrupole formalism, we get:

$$\begin{bmatrix}\theta_a \\ \Phi_a\end{bmatrix} = \begin{bmatrix}1 & R \\ 0 & 1\end{bmatrix} \times \begin{bmatrix}\theta_s \\ \dfrac{\theta_s}{Z}\end{bmatrix} = \begin{bmatrix}\theta_s\left(1+\dfrac{R}{Z}\right) \\ \dfrac{\theta_s}{Z}\end{bmatrix}$$

hence:

$$\theta_a = \theta_s\left(1 + \frac{R}{Z}\right) \qquad [3.46]$$

and:

$$\Phi_a = \frac{\theta_s}{Z} = \frac{\theta_a}{R+Z} \qquad [3.47]$$

with:

$\theta_a = L(T_a - T_i)$, where T_a is the room air temperature;

$\Phi_a = \frac{\varphi_0}{p}$; $R = \frac{1}{hS}$; $Z = \frac{1}{ES\sqrt{p}}$;

$\theta_a = \theta_s\left(1 + \frac{E}{h}\sqrt{p}\right)$;

$\frac{\varphi_0}{p} = ES\sqrt{p}\theta_s$;

$\theta_a = \frac{\varphi_0}{ES\sqrt{p}p}\left(1 + \frac{E}{h}\sqrt{p}\right) = \frac{\varphi_0}{ES\sqrt{p}p} + \frac{\varphi_0}{hSp}$;

$T_a - T_i = \frac{\varphi_0}{hS} + 2\frac{\varphi_0}{\sqrt{\pi}ES}\sqrt{t}$;

$E = \frac{\lambda}{\sqrt{a}} = \frac{1.8}{\sqrt{10^{-6}}} = 1{,}800\ \mathrm{W\,m^{-2}K^{-1}s^{1/2}}$;

$$t = \left(\frac{T_a - T_i - \dfrac{\varphi_0}{hS}}{2\dfrac{\varphi_0}{\sqrt{\pi}ES}}\right)^2 = \left(\frac{10 - \dfrac{1500}{8 \times 30}}{2 \times \dfrac{1500}{\sqrt{\pi} \times 30 \times 1800}}\right)^2 = 14{,}314\ \mathrm{s}.$$

The ground temperature is given by: $\varphi_0 = hS(T_a - T_s)$, hence: $T_s = T_a - \frac{\varphi_0}{hS} =$ 15.75°C.

The energy expended has the following value: $Q = \varphi_0 t = \frac{1{,}500}{1{,}000} \times \frac{14{,}314}{3{,}600} =$ 6 kWh.

The problem can also be solved by using the following equivalent electrical diagram:

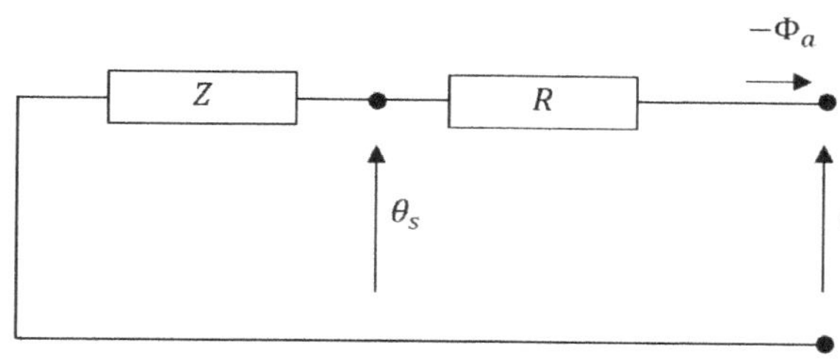

where R is the convection resistance and Z is the impedance of the semi-infinite medium. We can deduce:

$$\theta_a = (R + Z)\Phi_a$$

$$\theta_s = Z\Phi_a$$

with: $\Phi_a = \frac{\varphi_0}{p}$; $Z = \frac{1}{ES\sqrt{p}}$; $R = \frac{1}{hS}$ and we find relations [3.46] and [3.47].

2) The quadrupole equation is written as:

$$\begin{bmatrix} \theta_a \\ \Phi_a \end{bmatrix} = \begin{bmatrix} 1 & R \\ 0 & 1 \end{bmatrix} \times \begin{bmatrix} \theta_s \\ \frac{\theta_s}{Z} \end{bmatrix} = \begin{bmatrix} \theta_s \left(1 + \frac{R}{Z}\right) \\ \frac{\theta_s}{Z} \end{bmatrix}$$

Equations [3.46] and [3.47] remain valid but with: $\theta_a = \frac{T_a - T_i}{p}$, where T_a is the air temperature of the room:

$$R = \frac{1}{hS};\ Z = \frac{1}{ES\sqrt{p}}$$

From this, we deduce: $\Phi_a = \frac{\theta_s}{Z} = ES\sqrt{p}\theta_s = \frac{ES\sqrt{p}}{1+\frac{E}{h}\sqrt{p}} \frac{T_a - T_i}{p}$.

However, according to section A.7 of the Appendix: $L^{-1}\left[\frac{b}{p(b+\sqrt{p})}\right] = 1 - exp(b^2t)erfc(b\sqrt{t})$.

Hence: $T_s(t) = T_a - (T_a - T_i)exp\left(\frac{ah^2t}{\lambda^2}\right) erfc\left(\frac{h\sqrt{at}}{\lambda}\right)$

$$t = 48 \times 3{,}600 = 172{,}800 \text{ s}$$

$$T_s(t) = 22 - (22 - 12)\, exp\left(\frac{10^{-6} \times 8^2 \times 172{,}800}{1.8^2}\right) erfc\left(\frac{8 \times \sqrt{10^{-6} \times 172{,}800}}{1.8}\right) = 19.3°\text{C}$$

The lost heat flow rate is:

$$\varphi_0 = hS(T_a - T_s) = 8 \times 30 \times (22 - 19.3) = 654 \text{ W}$$

We could have used the same equivalent electrical diagram as before, but now with:

$$\theta_a = (R + Z)\Phi_a$$

$$\theta_s = Z\Phi_a$$

and $\theta_a = \frac{T_a - T_i}{p}$; $Z = \frac{1}{ES\sqrt{p}}$; $R = \frac{1}{hS}$ and we obtain the same result.

3) After one week:

$$t = 7 \times 24 \times 3{,}600 = 604{,}800 \text{ s}$$

$$T_s(t) = 22 - (22 - 12)\, exp\left(\frac{10^{-6} \times 8^2 \times 604{,}800}{1.8^2}\right) erfc\left(\frac{8 \times \sqrt{10^{-6} \times 604{,}800}}{1.8}\right) = 20.4°\text{C}$$

The lost heat flow rate is now only: $\varphi_0 = h\,S(T_a - T_s) = 8 \times 30 \times (22 - 20.4) = 377$ W.

With the assumption of a 1D transfer, we find that the heat flow rate φ_0 decreases over time and tends to zero.

4) The calculation is performed by treating the ground surface as a disk and assuming a uniform ground temperature.

The problem is identical to the previous case by taking $R = 0$ and replacing the impedance of the semi-infinite medium Z with the constriction impedance Z_c given in section A.13 of the Appendix by the approximation:

$$Z_c = \frac{1}{\frac{4}{\sqrt{\pi}}\lambda\sqrt{S} + ES\sqrt{p}}$$

Equations [3.46] and [3.47] remain valid by replacing Z with Z_c, therefore:

$$\Phi_a = \frac{\theta_s}{Z_c} = \frac{\theta_a}{p}\frac{1}{(Z_c + R)} = \frac{\theta_a}{p}\frac{1}{\frac{1}{\frac{4}{\sqrt{\pi}}\lambda\sqrt{S} + ES\sqrt{p}} + \frac{1}{hS}} = \frac{\theta_a}{p}\frac{hS\left(\frac{4}{\sqrt{\pi}}\lambda\sqrt{S} + ES\sqrt{p}\right)}{hS + \frac{4}{\sqrt{\pi}}\lambda\sqrt{S} + ES\sqrt{p}}$$

$$= \frac{\theta_a}{p}\frac{hS\left(\frac{4}{ES\sqrt{\pi}}\lambda\sqrt{S} + \sqrt{p}\right)}{\frac{h}{E} + \frac{4}{ES\sqrt{\pi}}\lambda\sqrt{S} + \sqrt{p}} = \frac{hS\theta_a}{p}\frac{\frac{4}{ES\sqrt{\pi}}\lambda\sqrt{S} + \sqrt{p}}{\frac{h}{E} + \frac{4}{ES\sqrt{\pi}}\lambda\sqrt{S} + \sqrt{p}}$$

$$\Phi_a = hS\theta_a\frac{c + \sqrt{p}}{p\left(d + \sqrt{p}\right)}$$

with: $c = \frac{4}{ES\sqrt{\pi}}\lambda\sqrt{S}$ and: $d = \frac{h}{E} + \frac{4}{ES\sqrt{\pi}}\lambda\sqrt{S}$.

$$\varphi_a = hS(T_a - T_i)\left[\frac{c}{d} + \frac{d - c}{d}erfc\left(d\sqrt{t}\right)exp(d^2 t)\right]$$

After 48 hours:

$$c = \frac{4\lambda}{E\sqrt{S}\sqrt{\pi}}\lambda = \frac{4 \times 1.8}{1{,}800 \times \sqrt{30}\sqrt{\pi}} = 4.12 \times 10^{-3}$$

$$d = \frac{h}{E} + \frac{4\lambda}{E\sqrt{S}\sqrt{\pi}} = \frac{h}{E} + c = \frac{8}{1{,}800} + 4.12 \times 10^{-3} = 4.86 \times 10^{-3}$$

$$\varphi_a = 8 \times 30 \times 10$$

$$\left[\frac{4.12}{4.86} + \frac{4.86 - 4.12}{4.86}erfc\left(4.86 \times 10^{-3}\sqrt{48 \times 3{,}600}\right)exp(4.86^2 \times 10^{-6} \times 48 \times 3{,}600)\right]$$

$\varphi_a = 760$ W instead of 654 W if we neglect 3D transfers.

In the steady state:

$$\Phi_a = \frac{\theta_s}{Z_c} = \frac{\theta_a}{p}\frac{1}{(Z_c + R)} = \frac{\theta_a}{p}\frac{1}{\frac{1}{\frac{4}{\sqrt{\pi}}\lambda\sqrt{S}} + \frac{1}{hS}}$$

hence: $\varphi = \frac{T_a - T_i}{\frac{1}{\frac{4}{\sqrt{\pi}}\lambda\sqrt{S}} + \frac{1}{hS}} = \frac{10}{\frac{1}{4\times 1.8\times\sqrt{\frac{30}{\pi}}} + \frac{1}{8\times 30}} = 204$ W.

5) In the steady state, assuming the indoor T_s floor temperature is uniform, Laraqi's formula (established in cylindrical coordinates) can be used to calculate the thermal resistance between the indoor floor surface and the floor surface exterior:

$$R_c = \frac{\sqrt{k}}{2\,\lambda r_1 ln\left(\frac{1+\sqrt{k}}{1-\sqrt{k}}\right)} \text{ with: } k = \frac{r_0}{r_1}$$

We can consider that the vertical flow is zero at the level of the thermal insulation and therefore that:

$$r_0 = \sqrt{\frac{S}{\pi}} \text{ and that } r_1 = r_0 + e_i$$

where e_i is the thickness of the insulation.

So we have:

$$\varphi = \frac{T_a - T_i}{R_c + \frac{1}{hS}} = \frac{T_a - T_i}{\frac{\sqrt{k}}{2\,\lambda r_1 ln\left(\frac{1+\sqrt{k}}{1-\sqrt{k}}\right)} + \frac{1}{hS}}$$

$$r_0 = \sqrt{\frac{S}{\pi}} = \sqrt{\frac{30}{\pi}} = 3.09 \text{ m}$$

$$r_1 = r_0 + e_i = 3.09 + 0.1 = 3.19 \text{ m}$$

$$\sqrt{k} = \sqrt{\frac{r_0}{r_1}} = \sqrt{\frac{3.09}{3.19}} = 0.984$$

$$\varphi = \frac{22 - 12}{\frac{0.984}{2 \times 1.8 \times 3.09\, ln\left(\frac{1 + 0.984}{1 - 0.984}\right)} + \frac{1}{8 \times 30}} = 444 \text{ W}$$

The losses have doubled compared to the previous calculation, so the losses from the floor to the outside air are of the same order of magnitude as the losses to the ground in the case considered.

3.3.21. *Periodic temperature variation in an unconditioned room*

We are in a tropical climate where the temperature varies between 18°C (at 4 a.m.) and 36°C (at 4 p.m.). A non-air-conditioned classroom, where lessons take place from 7 a.m. to 2 p.m., has to be built with earth brick walls of thickness e_1. The ceiling is assumed to be perfectly insulated. The floor is made of a perfectly insulated concrete screed with a thickness of $e_2 = 6$ cm (poured over polystyrene). The floor area is $S_2 = 8 \times 4$ m^2, and the height of the walls is $h = 3$ m.

Data: bricks: $\lambda_1 = 1.5 \text{ W m}^{-1}\text{K}^{-1}$; $\rho_1 c_1 = 2 \times 10^6 \text{J m}^{-3}\text{K}^{-1}$;

concrete: $\lambda_2 = 1.75 \text{ W m}^{-1}\text{K}^{-1}$; $\rho_2 c_2 = 2 \times 10^6 \text{ J m}^{-3}\text{K}^{-1}$;

polystyrene: $\lambda_3 = 0.03 \text{ W m}^{-1}\text{K}^{-1}$; $\rho_3 c_3 = 5 \times 10^4 \text{ J m}^{-3}\text{K}^{-1}$;

external and internal exchange coefficients: $h_e = 10 \text{ W m}^{-1}\text{K}^{-1}$;

$h_i = 5 \text{ W m}^{-1}\text{K}^{-1}$.

1) Calculate the amplitude of the indoor air temperature variation and its phase shift with respect to the outdoor air temperature for the following three wall thicknesses: 10 cm, 20 cm and 30 cm. Plot on the same graph the three curves of the evolution of the interior temperature over one day. Comment.

2) We want to insulate the wall with a thickness of 5 cm of polystyrene. On which side of the wall is it best to place it? Calculate the amplitude and the phase shift for a wall thickness $e_1 = 20$ cm with the insulation placed inside then outside the wall. Conclude on the interest of isolating in this case.

Corrected

1) It is assumed that the variation in the outside temperature is sinusoidal in the form:

$$T_a(t) = T_e + T_0 cos(\omega t)$$

with: $T_e = 27°\text{C}$; $T_0 = 9°\text{C}$ and $\omega = \frac{2\pi}{24 \times 3600} = 7.272 \times 10^{-5}$.

We set: $\overline{T}(x,t) = T(x,t) - 27°\text{C} = f(x) exp(i\omega t)$.

We write the quadrupole relation between the outside air and the inside air:

$$\begin{bmatrix} f_e \\ F_e \end{bmatrix} = \begin{bmatrix} 1 & R_{ce} \\ 0 & 1 \end{bmatrix} \begin{bmatrix} A_m & B_m \\ C_m & D_m \end{bmatrix} \begin{bmatrix} 1 & R_{ci} \\ 0 & 1 \end{bmatrix} \begin{bmatrix} f_i \\ F_i \end{bmatrix} = \begin{bmatrix} A & B \\ C & D \end{bmatrix} \begin{bmatrix} f_i \\ F_i \end{bmatrix}$$

where $\begin{bmatrix} A_m & B_m \\ C_m & D_m \end{bmatrix}$ is the quadrupole of the walls with: $R_{ce} = \frac{1}{h_e S_1}$; $R_{ci} = \frac{1}{h_i S_1}$, where: $S_1 = h \times perimeter$.

We write the quadrupole relationship between indoor air and soil:

$$\begin{bmatrix} f_i \\ F_i \end{bmatrix} = \begin{bmatrix} A_s & B_s \\ C_s & D_s \end{bmatrix} \begin{bmatrix} f_s \\ 0 \end{bmatrix}$$

where $\begin{bmatrix} A_s & B_s \\ C_s & D_s \end{bmatrix}$ is the quadrupole of the slab.

We deduce: $F_i = \frac{C_s}{A_s} f_i$

and: $f_i = f_e \frac{1}{A + B\frac{C_s}{A_s}}$.

We calculate the matrix product $M = \begin{bmatrix} A & B \\ C & D \end{bmatrix} = \begin{bmatrix} 1 & R_{ce} \\ 0 & 1 \end{bmatrix} \begin{bmatrix} A_m & B_m \\ C_m & D_m \end{bmatrix} \begin{bmatrix} 1 & R_{ci} \\ 0 & 1 \end{bmatrix}$ using Matlab with:

$$\begin{bmatrix} A_m & B_m \\ C_m & D_m \end{bmatrix} = \begin{bmatrix} cosh(q_1 e_1) & \frac{1}{\lambda_1 q_1 S_1} sinh(q_1 e_1) \\ \lambda_1 q_1 S_1 sinh(q_1 e_1) & cosh(q_1 e_1) \end{bmatrix}$$

We calculate $A_s = cosh(q_2 e_2)$ and $C_s = \lambda_2 q_2 S_2 sinh(q_2 e_2)$.

We extract the value of the complex numbers $A = M(1,1)$ and $B = M(1,2)$, then we calculate the modulus and the argument of $f_i = f_e \frac{1}{A + B\frac{C_s}{A_s}}$ using the appropriate Matlab functions (abs and angle), below is an example of a program:

```
TO= 9;
S1= 72; S2 =32; % Wall and floor area
he =10;hi=5; % Transfer coefficients
w =2*pi/(24*3600); % Pulsation
Rce =1/ he /S 1; % External resistance
R ci =1/hi/S 1; % Internal resistance
e 1=0.2;lam1=1.5;roc1=2e6;a1=lam1/roc1; % Wall properties
q1=(i*w/a 1)^ 0.5;
```

```
e2=0. 06;lam 2=1.75;roc2=2e6;a2=lam2/roc2; % Slab properties
q2=(i*w/a 2)^ 0.5;
Am=cosh(q1*e1 );Bm
=(sinh(q1*e1))/(lam1*q1*S1);Cm=lam1*q1*S1*sinh(q1*e1);Dm=Am;
As= cosh (q2*e2 );Cs =lam2*q2*S2* sinh (q2*e2);
Me=[1,Rce ;0.1];Mm=[Am,Bm;Cm,Dm ];Mi=[1,Rci;0.1];
M=Me*Mm*Mi;
A=M(1.1 );B =M(1.2);
f=T0* abs (1/(A+B*Cs/ As ))
psi=angle(1/(A+B*Cs/ As ))
```

and we get the following results:

e_1**(cm)**	10	20	30
$\vert f\vert$**(°C)**	2.66	1.23	0.61
ψ**(rd)**	-1.83	-2.51	-3.19

It can be seen that between 7 a.m. and 2 p.m., the interior temperature is minimal for a wall thickness of 10 cm. Note, however, that these results would be modified if the incident solar flux on the wall were taken into account, which would penalize the least insulating wall, namely the one 10 cm thick.

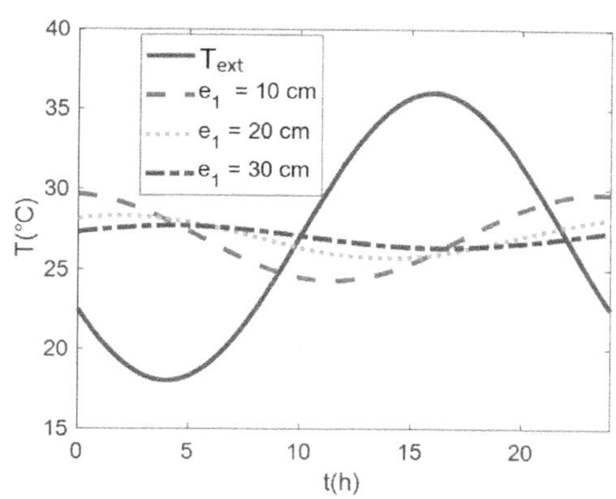

2) The quadrupole relationship between outdoor air and indoor air is modified as follows:

– in case (1) where the insulation is on the outside, the calculation is identical, replacing the quadrupole matrix of the wall with:

$$\begin{bmatrix} A_i & B_i \\ C_i & D_i \end{bmatrix} \begin{bmatrix} A_m & B_m \\ C_m & D_m \end{bmatrix}$$

where $\begin{bmatrix} A_i & B_i \\ C_i & D_i \end{bmatrix}$ is the quadrupole matrix of the insulator;

– in case (2) where the insulation inside, the calculation is identical by replacing the quadripolar matrix of the wall with:

$$\begin{bmatrix} A_m & B_m \\ C_m & D_m \end{bmatrix} \begin{bmatrix} A_i & B_i \\ C_i & D_i \end{bmatrix}$$

In both cases, we apply the calculation procedure in Matlab described in the previous question and we obtain:

– case (1):

$$|f| = 0.10°\text{C}; \ \psi = -2.85 \text{ rd}$$

– case (2):

$$|f| = 0.25°\text{C}; \ \psi = -3.27 \text{ rd}$$

The position of the insulation has little influence, and the temperature inside is almost constant and equal to the average temperature of the outside air. The insulation avoids a peak in interior temperature.

3.3.22. *Periodic flow variation in an air-conditioned room*

We study the same room as that defined in Exercise 3.3.21 subject to the same external conditions, but the room is now air-conditioned to impose a constant indoor $T_i = 22°\text{C}$ temperature.

Calculate for a wall thickness $e_1 = 20$ cm the heat flow entering the room in the following three cases:

– uninsulated wall;

– wall with 5 cm of polystyrene inside;

– wall with 5 cm of polystyrene on the outside.

Comment on the results.

Corrected

It is assumed that the variation of the outside temperature is sinusoidal of the form:

$$T_{ext}(t) = T_e + T_0 cos(\omega t)$$

with $T_e = 27°C$; $T_0 = 9°C$ and $\omega = \frac{2\pi}{24 \times 3600} = 7.272 \times 10^{-5}$.

The indoor temperature is fixed:

$$T_{int}(t) = T_i$$

The problem being linear, we can break it up into a sum of a problem in the steady state and a problem in the established periodic regime and calculate the two corresponding heat fluxes:

– steady-state regime: $T_{ext}(t) = T_e$ and $T_{int}(t) = T_i$;

– established periodic regime: $T_{ext}(t) = T_0 cos(\omega t)$ and $T_{int}(t) = 0$.

In the steady-state regime:

$$\varphi_1 = \frac{T_e - T_i}{R_{ce} + R_m + R_i} \quad \text{(a)}$$

with: $R_{ce} = \frac{1}{h_e S_1}$; $R_{ci} = \frac{1}{h_i S_1}$; $R_m = \frac{e_1}{\lambda_1 S_1}$.

In the established periodic regime: we write the quadrupole relation between the outside air and the inside air (same method as for Exercise 3.3.21):

$$\begin{bmatrix} f_e \\ F_e \end{bmatrix} = \begin{bmatrix} 1 & R_{ce} \\ 0 & 1 \end{bmatrix} \begin{bmatrix} A_m & B_m \\ C_m & D_m \end{bmatrix} \begin{bmatrix} 1 & R_{ci} \\ 0 & 1 \end{bmatrix} \begin{bmatrix} 0 \\ F_i \end{bmatrix} = \begin{bmatrix} A & B \\ C & D \end{bmatrix} \begin{bmatrix} 0 \\ F_i \end{bmatrix} \quad \text{(b)}$$

with: $R_{ce} = \frac{1}{h_e S_1}$; $R_{ci} = \frac{1}{h_i S_1}$, where: $S_1 = h \times perimeter$.

And: $\begin{bmatrix} A_m & B_m \\ C_m & D_m \end{bmatrix} = \begin{bmatrix} cosh(q_1 e_1) & \frac{1}{\lambda_1 q_1 S_1} sinh(q_1 e_1) \\ \lambda_1 q_1 S_1 sinh(q_1 e_1) & cosh(q_1 e_1) \end{bmatrix}$.

We deduce: $f_e = B F_i$ with $f_e = T_0$, hence:

$$F_i = \frac{T_0}{B} \quad \text{(c)}$$

$\varphi_2 = |F_i|$ with a phase difference ψ between the flow and the outside temperature.

The total incoming flow is finally written as:

$$\varphi = \frac{T_e - T_i}{R_{ce} + R_m + R_i} + \left|\frac{T_0}{B}\right| \cos(\omega t + \psi) \quad \text{(d)}$$

In the case of an insulated wall, R_m is written as: $R_m = \frac{e_1}{\lambda_1 S_1} + \frac{e_3}{\lambda_3 S_1}$.

If the insulation is inside, we replace in relation (a) the matrix $\begin{bmatrix} A_m & B_m \\ C_m & D_m \end{bmatrix}$ by the product $\begin{bmatrix} A_m & B_m \\ C_m & D_m \end{bmatrix}\begin{bmatrix} A_i & B_i \\ C_i & D_i \end{bmatrix}$.

If it is outside, the matrix $\begin{bmatrix} A_m & B_m \\ C_m & D_m \end{bmatrix}$ is replaced by the product $\begin{bmatrix} A_i & B_i \\ C_i & D_i \end{bmatrix}\begin{bmatrix} A_m & B_m \\ C_m & D_m \end{bmatrix}$.

We calculate the matrix product $M = \begin{bmatrix} A & B \\ C & D \end{bmatrix} = \begin{bmatrix} 1 & R_{ce} \\ 0 & 1 \end{bmatrix}\begin{bmatrix} A_m & B_m \\ C_m & D_m \end{bmatrix}\begin{bmatrix} 1 & R_{ci} \\ 0 & 1 \end{bmatrix}$ in Matlab as in Exercise 3.3.21, and we calculate $B = M(1,2)$ and deduce F_i by relation (c), then φ by relation (d) using the same type of program as that given in Exercise 3.3.21.

We thus obtain $\varphi = \varphi_1 + \varphi_2 cos(\omega t + \psi)$ with:

	φ_1**(W)**	φ_2**(W)**	ψ**(rd)**
Without insulation	831	559	-1.69
Interior insulation	171	73	-1.93
External insulation	171	46	-2.05

Insulation with 5 cm of polystyrene allows us to divide by almost 5 the heat input and therefore the air-conditioning needs. Insulation placed on the outside allows us to reduce peak demand and would therefore a priori be the best solution.

4

Convective Heat Transfer

4.1. Reminders on dimensional analysis

4.1.1. *Fundamental dimensions*

Physical quantities can be expressed in terms of a limited number of fundamental dimensions.

EXAMPLES.– *Velocity:* $L.T^{-1}$. T^{-1}; *dynamic viscosity:* $M.L^{-1}.T^{-1}$; *force:* $M.L.T^{-2}$.

On these examples, we see that the number of fundamental dimensions is 3: mass M, length L, and time T.

These three fundamental dimensions are not always sufficient. For heat transfer problems, it is necessary to add a fourth dimension: temperature θ.

The method of dimensional analysis, which is based on the principle of dimensional homogeneity of the terms of an equation, is known as the Vaschy–Buckingam theorem or the grouping theorem π.

4.1.2. *Principle of the method*

If we can mathematically represent a physical law by expressing a physical variable G_1 as a function of a certain number of other independent physical variables, $G_2, \ldots, G_n$, i.e., if $G_1 = f(G_2, \ldots, G_n)$ or $f(G_1, G_2, \ldots, G_n) = 0$, the problem can be simplified in the following way:

– We write for each variable G_i, the dimensional equation according to the fundamental dimensions. We then have n equations requiring p fundamental dimensions to characterize all the physical quantities.

– We take p from these n equations which we consider as basic equations. Although the choice of equations taken is arbitrary, each fundamental dimension must appear at least once in all of the p equations. The dimensional equations must be linearly independent.

– The $(n-p)$ remaining equations are then presented in the form of $(n-p)$ ratios without dimensions called π groupings which are "reduced variables". We then obtain a reduced equation:

$$g(\pi_1, \pi_2, \dots \pi_{n-p}) = 0$$

A π grouping is the ratio of a dimension equation of a physical quantity not belonging to the set of basic equations, to the product of the basic equations, each of them being raised to a certain power:

$$\pi_i = \frac{[G_i]}{[G_1]^{a_i}[G_2]^{b_i}\dots[G_p]^{e_i}}$$

For each fundamental dimension M, L, T, θ appearing in the denominator, the sum of the exponents is made and then identified with the exponent of the same dimension appearing in the physical quantity of the numerator. We thus obtain a linear system of p equations whose resolution allows us to determine the p exponents of the basic equations of the denominator.

It is then sufficient to write the ratio π as a function of the physical quantities attached to the initial dimension equations.

4.1.3. *Application example*

Let consider a fluid in forced circulation in a cylindrical pipe for which it is proposed to determine the convection coefficient h relating to the fluid-wall heat transfer which corresponds to a forced convection.

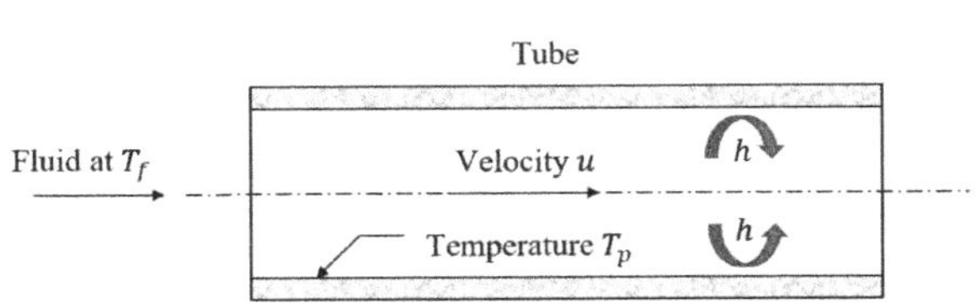

Figure 4.1. *Diagram of the studied configuration. For a color version of this figure, see www.iste.co.uk/jannot/heattransfer1.zip*

4.1.3.1. *Determination of physical quantities*

It is necessary to determine all the parameters on which the exchange coefficient h depends, they are in this case:

– the characteristics of the fluid:

- λ, thermal conductivity,
- c, specific heat,
- ρ, density,
- μ, dynamic viscosity;

– flow characteristics: u, average fluid velocity;

– the geometry of the exchange surface: D, pipe diameter.

Hence: $f(\lambda, c, \rho, \mu, u, D, h) = 0$.

4.1.3.2. *Dimensional equation of each quantity*

It is then necessary to write the equation with the fundamental dimensions M, L, T, θ of each quantity, which is written here as:

– λ: $M.T^{-3}.L.\theta^{-1}$;

– c: $T^{-2}.L^2.\theta^{-1}$;

– ρ: $M.L^{-3}$;

– μ: $M.T^{-1}.L^{-1}$;

– u: $T^{-1}.L$;

– D: L;

– h: $M.T^{-3}.\theta^{-1}$.

4.1.3.3. *Determination of π groupings*

It is now necessary to choose four basic equations so that the four fundamental dimensions appear at least once in all the equations.

Take for example: λ, ρ, D, u, there remain h, c and μ.

We then write the three dimensionless ratios corresponding to these variables in the form:

$$\pi_1 = \frac{h}{\lambda^{a_1}\rho^{b_1}D^{c_1}u^{d_1}};\ \pi_2 = \frac{c}{\lambda^{a_2}\rho^{b_2}D^{c_2}u^{d_2}};\ \pi_3 = \frac{\mu}{\lambda^{a_3}\rho^{b_3}D^{c_3}u^{d_3}}$$

For each ratio π, we replace the physical quantities by their dimensional equations, which gives for example for π_1:

$$[\pi_1] = \frac{M.T^{-3}.\theta^{-1}}{[M.T^{-3}.L.\theta^{-1}]^{a_1}[M.L^{-3}]^{b_1}[L]^{c_1}[L.T^{-1}]^{d_1}}$$

For each fundamental dimension, we identify the power exponents between numerator and denominator relative to the same dimension, which leads to the system:

– T: $-3 = -3a_1 - d_1$;

– L: $0 = a_1 - 3b_1 + c_1 + d_1$;

– θ: $-1 = -a_1$;

– M: $1 = a_1 + b_1$.

The ratio $\pi_1$is therefore written as $\pi_1 = \frac{hD}{\lambda}$; we obtain in the same way: $\pi_2 = \frac{\rho u D c}{\lambda}$ and $\pi_3 = \frac{\mu}{\rho u D}$.

The Vaschy–Buckingam theorem allows us to affirm that the equation: $f(\lambda, c, \rho, \mu, u, D, h) = 0$ between seven variables can be expressed using the three dimensionless numbers π_1, π_2 and π_3 in the form:

$$f(\pi_1, \pi_2, \pi_3) = 0 \text{ or } \pi_1 = f(\pi_2, \pi_3)$$

4.1.3.4. *Physical meaning of these groupings*

– $\pi_1 = \frac{hD}{\lambda}$ is the Nusselt number, it can also be written as: $Nu = \frac{\frac{D}{\lambda}}{\frac{1}{h}}$.

It is therefore the ratio of the thermal resistance of conduction by the thermal resistance of convection. It is higher as convection is predominant over conduction. It characterizes the heat transfer type.

– $\pi_3 = \frac{\mu}{\rho u D} = \frac{1}{Re}$, is the inverse of the Reynolds number which characterizes the flow regime in the pipe.

– $\pi_2 = \frac{\rho u D c}{\lambda}$, is the Péclet number. We prefer to use a new dimensional number, the product $\pi_2 \times \pi_3 = \frac{c\mu}{\lambda}$ called the Prandtl number. This number can be calculated for a given fluid independently of the experimental conditions (it only depends on the temperature) and characterizes the influence of the nature of the fluid on the heat transfer by convection.

We therefore prefer to look for an equation in the form:

$$Nu = f(Re, Pr) \qquad [4.1]$$

4.1.4. *Advantages of using reduced quantities*

They essentially concern the representation, comparison and research of experimental results:

– The representation of the experimental results is simplified, we can have a curve linking two variables or an abacus linking three reduced variables instead of an equation linking $(3 + p)$ parameters.

– The comparison of experimental results is also very quick and easy, regardless of the researcher, even though the system of units used is different since the reduced quantities are dimensionless.

– The search for experimental results is facilitated and organized: if it is sufficient to draw a curve between two reduced variables, then it is sufficient to carry out a single series of experiments.

REMARK.– It must, however, be well understood that the method of dimensional analysis which provides the reduced quantities does not give the form of the equation which links them, the search for this relationship is the subject of the analysis of the experimental results.

Group			
$Re = \frac{\rho u D}{\mu}$	Reynolds number	$Ma = \frac{h}{\rho u c}$	Margoulis number
$Pr = \frac{c\mu}{\lambda}$	Prandtl number	$Gr = \frac{\beta g \Delta T \rho^2 L^3}{\mu^2}$	Grashof number
$Nu = \frac{hD}{\lambda}$	Nusselt number	$Ra = \frac{c\beta g \Delta T \rho^2 L^3}{\lambda\mu}$	Rayleigh number
$Pe = \frac{\rho u D c}{\lambda}$	Peclet number		

Table 4.1. *Some dimensionless numbers*

4.2. Convection without phase change

4.2.1. *Generalities and definitions*

Heat transfer that occurs simultaneously with flow movement is called convective heat transfer. This mode of heat exchange exists within fluid media in which it is generally preponderant.

4.2.1.1. *Natural and forced convection*

Depending on the nature of the mechanism that causes the movement of the fluid, we distinguish:

– *natural or free convection*: the fluid is set in motion under the sole effect of differences in density resulting from temperature differences at the boundaries and from a field of external forces (gravity);

– *forced convection*: fluid movement is induced by a cause independent of temperature differences (pump, fan).

The study of heat transfer by convection enables in particular determining the heat transfer coefficient occurring between a fluid and a wall.

4.2.1.2. *Flow regime*

Given the link between flow movement and heat transfer, it is necessary to take the flow regime into account. Consider as an example the flow of a fluid in a pipe.

In laminar regime, the flow takes place in practically independent layers. Between two adjacent fluid layers, the heat exchange therefore takes place:

– by conduction only if we consider a direction normal to the fluid velocity vector;

– by convection and conduction (negligible) if we consider a direction parallel to the fluid velocity vector.

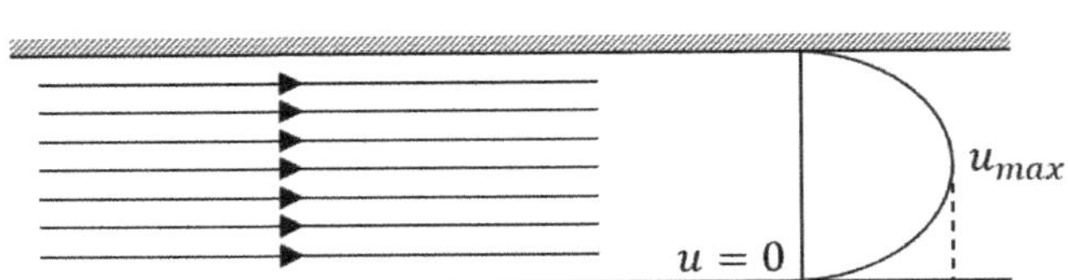

Figure 4.2. *Schematization of a laminar flow*

In *turbulent regime*, the flow is not unidirectional.

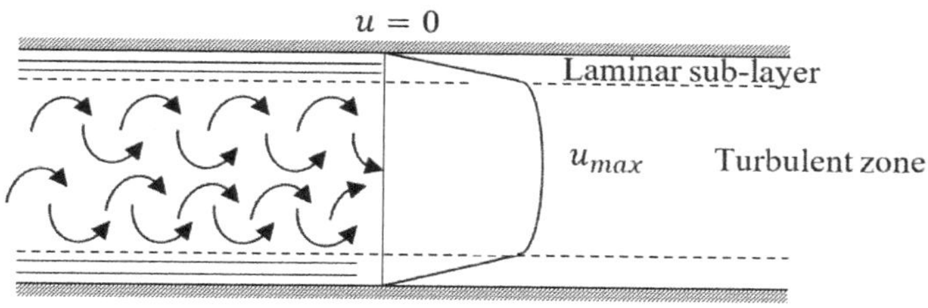

Figure 4.3. *Schematization of a turbulent flow*

The heat exchange in the turbulent zone takes place by convection and conduction in all directions. It is verified that molecular conduction is generally negligible compared to convection and "turbulent diffusion" (mixing of the fluid due to turbulent agitation) outside the laminar sub-layer.

4.2.2. *Expression of heat flow rate*

4.2.2.1. *Reynolds analogy*

Just as at the molecular level, the viscosity of gases is explained by the transmission of the momentum of molecules during intermolecular collisions, the transmission of heat is explained by the transmission of kinetic energy during these same collisions.

This intimate link between the phenomena of viscosity and heat transfer leads to the Reynolds analogy: in a fluid flow with heat transfer, the velocity profile and the temperature profile are linked by a similarity relation schematized in Figure 4.4. This similarity will be demonstrated later in the case of a flow on a heated flat plate.

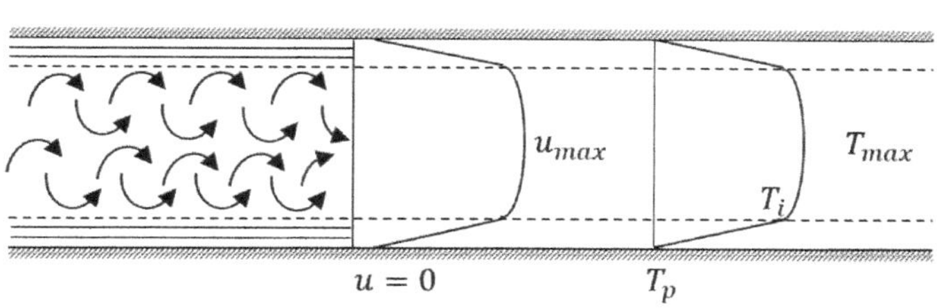

Figure 4.4. *Representation of the Reynolds analogy in the case of turbulent flow in a tube*

4.2.2.2. *Dynamic and thermal boundary layers*

Whatever the flow regime, remains at the wall a dynamic boundary layer in which the flow is laminar and whose thickness is all the more reduced as the Reynolds number is high. The thickness of this boundary layer varies according to many parameters: nature of the fluid, temperature, wall roughness, etc.

The Reynolds analogy shows that the thermal gradient is particularly important in the vicinity of the wall, in a thermal boundary layer that develops analogously to the dynamic boundary layer. Whatever the flow regime of the fluid, it is considered that the thermal resistance is entirely located in this thermal boundary layer which acts as an insulator.

This corresponds to the Prandtl model shown in Figure 4.5 as an example for the turbulent flow of a fluid in a pipe.

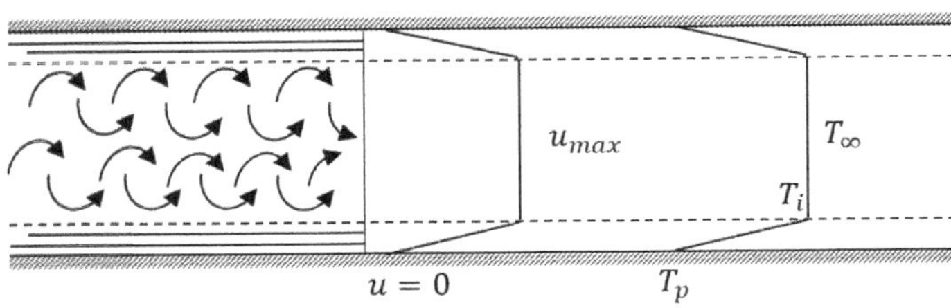

Figure 4.5. *Representation of the Prandtl model for turbulent flow in a pipe*

4.2.2.3. *Expression of flow*

Whatever the type of convection (free or forced) and the flow regime of the fluid (laminar or turbulent), the heat flow rate φ is given by the equation known as Newton's law:

$$\varphi = hS\Delta T \qquad [4.2]$$

where ΔT is the temperature difference between the wall and the fluid.

This expression is an approximation which in fact replaces the resolution of a complex fluid mechanics problem (see section A.16 of the Appendix).

The major problem to solve before the calculation of the heat flow rate consists of determining the convective heat transfer coefficient h which depends on a large number of parameters: characteristics of the fluid, flow, temperature and shape of the exchange surface.

Table 4.2 gives the order of magnitude of the convective heat transfer coefficient for different configurations.

Configuration	h (W m^{-2}K^{-1})
Natural convection	
In a gas	2–10
In a liquid	100–1,000
Forced convection	
In a gas	10–200
In a liquid	100–5,000
Boiling water	
In a container	2,500–35,000
Flowing in a tube	5,000–100,000
Water condensation at 1 atm	
On a vertical surface	1,000–11,000
Outside horizontal tubes	10,000–25,000

Table 4.2. *Order of magnitude of the convective heat transfer coefficient*

4.2.3. *Calculation of heat flow rate in forced convection*

4.2.3.1. *Analytical calculation*

In certain simple cases, a theoretical calculation leads to an analytical expression of the heat flow exchanged by convection between a fluid and a wall. We will treat here as an example the classical case of the laminar flow in steady state of a fluid with constant physical properties at the temperature T_∞, on a flat wall of length L maintained at a temperature T_p (see Figure 4.6).

It can be seen that the velocity of the fluid evolves from a zero value at the wall to a value close to u_∞ in a zone of thickness $\delta(x)$ called the dynamic boundary layer. In the same way, the temperature of the fluid evolves from the value T_p at the wall to a value close to T_∞ in a zone of thickness $\Delta(x)$ called the thermal boundary layer.

The mass conservation equation is written in integral form (see section A.16 of the Appendix):

$$\int_\Lambda \frac{\partial \rho}{\partial t} dv + \int_\Sigma \rho \vec{V}.\vec{n} dS = 0$$

where $\vec{n}$ is the exterior normal to Σ, and v *is* the volume.

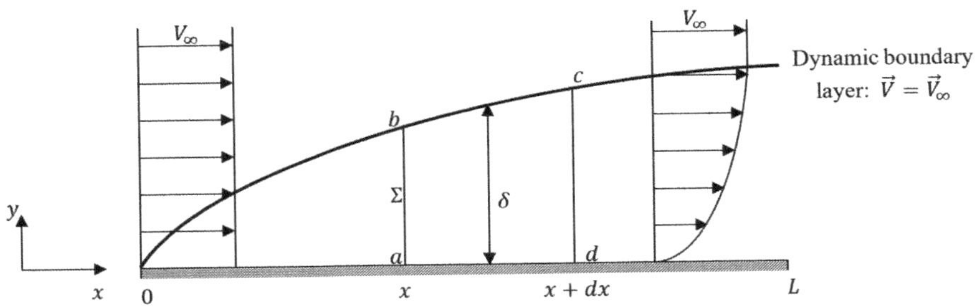

Figure 4.6. *Schematization of the development of a dynamic boundary layer on a flat plate*

In steady state: $\frac{\partial \rho}{\partial t} = 0$. Let us apply this equation to the volume $[abcd]$ represented in Figure 4.6:

$$\int_\Sigma \rho \vec{V}.\vec{n}dS = \underbrace{-\rho\left(\int_0^\delta udy\right)_x}_{\text{mass flow exiting by } ab} + \underbrace{\rho\left(\int_0^\delta udy\right)_{x+dx}}_{\text{mass flow exiting by } cd} + \underbrace{\int_{bc} \rho\overrightarrow{V_\infty}.\vec{n}dS}_{\text{mass flow exiting by } bc}$$

We deduce: $\int_{bc} \rho\overrightarrow{V_\infty}.\vec{n}dS = -\rho\frac{d}{dx}\left(\int_0^\delta udy\right)dx$.

The momentum conservation equation in steady state (Euler's theorem, see section A.16 of the Appendix) is written as:

$$\int_\Sigma \rho\vec{V}(\vec{V}.\vec{n})dS = \int_\Lambda \rho\vec{f}dv + \int_\Sigma \overrightarrow{F_p}dS = \int_\Sigma \overrightarrow{F_p}dS$$

because $f = 0$ (no volume force due to an external field).

where $\overrightarrow{F_p}$ are the external forces (per unit area) exerted by contact on the faces of the surface Σ bounding the volume Λ.

Let us apply this equation to the volume $[abcd]$:

$$\int_\Sigma \rho\vec{V}(\vec{V}.\vec{n})dS = -\rho\left(\int_0^\delta \vec{V}udy\right)_x + \rho\left(\int_0^\delta \vec{V}udy\right)_{x+dx} + \int_{bc} \rho\overrightarrow{V_\infty}(\overrightarrow{V_\infty}.\vec{n})dS$$

(per unit width).

Or, in projection along Ox:

$$\int_\Sigma \rho\vec{V}(\vec{V}.\vec{n})dS = -\rho\left(\int_0^\delta u^2dy\right)_x + \rho\left(\int_0^\delta u^2dy\right)_{x+dx} + \int_{bc}\rho u_\infty(\overrightarrow{V_\infty}.\vec{n})dS$$

$$\int_\Sigma \rho\vec{V}(\vec{V}.\vec{n})dS = -\rho\left(\int_0^\delta u^2dy\right)_x + \rho\left(\int_0^\delta u^2dy\right)_{x+dx} + \rho u_\infty\int_{bc}(\overrightarrow{V_\infty}.\vec{n})dS$$

Hence, using the mass conservation equation:

$$\int_\Sigma \rho\vec{V}(\vec{V}.\vec{n})dS = \rho\frac{d}{dx}\left(\int_0^\delta u^2dy\right)dx \quad - \rho u_\infty\frac{d}{dx}\left(\int_0^\delta udy\right)dx$$

Let us analyze the forces along Ox:

– on $[ad]$ the parietal friction is exerted;

– on $[bc]$ as the velocity profile is uniform, there is no friction;

– there are no pressure forces since the pressure is uniform in the flow.

We can therefore write:

$$\int_\Sigma \overrightarrow{F_p}dS = -\tau_p dx$$

hence:

$-\tau_p dx = \rho\frac{d}{dx}\left(\int_0^\delta u^2dy\right)dx \quad - \rho u_\infty\frac{d}{dx}\left(\int_0^\delta udy\right)dx$ because $\int_\Sigma \rho\vec{V}(\vec{V}.\vec{n})dS = \int_\Sigma \overrightarrow{F_p}dS$.

We deduce:

$$\frac{\tau_p}{\rho u_\infty{}^2} = \frac{d}{dx}\left[\delta\int_0^1\frac{u}{u_\infty}\left(1-\frac{u}{u_\infty}\right)d\left(\frac{y}{\delta}\right)\right] \qquad \text{(a)}$$

We seek a priori the velocity in the simple form of a parabolic profile: $\frac{u}{u_\infty} = a + b\left(\frac{y}{\delta}\right) + c\left(\frac{y}{\delta}\right)^2$.

The velocity is zero at the wall: $u(y=0)=0$.

The continuity of velocity and friction at the border of the boundary layer imposes the following conditions:

$$u(y=\delta) = u_\infty$$

$$\mu\left(\frac{du}{dy}\right)_{y=\delta} = 0$$

We deduce that:

$$\frac{u}{u_\infty} = \frac{y}{\delta}\left(2 - \frac{y}{\delta}\right) \qquad \text{(b)}$$

and $\tau_p = \mu\left(\frac{\partial V}{\partial y}\right)_{y=0} = 2\mu\frac{u_\infty}{\delta}$.

Equations (a) and (b) lead to: $\frac{2\mu}{\rho u_\infty \delta} = \frac{2}{15}\frac{d\delta}{dx}$.

Then, by integration: $\delta^2 = 30\frac{\mu x}{\rho u_\infty}$ or again: $\left(\frac{\delta}{x}\right)^2 = 30\frac{\mu}{\rho u_\infty x} = \frac{30}{Re_x}$.

$$\tau_p = 2\mu\frac{u_\infty}{\delta} = 2\mu u_\infty\sqrt{\frac{\rho u_\infty}{30\mu x}}$$

We deduce the expression of the coefficient of friction: $C_{f_x} = \frac{\tau_p}{\frac{1}{2}\rho u_\infty^2} = \frac{2\mu u_\infty\sqrt{\frac{\rho u_\infty}{30\mu x}}}{\frac{1}{2}\rho u_\infty^2} = \frac{0.73}{Re_x^{0.5}}$.

A refined analysis (local equations and no assumptions on the shape of the velocity profile) would lead to a constant of 0.664 instead of 0.73.

Now let us calculate the temperature profile. At constant pressure, the enthalpy change of a system is equal to the heat supplied to that system.

By applying this principle to a volume (Λ) of surface (Σ) and by neglecting the viscous dissipation (internal source of heat corresponding to the degradation of mechanical energy into heat), we get:

$$\int_\Lambda \frac{\partial}{\partial t}(\rho H)dv + \int_\Sigma \left(\rho H\vec{V} + \vec{q}\right).\vec{n}dS = 0$$

with:

– H, the mass enthalpy of the fluid;

– $\vec{q}$, the heat flux vector;

– $\vec{n}$, the normal exterior to Σ;

– v, the volume.

Let us apply this steady-state equation to the volume $[a'b'c'd']$ shown in Figure 4.7 for a fluid such that $H = c_p(T - T_0)$ (where the pressure is assumed to be constant).

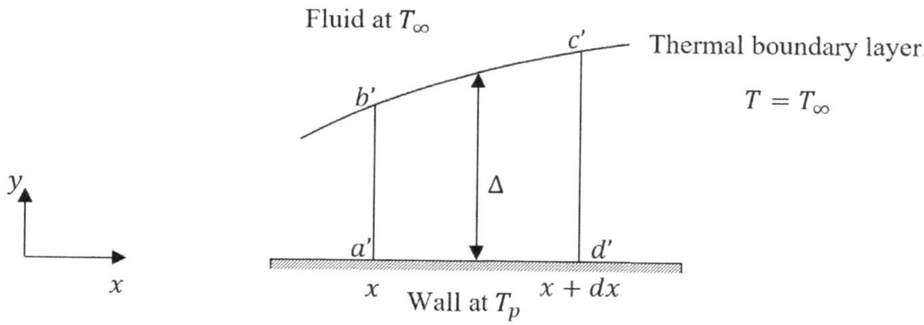

Figure 4.7. *Schematization of the thermal boundary layer on a flat plate*

The conductive heat flux is zero on the surface $[b'c']$ since outside the thermal boundary layer the temperature is uniform and is equal to T_∞. However, the longitudinal heat flow rate (following Ox) is neglected in front of the transverse heat flow rate (following Oy), the temperature varying much more quickly in the direction Oy than in the direction Ox (boundary layer hypothesis). We then get:

$$-\left(\rho c \int_0^{\Delta} uT dy\right)_x + \left(\rho c \int_0^{\Delta} uT dy\right)_{x+dx} + \rho c T_\infty \int_{b'c'} u dS - q_p dx = 0$$

where: $q_p = -\lambda \left(\frac{\partial T}{\partial y}\right)_{y=0}$ is the heat flux exchanged at the wall (positive if entering the volume $[a'b'c'd']$).

By applying conservation of mass to volume $[a'b'c'd']$, we get:

$$\int_{b'c'} \rho \vec{V}.\vec{n} dS = \rho \int_{b'c'} u dS = -\rho \frac{d}{dx}\left(\int_0^{\Delta} u dy\right)$$

hence:

$$\frac{d}{dx}\left(\rho c \int_0^{\Delta} uT dy\right) dx - \rho c T_\infty \frac{d}{dx}\left(\int_0^{\Delta} u dy\right) dx - q_p dx = 0.$$

Thus:

$$\frac{q_p}{\rho c u_\infty (T_p - T_\infty)} = \frac{d}{dx}\left[\Delta \int_0^1 \frac{u}{u_\infty}\left(1 - \frac{T - T_p}{T_\infty - T_p}\right) d\left(\frac{y}{\Delta}\right)\right] \qquad \text{(c)}$$

We seek a priori the temperature in the form:

$$\frac{T-T_p}{T_\infty-T_p} = a + b\left(\frac{y}{\Delta}\right) + c\left(\frac{y}{\Delta}\right)^2.$$

The continuity of temperature and heat flow imposes the following conditions:

$$\left(\frac{T-T_p}{T_\infty-T_p}\right)_{y=0} = 0$$

$$\left(\frac{T-T_p}{T_\infty-T_p}\right)_{y=\Delta} = 1$$

$$\lambda\left(\frac{dT}{dy}\right)_{y=\Delta} = 0$$

We deduce that:

$$\frac{T-T_p}{T_\infty-T_p} = \frac{y}{\Delta}\left(2-\frac{y}{\Delta}\right) \qquad \text{(d)}$$

and $q_p = -\lambda\left(\frac{\partial T}{\partial y}\right)_{y=0} = 2\lambda\frac{T_p-T_\infty}{\Delta}$.

Equations (c) and (d) allow us to write:

$$\frac{q_p}{\rho c u_\infty(T_p-T_\infty)} = \frac{d}{dx}\left\{\Delta\int_0^1 \frac{y}{\delta}\left(2-\frac{y}{\delta}\right)\left[1-\frac{y}{\Delta}\left(-2\frac{y}{\Delta}\right)\right] d\left(\frac{y}{\Delta}\right)\right\}.$$

We consider the case where $\Delta \leq \delta$ and we assume that $r = \frac{\Delta}{\delta} \leq 1$ remains constant.

The previous equation then becomes: $\Delta\frac{d\Delta}{dx} = \frac{12\lambda}{\rho c u_\infty r\left(1-\frac{r}{5}\right)}$.

Then, after integration: $\left(\frac{\Delta}{x}\right)^2 = \frac{24}{Re_x Pr}\frac{1}{r\left(1-\frac{r}{5}\right)}$, where $Re_x = \frac{\rho u_\infty x}{\mu}$.

In addition: $\left(\frac{\delta}{x}\right)^2 = \frac{30}{Re_x}$.

r is therefore the solution of the equation:

$$r^2 = \frac{4}{5\,Pr\left(1-\frac{r}{5}\right)r} \qquad \text{(e)}$$

In the case $Pr = 1$, the solution of equation (e) is $r = 1$, the dynamic and thermal boundary layers have the same thickness, and there is a complete analogy between heat and momentum transfers. This is the case for gases for which $Pr \approx 1$.

The case $r < 1$ corresponds to the case $Pr > 1$, it is the case of water, for example ($Pr \approx 7$). An approximate solution of equation (e) is then: $\frac{\Delta}{\delta} \approx Pr^{-\frac{1}{3}}$.

The heat flow density at the wall is written as:

$$q_p = h(T_p - T_\infty) = 2\frac{\lambda}{\Delta}(T_p - T_\infty).$$

And: $Nu_x = \frac{hx}{\lambda} = \frac{2x}{\delta}\frac{\delta}{\Delta} = 0.36\, Re_x^{\frac{1}{2}} Pr^{\frac{1}{3}}$.

A more precise calculation would lead to:

$$Nu_x = 0.332\, Re_x^{\frac{1}{2}}\, Pr^{\frac{1}{3}} \qquad [4.3]$$

The global flow is obtained by integration of q_p between $x = 0$ and $x = L$, and we deduce the average Nusselt number on the surface of length L:

$$\overline{Nu_L} = 0.664\, Re_L^{\frac{1}{2}}\, Pr^{\frac{1}{3}} \qquad [4.4]$$

4.2.3.2. *Approximate calculation*

In more complex cases where an analytical solution cannot be established, correlations deduced from experiments or numerical calculations are used.

The application of dimensional analysis shows that the equation linking the heat flow rate transferred by convection to the variables on which it depends can be sought in the form of an equation between three dimensionless numbers:

$$Nu = f(Re, Pr) \qquad [4.5]$$

Defined by:

– $Re = \frac{\rho u D}{\mu}$, the Reynolds number;

– $Pr = \frac{c\mu}{\lambda}$, the Prandtl number;

– $Nu = \frac{hD}{\lambda}$, the Nusselt number.

where D is the characteristic dimension of the geometry considered which will be, for example, the hydraulic diameter 4 × cross-sectional area of the flow/wetted perimeter of the cross-section $D_h = \frac{4 \times Passage\ Section}{Perimeter}$ for a flow in a pipe (equal to the internal diameter for a cylindrical pipe), the external diameter for an external flow perpendicular to a tube, the length for free surface flow on a plate.

The calculation of a heat flow rate transmitted by forced convection is therefore carried out as follows:

1) calculation of dimensionless Reynolds and Prandtl numbers;

2) depending on the value of Re and the configuration → choice of the correlation (function f in equation [4.5]);

3) calculation of Nu by application of this correlation;

4) calculation of $h = \frac{\lambda Nu}{D}$ and of $\varphi = hS\,\Delta T$.

REMARK.– For external flow, the temperature of the fluid is taken away from the wall and $\Delta T = T_p - T_\infty$. For an internal flow (in a pipe, for example), $\Delta T = T_p - T_m$, where T_m is the average temperature of the fluid.

For forced convection, the main correlations are given in section A.17 of the Appendix. The properties of the fluid (c, ρ, λ, μ) are calculated at an average temperature called the film temperature: $T_f = \frac{T_p + T_\infty}{2}$ or $T_f = \frac{T_p + T_m}{2}$.

4.2.4. *Calculation of heat flow rate in natural convection*

4.2.4.1. *Mechanism of natural convection*

Consider a fluid at rest in contact with a flat wall at temperature T_0. If the wall is brought to a temperature $T = T_0 + \Delta T$, the fluid in contact with the wall will heat up by conduction and the mass of the unit volume will change from ρ_0 to $\rho_0 - \Delta\rho$ (see Figure 4.8).

It will therefore be subject to an upward force $\vec{f} = -\Delta\rho\vec{g}$. The fundamental principle of dynamics allows us to evaluate the acceleration of the fluid: for a unit volume: $m = \rho$ hence: $\Delta\rho g = \rho\gamma$ and $\gamma = \frac{\Delta\rho}{\rho} g$.

By introducing the coefficient of volume expansion β of the fluid defined by: $\beta = -\frac{1}{\rho}\left(\frac{\Delta\rho}{\Delta T}\right)_P$, we get:

$$\gamma = -\beta g \Delta T$$

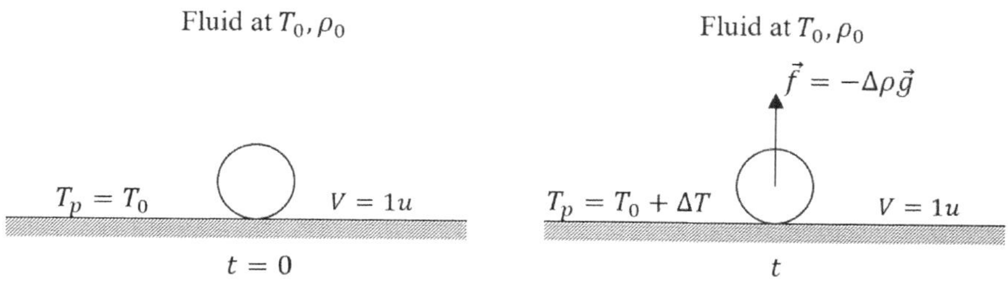

Figure 4.8. *Representation of the natural convection mechanism*

$\beta g \Delta T$ is therefore the modulus of the acceleration produced by the thermal expansion due to the variation ΔT of the temperature T_0. This movement of the fluid induced by density differences resulting from temperature gradients will give rise to convection currents.

In the case of heat transfer by natural convection along a flat plate, the convection coefficient depends on the characteristics of the fluid: $\lambda, \rho, \mu, c, \beta, g$, on the wall characterized by the length L and on the temperature difference ΔT, which can be expressed by an equation of the type:

$$f(\lambda, \rho, \mu, c, \beta, g, L, \Delta T, h) = 0$$

In the system M, L, T, θ, this equation between nine quantities is reduced to an equation between five dimensionless numbers, a refined analysis of the five numbers shows that the three quantities β, g and ΔT always appear in the form of the product $\beta g \Delta T (\gamma$ acceleration), which we can translate into an equation of the type:

$$f(\lambda, \rho, \mu, c, \gamma, L, h) = 0$$

This equation between seven quantities is reduced to an equation between three-dimensional numbers

$$Nu = f(Gr, Pr) \quad [4.6]$$

defined by:

– $Gr = \frac{\beta g \Delta T \rho^2 L^3}{\mu^2}$, the Grashof number;

– $Pr = \frac{c\mu}{\lambda}$, the Prandtl number;

– $Nu = \frac{hL}{\lambda}$, the Nusselt number.

4.2.4.2. *Physical meaning of the Grashof number*

When the unit mass of the fluid, subjected to the acceleration $\beta g \Delta T$ undergoes a variation of altitude L, the conservation of energy makes it possible to write: $\frac{u^2}{2} = \beta g \Delta T L$.

$\frac{u^2}{2}$ represents the change in kinetic energy and $\beta g \Delta T L$ the change in potential energy.

We can therefore put the Grashof number in the form:

$$Gr = \frac{1}{2}\left(\frac{uL\rho}{\mu}\right)^2$$

It is therefore proportional to the square of a Reynolds number characterizing the flow. In practice, in natural convection, the current that arises remains laminar until the Grashof number reaches a value of approximately 10^9.

4.2.4.3. *Calculation of heat flow in natural convection*

Application of dimensional analysis shows that the equation linking the heat flow rate transferred by convection to the variables on which it depends can be sought in the form of an equation between three dimensionless numbers: $Nu = f(Gr, Pr)$.

The calculation of the heat flow rate transmitted by natural convection is therefore carried out as follows:

1) calculation of dimensionless Grashof and Prandtl numbers;

2) depending on the value of Gr and of the configuration $\rightarrow$ choice of the correlation;

3) calculation of Nu by application of this correlation;

4) calculation of $h = \frac{\lambda Nu}{L}$ and of $\varphi = hS(T_p - T_\infty)$.

For natural convection, the main correlations are given in section A.18 of the Appendix. Fluid properties (c, ρ, λ, μ) are calculated at mean film temperature as in forced convection.

4.3. Convection with phase change

4.3.1. *Condensation*

4.3.1.1. *Phenomena*

The heat exchanges between a vapor condensing on a wall and the wall itself are linked to the types of condensation which essentially depend on liquid–wall interactions.

If the liquid does not wet the surface, it then forms, at certain points, droplets of liquid which run off along the wall. This type of condensation can only be observed if the wall has a smooth and clean surface. In the case of drop condensation, the liquid that does not form a continuous film on the wall offers negligible thermal resistance.

However, the type of condensation generally encountered in practice is film condensation: the wall is isolated from the vapor by a continuous film of liquid which acts as a thermal insulator between the wall and lowers the value of the convective heat transfer coefficient h compared to drop condensation.

4.3.1.2. *Value of coefficient h for film condensation*

Nusselt's theory, established in 1916, analytically relates the transfer coefficient h to the various physical parameters involved in the film condensation of a fluid on a wall.

4.3.1.2.1. Vertical wall

Hypotheses:

– laminar flow of the film;

– constant wall temperature;

– constant temperature gradient in the film;

– large curvature radius of the condensate film.

Note T_g the saturation temperature (dew) of the vapor and T_p ($<T_g$) the constant temperature of the vertical wall (see Figure 4.9).

The forces acting on the system consisting of the liquid with a thickness dx between y and δ, and with unit length along Oz (grey surface) are:

– the gravity force: $\rho_l g(\delta - y)dx$;

– the force due to displaced water vapor: $\rho_v g(\delta - y)dx$;

– the viscous friction force: $\mu_l \frac{du}{dy} dx$ (Newtonian fluid hypothesis).

The balance of forces is written as: $\rho_l g(\delta - y)dx = \mu_l \frac{du}{dy} dx + \rho_v g(\delta - y)dx$.

By integrating the previous equation between $y = 0$ and $y = \delta$ with the boundary condition $u = 0$ at $y = 0$, we obtain:

$$u = \frac{(\rho_l - \rho_v)g}{\mu_l}\left(\delta y - \frac{1}{2}y^2\right)$$

The mass flow rate of condensed liquid at a height x (per unit length according to Oz) is given by:

$$\dot{m} = \int_0^\delta \rho_l \left[\frac{(\rho_l - \rho_v)g}{\mu_l}\left(\delta y - \frac{1}{2}y^2\right)\right] dy = \rho_l \frac{(\rho_l - \rho_v)g\delta^3}{3\mu_l}$$

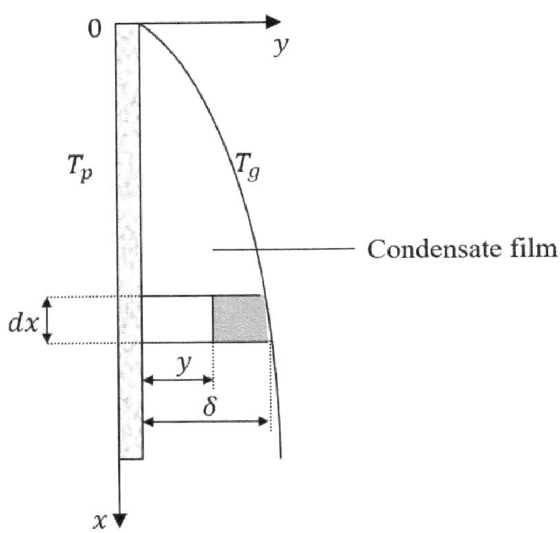

Figure 4.9. *Schematization of condensation on a vertical wall*

The heat flow rate transferred by the condensate to the wall over the height dx is written as:

$$\varphi_x = \lambda dx \left(\frac{\partial T}{\partial y}\right)_{y=0} = \lambda_l dx \frac{T_g - T_p}{\delta}$$

Between the heights x and $x + dx$, the thickness of the liquid film changes from δ to $\delta + d\delta$ due to condensation at height dx. The amount of condensed vapor between x and $x + dx$ is written as:

$$\frac{d}{dx}\left[\rho_l \frac{(\rho_l-\rho_v)g\delta^3}{3\mu_l}\right] dx = \frac{d}{d\delta}\left[\rho_l \frac{(\rho_l-\rho_v)g\delta^3}{3\mu_l}\right]\frac{d\delta}{dx} dx = \rho_l \frac{(\rho_l-\rho_v)g\delta^2 d\delta}{\mu_l}$$

The heat flow rate transferred by the condensate to the wall must be equal to the latent heat of condensation released by the quantity of steam calculated above, i.e.:

$$\rho_l \frac{(\rho_l-\rho_v)g\delta^2 d\delta}{\mu_l} \Delta H = \lambda_l dx \frac{T_g-T_p}{\delta}$$

Integration of this equation with the boundary condition $\delta = 0$ at $x = 0$ leads to:

$$\delta = \left[\frac{4\mu_l\lambda_l x(T_g - T_p)}{g\Delta H\rho_l(\rho_l - \rho_v)}\right]^{\frac{1}{4}}$$

The local convective heat transfer coefficient (at position x) verifies: $h_x dx(T_g - T_p) = \lambda_l dx \frac{T_g-T_p}{\delta}$.

Hence:

$$h_x = \frac{\lambda_l}{\delta}.$$

Thus:

$$h_x = \left[\frac{g\Delta H\rho_l(\rho_l-\rho_v)\lambda_l{}^3}{4\mu_l x(T_g-T_p)}\right]^{\frac{1}{4}}.$$

The average transfer coefficient h_v is obtained by integrating the local coefficient over the height L of the condensing surface: $h_v = \frac{1}{L}\int_0^L h_x dx$.

Thus finally:

$$h_v = \frac{2\sqrt{2}}{3}\left[\frac{g\Delta H\rho_l(\rho_l - \rho_v)\lambda_l{}^3}{\mu_l\, L\, \Delta T}\right]^{\frac{1}{4}} \quad [4.7]$$

with:

– ΔH, the latent heat of condensation (J kg^{-1});

– ΔT, the difference between the dew temperature of the vapor and the temperature of the wall (K);

– L, the height of the wall (m);

– condition of validity: $Re < 2{,}100$.

Consider, for example, the case of a vertical tube of outside diameter D_e.

Given $\dot{m}$ the mass flow rate of condensate and S the flow section of the liquid film, the hydraulic diameter D_h of the film is defined by:

$$D_h = 4\frac{flow\ section}{perimeter} = 4\frac{S}{\pi D_e} \text{ hence } Re = \frac{\rho_l u D_h}{\mu_l} = \frac{\rho_l u}{\mu_l} 4\frac{S}{\pi D_e} = \frac{4\dot{m}}{\mu_l \pi D_e}$$

The condition of validity is therefore written in this case: $\frac{4\dot{m}}{\mu_l \pi D_e} < 2{,}100$.

REMARK.– The physical quantities relating to the liquid are evaluated at the film temperature defined by Drew's formula: $T_f = \frac{3T_p + T_v}{4}$.

4.3.1.2.2. Horizontal tube

An average value of h for a horizontal tube of outside diameter D_e can be calculated by:

$$h_h = 0.725\left[\frac{g\Delta H \rho_l(\rho_l - \rho_v)\lambda_l{}^3}{\mu_l D_e \Delta T}\right]^{\frac{1}{4}} \qquad [4.8]$$

with the validity condition: $Re = \frac{4\dot{m}}{\mu_l \pi D_e} < 2{,}100$.

4.3.1.2.3. Comparison between a horizontal and vertical tube

If we note L_v the length of the vertical tube and D_{e_h} the outside diameter of the horizontal tube, the ratio of the two expressions of h leads to: $\frac{h_h}{h_v} = 0.769\left(\frac{L_v}{D_{e_h}}\right)^{\frac{1}{4}}$.

Hence $h_h > h_v$ if $L_v > 2.86\ D_{e_h}$, which is almost always the case. Under the same temperature conditions, the transfer coefficient is higher on a horizontal tube than on a vertical tube.

In the case of tube bundle condensers, the tubes not being all in the same horizontal plane, the liquid falling from a tube will "thicken" the film that exists on the tube located below it, so that the heat transfer coefficient h is lower on the lower tubes.

Taking into account the recycling of the condensate on the lower tubes, Nusselt proposed the following equation to calculate the average value of h for a set of N tubes located in the same vertical plane:

$$\overline{h_h} = 0.725\left[\frac{g\Delta H\rho_l(\rho_l-\rho_v){\lambda_l}^3}{N\,\mu_l D_e\Delta T}\right]^{\frac{1}{4}} \qquad [4.9]$$

Experience has shown that this theoretical formula gives values of h lower than those determined experimentally and that the value of h given by formula [4.9] should be multiplied by a corrective factor according to the following formula:

$$\overline{h_{h\,corrected}} = \overline{h_h}\left[1+0.2(N-1)\frac{c_p\Delta T}{\Delta H}\right] \qquad [4.10]$$

4.3.2. *Boiling*

4.3.2.1. *Formation of droplets and bubbles*

The equilibrium pressure of a liquid droplet of radius r_g in its vapor is of the form:

$$p_{rg}(T) = p_s(T) + \frac{2\sigma}{r_g}$$

where σ is the surface tension of the liquid (see Figure 4.10). If $r_g \to 0$, then $p_{rg} \to \infty$, so that a droplet could theoretically not originate in a vapor which is a continuous medium.

During the cooling of a vapor at constant pressure, condensation will therefore be initiated on "germs" of very small diameters (dust in suspension in the atmosphere, for example) at a temperature T_g below the saturation temperature $T_s(p)$. The development of condensation will then have the effect of increasing the size of the droplets and reducing the difference between T_g and $T_s(p)$.

Similarly, when a liquid is heated, it is supposed that on the hot walls on which the boiling takes place there are discontinuities (small cavities containing air) which serve as "germs" favoring the birth of small-diameter bubbles at a temperature T_b above the saturation temperature $T_s(p)$. The development of boiling will then have the effect of increasing the size of the bubbles and reducing the difference between T_b and $T_s(p)$.

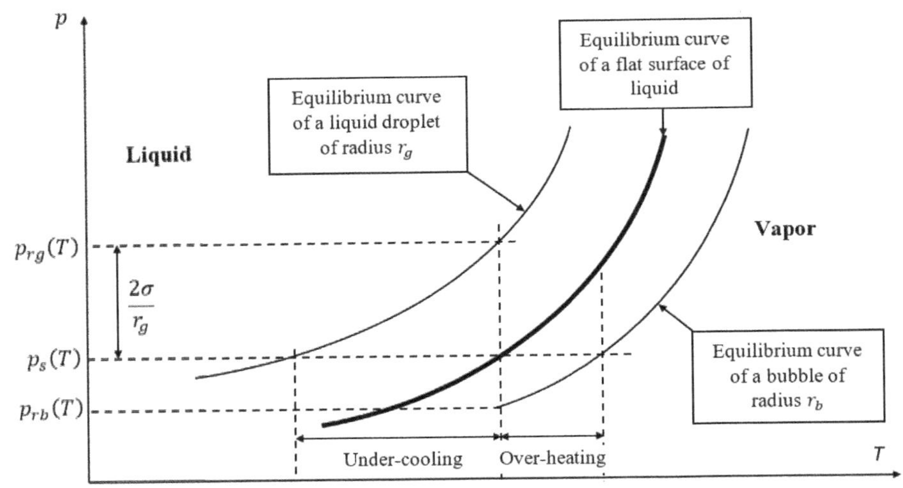

Figure 4.10. *General liquid–vapor equilibrium diagram*

4.3.2.2. *The different boiling regimes*

The variations of the heat transfer coefficient h as a function of the temperature difference $T_p - T_s(p)$, where T_p is the temperature of the heated wall, have the same shape for a large number of liquids, they are represented by the graph of Nukiyama (see Figure 4.11).

4.3.2.3. *Area AB*

Although $T_p > T_s(p)$, there are no bubbles yet. Liquid-wall exchange takes place by natural convection and obeys Newton's law: $\varphi = hS(T_p - T_\infty)$, h is calculated by the correlations concerning natural convection (see section A.18 of the Appendix). Evaporation occurs on the flat, free surface of the liquid in contact with the air.

4.3.2.4. *Area BC*

The bubbles rise in a column from isolated points on the wall: the "sites" with a frequency of around 100 per second. Then, the bubbles become increasingly numerous and isolate the wall with an almost continuous layer of vapor. Heat removal is mainly in the form of latent heat of vaporization. This is the *nucleate boiling* zone.

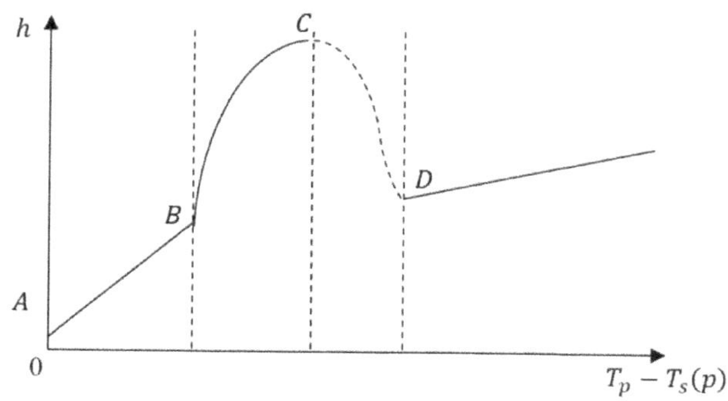

Figure 4.11. *Schematic representation of the Nukiyama graph*

The heat flux density ϕ transferred in this area can be calculated by the following formula (Rosenhow et al. 1985):

$$\frac{c_l \Delta T}{\Delta H \, (Pr_l)^s} = C \left[\frac{\phi}{\mu_l \Delta H} \sqrt{\frac{\sigma}{g(\rho_l - \rho_v)}} \right]^{0.33} \qquad [4.11]$$

where:

– c_l, the specific heat of the liquid;

– ΔT, the temperature difference $T_p - T_s(p)$;

– ΔH, the latent heat of vaporization;

– Pr_l, the Prandtl number of the liquid at saturation;

– σ, the surface tension (value for water in Table 4.3);

– g, the acceleration due to gravity;

– ρ_l, the liquid density;

– ρ_v, the vapor density;

– C, the constant determined experimentally (see values in Table 4.4);

– s, $s = 1$ for water, $s = 1.7$ for other liquids.

Saturation temperature	**°C**	0	15.6	37.8	60	93.3	100	160	226.7	293.3	360	374.1
Surface tension	**10^{-3} N m^{-1}**	75.6	73.3	69.8	66.0	60.1	58.8	46.1	32.0	16.2	1.46	0

Table 4.3. *Surface tension value σ for water (from Holman (1990))*

Configuration	***C***	**Configuration**	***C***
Water–copper	0.013	Benzene–chromium	0.010
Water–platinum	0.013	Ethyl alcohol–chromium	0.027
Water–brass	0.006	n-Pentane–chromium	0.015
Water–copper polished with emery	0.0128	n-Pentane–copper polished with emery	0.0154
Water–polished stainless steel	0.0080	n-Pentane–nickel polished with emery	0.0127
Carbon–copper tetrachloride	0.013	Isopropyl alcohol–copper	0.00225
Carbon tetrachloride–polished copper	0.007	Alcohol n-butyl–copper	0.00305

Table 4.4. *Values of the constant C for various fluid/heating surface configurations (from Holman (1990))*

4.3.2.5. *Point C*

The layer of vapor completely isolates the wall from the liquid and heat can no longer be transmitted except through the vapor of very low thermal conductivity. The sudden increase in the thermal resistance causes a sudden increase in the temperature of the heating wall to a level which will make it possible to evacuate the flow supplied to the wall both by conduction-convection and by radiation. Suddenly moving from point C to point D, where the temperature is well over 1,000°C, the wall has melted in most cases, which is why point C is called the exhaustion burn-out point.

The determination of the burn-out point is crucial in the study of boiling for obvious safety reasons. The correlation most used to determine this burn-out heat flux is the following (Zuber 1958):

$$\phi_{bo} = \frac{\pi\rho_v}{24}\Delta H\left[\frac{\sigma g(\rho_l - \rho_v)}{{\rho_v}^2}\right]^{\frac{1}{4}}\left(1+\frac{\rho_v}{\rho_l}\right)^{\frac{1}{2}} \qquad [4.12]$$

4.3.2.6. *Area CD*

The CD area is an unstable area.

4.3.2.7. *Area DE*

The *film boiling* area in which the transfer of heat from the wall to the liquid takes place by conduction and by radiation through the continuous vapor layer. Heat transfer coefficients can be calculated by (Bromley 1950):

– conduction:

$$h_c = 0.62\left[\frac{\lambda_v{}^3\rho_v(\rho_l-\rho_v)g(\Delta H+0.4c_v\Delta T)}{D\,\mu_v\Delta T}\right]^{\frac{1}{4}} \quad [4.13]$$

where D is the diameter of the heating tube;

– radiation:

$$h_r = \frac{\sigma\varepsilon\left[T_p{}^4-\left(T_s(p)\right)^4\right]}{T_p-T_s(p)} \quad [4.14]$$

– global:

$$h = h_c\left(\frac{h_c}{h}\right)^{\frac{1}{3}} + h_r \quad [4.15]$$

The empirical equation [4.15] requires the use of an iterative method to calculate the global coefficient h.

4.3.2.8. *Interest of heat transfer by boiling*

In addition to water steam generators widely used in the food and textile industries, this type of transfer is used for the extraction of very high calorific powers from very small surfaces (cooling of nuclear reactor cores, rocket engines) due to high values of the transfer coefficients, of the order of 100,000 W m^{-2} K^{-1}.

4.4. Corrected exercises

4.4.1. *Forced convection in and around a tube*

A stainless steel pipe $\lambda_{stainless\ steel} = 14\ \mathrm{W\ m^{-1}K^{-1}}$, with an inside diameter $d_i = 30$ mm and an outside diameter, $d_i = 40$ mm is traversed by a flow of water at an average velocity of 3 m s^{-1} and at a temperature $T_i = 70°C$. Its inner wall is covered with a thin deposit ($e = 0.5$ mm) of material whose conductivity is $\lambda_{deposit} = 0.06\ \text{W m}^{-1}\text{K}^{-1}$.

Outside the steel layer, an air flow circulates normally to the surface at a velocity of 8 m s^{-1} and at a temperature of $T_e = 15°C$.

1) Calculate the internal and external exchange coefficients h_i and h_e using classical correlations.

2) Calculate each thermal resistance between the liquid and the air and give the equivalent electrical diagram.

3) Calculate the temperature on the outer side of the deposit and on the inner and outer sides of the steel tube.

Data: for air: $\lambda = 0.026\ \text{W m}^{-1}\text{K}^{-1}$; $\nu = 1.57 \times 10^{-5}\text{m}^2\text{s}^{-1}$; $Pr = 0.7$, for water: $\lambda = 0.65\ \text{W m}^{-1}\text{K}^{-1}$; $\nu = 0.048 \times 10^{-5}\text{m}^2\text{s}^{-1}$; $Pr = 3$.

Corrected

1) $d_0 = d_i - 2e = 30 - 2 \times 0.5 = 29$ mm.

$$Re_i = \frac{u_i d_0}{\nu_i} = \frac{3\times0.029}{0.048.10^{-5}} = 181\ 250\text{, the flow is turbulent}$$

$$Nu_i = 0.023 Re_i^{0.8} Pr_i^{0.3} = 0.023 \times 181\ 250^{0.8} \times 3^{0.3} = 214.62$$

$$h_i = \frac{\lambda_i Nu_i}{d_0} = \frac{0.65\times514.62}{0.029} = 11\ 534\ \text{W m}^{-2}\text{K}^{-1}$$

$$Re_e = \frac{u_e d_e}{\nu_e} = \frac{8\times0.04}{1.57.10^{-5}} = 20{,}382$$

$$Nu_e = C Re_e^{\ n} Pr_e^{\ 1/3} = 0.193 \times 20\ 383^{0.618} \times 0.7^{1/3} = 78.89$$

$$h_e = \frac{\lambda_e Nu_e}{d_e} = \frac{0.026\times78.89}{0.04} = 51.3\ \text{W m}^{-2}\text{K}^{-1}$$

2)

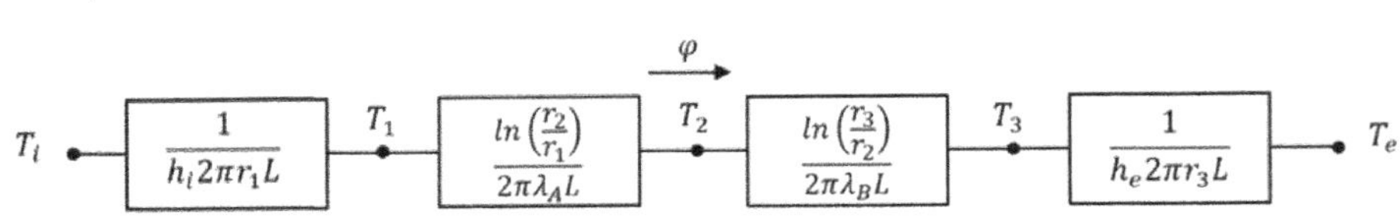

$R_1 = \frac{1}{h_i \pi d_O L} = \frac{1}{11{,}534 \times \pi \times 0.029} = 0.00095\ \mathrm{K\,W^{-1}\,m}$; $R_2 = \frac{ln\left(\frac{d_i}{d_O}\right)}{2\pi\lambda_{dépôt}L} = \frac{ln\left(\frac{0.03}{0.029}\right)}{2\pi \times 0.06} = 0.0899\ \mathrm{K\,W^{-1}\,m}$

$R_3 = \frac{ln\left(\frac{de}{d_{Oi}}\right)}{2\pi\lambda_{inox}L} = \frac{ln\left(\frac{0.04}{0.03}\right)}{2\pi \times 14} = 0.00327\ \mathrm{K\,W^{-1}\,m}$; $R_4 = \frac{1}{h_e \pi d_e L} = \frac{1}{51.3 \times \pi \times 0.04} = 0.1552\ \mathrm{K\,W^{-1}\,m}$

$$R = R_1 + R_2 + R_3 + R_4 = 0.2493\ \mathrm{K\,W^{-1}\,m}$$

3) $\varphi = \frac{T_i - T_e}{R} = \frac{55}{0.2493} = 220.6\ \mathrm{W\,m^{-1}}$.

$$T_1 = T_i - \varphi R_1 = 70 - 2\,206 \times 0.00095 = 69.8°\mathrm{C}$$

$$T_2 = T_i - \varphi(R_1 + R_2) = 70 - 2\,206 \times (0.00095 + 0.0899) = 50.0°\mathrm{C}$$

$$T_3 = T_i - \varphi(R_1 + R_2 + R_3) = 70 - 2\,206 \times (0.00095 + 0.0899 + 0.00327) = 49.2°\mathrm{C}$$

4.4.2. *Water flow in a heating tube*

A flow of water of 10 kg min^{-1} is heated from $T_e = 20°\mathrm{C}$ to $T_s = 90°\mathrm{C}$ during its passage through a tube with an inside diameter of 20 mm and a length of 3 m. The tube is equipped with a heating resistor that produces a uniform heat flow rate over its entire surface. It is considered first that the outer surface of the tube is perfectly insulated.

1) Calculate the power dissipated by the resistor and the heat flow rate per unit length of tube.

2) Calculate the convective heat transfer coefficient h_i between the water and the inner wall of the tube.

3) Show that the temperature difference ΔT between the inner wall of the tube and the water is constant and calculate the temperature of the tube T_p at its end on the side of the water outlet.

We no longer neglect in the following the external heat losses by convection and by radiation. The copper tube has a thickness of 2 mm, a thermal conductivity $\lambda_{Cu} = 400\ \mathrm{W\,m^{-1}K^{-1}}$ and an emissivity $\varepsilon = 0.9$. It is arranged horizontally in air at $T_e = 20°\mathrm{C}$.

4) Calculate the transfer coefficients h_{ce} by natural convection (the flow is laminar) and h_{re} by radiation on the outer wall of the tube. These values will be calculated at the inlet and at the outlet of the tube.

5) Calculate the water outlet temperature if the heat flow rate is maintained at the value calculated in 1. We can consider average values of h_{ce} and of h_{re} to carry out this calculation.

6) Calculate the temperature of the outer surface of the tube at its end on the water outlet side.

7) Compare with the initial value of 90°C (perfectly insulated tube) and show that this result was predictable.

Data for water: $\rho = 993 \text{ kg m}^{-3}$; $\mu = 6.83 \times 10^{-4} \text{ kg m}^{-1}\text{s}^{-1}$; $c = 4{,}174 \text{ J kg}^{-1}\text{K}^{-1}$; $\lambda = 0.63 \text{ W m}^{-1}\text{K}^{-1}$.

Corrected

1) $\varphi = \dot{m}c\,\Delta T = \frac{10}{60} \times 4\,174 \times 70 = 48\,697 \text{ W}.$

$$\phi = \frac{\varphi}{L} = 16\,232 \text{ W m}^{-1}$$

2) $Re = \frac{\rho u D}{\mu}.$

$$u = \frac{\dot{m}}{\rho \pi R_i{}^2} = \frac{10}{60 \times 993 \times \pi \times 0.01^2} = 0.534 \text{ m s}^{-1}$$

$$Re = \frac{993 \times 0.534 \times 0.02}{6.82.10^{-4}} = 15\,558;\ Pr = \frac{c\mu}{\lambda} = \frac{4{,}174 \times 6.82.10^{-4}}{0.63} = 4.519$$

$$Nu = 0.023\, Re^{0.8} Pr^{0.4} = 0.023 \times 15\,558^{0.8} \times 4.519^{0.4} = 94.9$$

$$h_i = \frac{\lambda Nu}{D_i} = \frac{0.63 \times 94.9}{0.02} = 2{,}989 \text{ W m}^{-2}\text{K}^{-1}$$

3) $\phi = h_i \pi D_i\, \Delta T$ so if ϕ is constant then ΔT is constant.

$$\Delta T = \frac{\phi}{h_i \pi D_i} = \frac{16{,}232}{2{,}989 \times \pi \times 0.02} = 86.4°\text{C}$$

$$T_{p_{x=L}} = 90 + 86.4 = 176.4°\text{C}$$

4) $h_{ce} = 1.26\left(\frac{T_t - T_a}{D_e}\right)^{0.25}$.

At the inlet: $h_{ce} = 1.26\left(\frac{86.4}{0.022}\right)^{0.25} = 10.0\ \mathrm{W\ m^{-2}\ K^{-1}}$.

At the outlet: $h_{ce} = 1.26\left(\frac{70+86.4}{0.022}\right)^{0.25} = 11.6\ \mathrm{W\ m^{-2}\ K^{-1}}$.

$$h_{re} = \sigma\varepsilon\left(T_t^{\ 2} + T_a^{\ 2}\right)\left(T_t + T_a\right)$$

At the inlet: $h_{ce} = 5.67.10^{-8} \times 0.9 \times ((293 + 86.4)^2 + 293^2)\ (293 + 86.4 + 293)$.

$$h_{re} = 7.88\ \mathrm{W\ m^{-2}\ K^{-1}}$$

At the outlet: $h_{ce} = 5.67.10^{-8} \times 0.9 \times ((363 + 86.4)^2 + 293^2)\ (363 + 86.4 + 293)$.

$$h_{re} = 10.9\ \mathrm{W\ m^{-2}\ K^{-1}}$$

Average values:

$h_{ce} = 10.8\ \mathrm{W\ m^{-2}\ K^{-1}}$ and $h_{re} = 9.4\ \mathrm{W\ m^{-2} K^{-1}}$; thus $h_e = 20.2\ \mathrm{W\ m^{-2}\ K^{-1}}$

5) The thermal resistance of the copper tube ($\lambda_{Cu} = 400\ \mathrm{W\ m^{-1} K^{-1}}$) is negligible compared to the equivalent thermal resistance (convection and radiation in parallel) between the tube and the outside; therefore the temperature difference between the internal and external surface of the tube is negligible, and we will call T_p the tube temperature.

Balances between x and $x + dx$:

– on the fluid portion: $\dot{m}cdT = h_i\pi D_i dx\left(T_p - T\right)$;

– on the piece of tube: $\phi = h_i\pi D_i\left(T_p - T\right) + h_e\pi D_e\left(T_p - T_a\right)$;

– thus: $\phi = \left(h_i\pi D_i + h_e\pi D_e\right)\left(T_p - T\right) + h_e\pi D_e\left(T - T_a\right)$.

We deduce: $T_{x=L} = T_a + \frac{\phi}{h_e\pi D_e}\left[1 - exp\left(-\frac{h_e\pi D_e L}{A}\right)\right]$.

where: $A = \dot{m}c\frac{h_i D_i + h_e D_e}{h_i D_i}$.

We find: $T_{x=L} = 89.3°C$, this value is very close to the value obtained with a perfectly insulated tube (90°C). This result was predictable because $h_i \gg h_e$.

4.4.3. *Air cooling in a duct*

A blast furnace is blown with hot air at an absolute pressure of 3×10^5 Pa and at a temperature $T_a = 1{,}092$ K.

Hot air circulates at high velocity $V = 34.7 \text{ m s}^{-1}$ in a horizontal supply pipe of internal diameter $D_i = 1.4$ m before being distributed in the nozzles of the blast furnace.

1) Calculate the exchange coefficient h_i between the air and the inner wall of the pipe. The choice of the temperature retained for the calculation of the properties of the air will be justified a posteriori.

2) The walls of the pipe are made of, going from the inside to the outside:

- a cylindrical layer of refractory bricks 0.2 m thick;
- a cylindrical layer of insulating bricks 0.2 m thick;
- a thin steel sheet.

Outside: on the one hand, the sheet metal exchanges heat by radiation with an environment at an average temperature of 20°C, and on the other hand, it exchanges heat by natural convection.

Give the equivalent electrical diagram.

3) Show that the natural convective flow outside the tube is turbulent and choose a simplified correlation to calculate the convective transfer coefficient $h_{e\,conv}$.

4) Give the literal expression of the overall thermal resistance between the hot air and the outside.

5) Write the system of equations to be solved and propose a simplified solution method by calculating the properties of the outside air at 293 K.

6) Calculate the heat flow rate and the temperatures at the interfaces. Were the assumptions for the calculation of the thermal properties of the air justified?

7) Calculate the temperature drop for 10 m of pipe.

Data:

– air at 1,092 K and 3.10^5 Pa: $\rho = 0.97 \text{ kg m}^{-3}$, $c = 1{,}165 \text{ J kg}^{-1}\text{K}^{-1}$, $\lambda = 0.0675 \text{ W m}^{-1}\text{K}^{-1}$, $\mu = 4.4 \times 10^{-5} \text{ kg m}^{-1}\text{s}^{-1}$;

– air at 293 K and 10^5 Pa: $\rho = 1.205 \text{ kg m}^{-3}$, $c = 1{,}000 \text{ J kg}^{-1}\text{K}^{-1}$, $\lambda = 0.0256 \text{ W m}^{-1}\text{K}^{-1}$, $\mu = 1.8 \times 10^{-5} \text{ kg m}^{-1}\text{s}^{-1}$;

– refractory bricks: $\lambda = 1.28 \text{ W m}^{-1}\text{K}^{-1}$; insulating bricks: $\lambda = 0.35 \text{ W m}^{-1}\text{K}^{-1}$.

Corrected

1) $Re_i = \frac{\rho V D_i}{\mu} = \frac{0.97\times 34.7\times 1.4}{4.4.10^{-5}} = 1.071 \times 10^6$.

$$Pr_i = \frac{c\mu}{\lambda} = \frac{1{,}165\times 4.4\times 10^{-5}}{0.0675} = 0.759.$$

$$Nu_i = 0.023\, Re_i^{\,0.8} Pr_i^{\,0.3} = 1{,}412.$$

$$h_i = \frac{\lambda Nu_i}{D_i} = \frac{0.675\times 1{,}412}{1.4} = 68.1 \text{ W m}^{-2}\text{ K}^{-1}.$$

2)

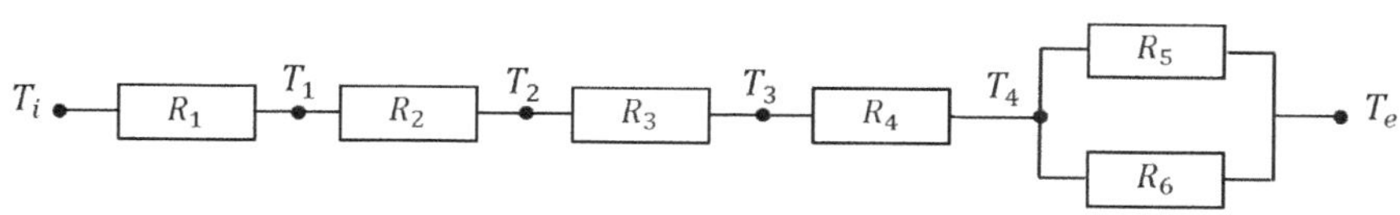

3) We have: $Gr = \frac{\beta g \Delta T \rho^2 D^3}{\mu^2} = \frac{9.81\times \Delta T\times 1.2^2\times 2.2^3}{293\times(1.8\times 10^{-5})^2} = 1.6 \times 10^9 \Delta T$ taking $\beta = \frac{1}{T}$.

$Gr > 10^9$, the flow is turbulent and we can retain: $h = 1.52\,(\Delta T)^{1/3}$ (see section A.18 of the Appendix).

4) Resistances are written as:

$$R_1 = \frac{1}{h_i 2\pi r_1};\ R_2 = \frac{ln\left(\frac{r_2}{r_1}\right)}{2\pi\lambda_1};\ R_3 = \frac{ln\left(\frac{r_3}{r_2}\right)}{2\pi\lambda_2};\ R_4 = \frac{ln\left(\frac{r_4}{r_3}\right)}{2\pi\lambda_3};\ R_5 = \frac{1}{h_{ce}2\pi r_4};\ R_6 = \frac{1}{h_{re}2\pi r_4}.$$

with:

$$h_{re} = \sigma\varepsilon\left(T_4^{\,2} + T_e^{\,2}\right)(T_4 + T_e) \text{ and } h_{ce} = 1.52\,(T_4 - T_e)^{1/3}.$$

5) $R = R_1 + R_2 + R_3 + R_4 + \frac{1}{\frac{1}{R_5}+\frac{1}{R_6}}$.

$$\varphi = \frac{T_i + T_e}{R}$$

$$T_4 = T_i - (R_1 + R_2 + R_3 + R_4)\varphi$$

6) Proposed resolution method:

- we set $T_4 = T_e + 50°C$;
- we calculate h_{re}, h_{ce}, R and φ;
- we recalculate T_4 by $T_4 = T_i - (R_1 + R_2 + R_3 + R_4)\varphi$;
- the calculation is repeated with this value of T_4 until convergence.

The steel sheet being thin and of high thermal conductivity, its thermal resistance has been neglected.

We find that two iterations are sufficient, and we end up with:

$$T_1 = 1{,}072.5 \text{ K}, T_2 = 890.3 \text{ K and } T_3 = T_4 = 358.3 \text{ K}$$

We obtain $T_1 \approx T_i$ validating the hypothesis $T_1 = T_i$.

7) The thermal balance on the air between x and $x + dx$ is written as:

$$\dot{m}cdT_i = -dx\frac{T_i - T_e}{R}$$

We deduce:

$$T_i(L) = T_a + [T_i(0) - T_a]exp\left(-\frac{L}{\dot{m}cR}\right) =$$

$$293 + (1{,}092 - 293)exp\left(-\frac{10}{0.97 \times 34.7 \times \frac{\pi \times 1.4^2}{4} \times 1{,}165 \times 0.137}\right)$$

$$T_i(L) = 1{,}091.0 \text{ K}$$

4.4.4. *Permeable-dynamic insulation of a house*

Consider a house with a square floor area of side a and height h. The exterior vertical walls consist of air-permeable walls of thickness L and thermal conductivity λ. To obtain good air quality in the house, it is necessary to renew the ambient air. In order to reduce heat losses through the walls, we plan to expel the volume of used air to the outside through the walls. The purpose of the exercise is to evaluate, under very simple assumptions, the benefit that can be derived from such so-called *permeable-dynamic insulation*.

1) If v_a denotes the air velocity per unit area of the wall, write in steady state the convection–diffusion equation describing the temperature field $T(x)$ through the wall (the air and the wall are locally at the same temperature).

2) Solve the previous equation assuming that the wall temperature is imposed inside ($x = 0$, $T = T_0$) and outside ($x = L$, $T = T_L$) of the house. Examine the cases where the Peclet number $Pe = \frac{(\rho c)_a v_a L}{L}$ tends to 0 or infinity.

3) If $T_0 > T_L$, calculate the heating power required to maintain the interior temperature at T_0 distinguishing the contributions due to air renewal and heat losses through the walls in the following two cases:

- if the used air is expelled outside through the vertical walls;
- if the used air is expelled directly outside.

In both cases, it will be assumed that the fresh air (replacing the extracted used air) is admitted at the temperature T_L and that the losses through the floor and through the ceiling are negligible.

4) What gain in heating can be expected from *permeable–dynamic insulation* by comparing it to the case where the air is expelled directly outside?

Numerical application: for air: $(\rho c)_a = 1{,}200\ \mathrm{J\,m^{-3}K^{-1}}$; for the wall: $\lambda = 0.5\ \mathrm{W\,m^{-1}K^{-1}}$and $L = 0.25$ m; we will consider a renewal of the volume of air every hour; for home $a = 14.4$ m; $h = 2.5$ m.

Corrected

1) The convective transport equation through the wall is written as:

$$(\rho c)_a v_a \frac{dT}{dx} - \lambda \frac{d^2T}{dx^2} = 0$$

2) The solution is elementary:

$$\frac{T-T_0}{T_L-T_0} = \frac{exp\left(Pe\frac{x}{L}\right)-1}{exp(Pe)-1} \text{ with } Pe = \frac{(\rho c)_a v_a L}{\lambda}$$

If $Pe \to 0$, $\frac{T-T_0}{T_L-T_0} \to \frac{x}{L}$ (pure conduction).

If $Pe \to \infty$, the temperature is equal everywhere to T_0 except in a thin boundary layer in the vicinity of $x = L$ in order to satisfy the boundary condition at $x = L$.

3) Calculation of the heating power required with permeable–dynamic insulation:

$$\varphi_{conduction} = -\lambda S\left(\frac{dT}{dx}\right)_{x=0} = \lambda S\frac{T_0-T_L}{L}\frac{Pe}{exp(Pe)-1}$$

$$\varphi_{renewal} = (\rho c)_a v_a (T_0 - T_L)S$$

$$\varphi_{heating_1} = (T_0 - T_L)S\left[\frac{\lambda}{L}\frac{Pe}{exp(Pe)-1} + (\rho c)_a v_a\right] = \frac{\lambda Pe}{L}\left[\frac{exp(Pe)}{exp(Pe)-1}\right](T_0 - T_L)S$$

Calculation of the heating power required with direct extraction:

$$\varphi_{conduction} = -\lambda S\left(\frac{dT}{dx}\right)_{x=0} = \lambda S\frac{T_0-T_L}{L}$$

$$\varphi_{renewal} = (\rho c)_a v_a (T_0 - T_L)S$$

$$\varphi_{heating_2} = \frac{\lambda}{L}(1 + Pe)(T_0 - T_L)S$$

4) $\text{gain} = \frac{\varphi_{heating_2}-\varphi_{heating_1}}{\varphi_{heating_2}} = \frac{\frac{\lambda}{L}+(\rho c)_a v_a-\frac{\lambda}{L}\frac{Pe}{exp(Pe)-1}-(\rho c)_a v_a}{\frac{\lambda}{L}+(\rho c)_a v_a}$.

$$\text{gain} = \frac{\frac{\lambda}{L}-\frac{\lambda}{L}\frac{Pe}{exp(Pe)-1}}{\frac{\lambda}{L}+(\rho c)_a v_a} = \frac{1-\frac{Pe}{exp(Pe)-1}}{1+Pe}$$

Numerical application:

– the air flow is: $\dot{m} = \frac{a^2h}{t_r}$ with $t_r = 3{,}600$ s (renewal of a volume of air every hour);

– the flow passes through the total surface $S = 4ah$.

Hence:

$$v_a = \frac{\dot{m}}{S} = \frac{a^2h}{t_r}\frac{1}{4ah} = \frac{a}{4t_r} = 10^{-3}\,\text{m s}^{-1}.$$

$$Pe = \frac{(\rho c)_a v_a L}{L} = \frac{1{,}200\times10^{-3}\times0.25}{0.5} = 0.6$$

$$gain = \frac{1-\frac{Pe}{exp(Pe)-1}}{1+Pe} = 17\%$$

4.4.5. Convection in a chimney

A vertical flue is considered, into which flue gas enters at a temperature $T_{fe} = 400°C$. The flue is a steel cylinder with an inside diameter of 30 cm, a thickness of 0.5 mm and a height of 3 m, its emissivity is equal to 0.9. Its outer surface is in contact with air at a constant temperature $T_a = 30°C$ and equal to that of the surrounding medium. The fumes circulation velocity inside the flue is $u = 3 \text{ m s}^{-1}$. The fumes will be considered to have the same thermal properties as air. We will make the initial hypothesis that the average temperature of the fumes is 365°C and we will check the validity of this hypothesis a posteriori at the end of the problem.

1) Calculate the internal heat transfer coefficient.

2) Calculate the global (convection + radiation) external heat transfer coefficient.

3) Calculate the global heat transfer coefficient between the fumes and the air.

4) Calculate the flue gas outlet temperature.

Corrected

1) The average flue gas temperature is: $\frac{400+330}{2} = 365°C$.

The average temperature of the flue verifies: $2\, h_e(365 - T_p) = h_e(T_p - 30)$, hence: $T_p = 250°C$.

The internal film temperature is equal to: $T_{f_i} = \frac{365+250}{2} = 300°C$.

Calculation of h_i:

$\rho = 0.616 \text{ kg m}^{-3}$; $c = 1{,}045 \text{ J kg}^{-1}\text{K}^{-1}$; $\lambda = 0.045 \text{ W m}^{-1}\text{K}^{-1}$; $\mu = 2.93 \times 10^{-5} \text{ Pa s}^{-1}$; $Pr = 0.68$

$$Re = \frac{\rho u D}{\mu} = \frac{0.616 \times 3 \times 0.03}{2.93 \times 10^{-5}} = 1{,}892$$

$$Nu = 0.023\, Re^{0.8} Pr^{0.3} = 8.57$$

$$h_i = \frac{\lambda Nu}{D} = \frac{0.0450 \times 8.57}{0.03} = 12.9 \text{ W m}^{-2}\text{K}^{-1}$$

2) The outer film temperature is equal to: $T_{fe} = \frac{250+30}{2} = 140°C$.

$\rho = 0.854 \text{ kg m}^{-3}$; $c = 1{,}016 \text{ Jkg}^{-1}\text{K}^{-1}$; $\lambda = 0.0345 \text{ W m}^{-1}\text{K}^{-1}$; $\mu = 2.34 \times 10^{-5} \text{ Pa s}^{-1}$; $Pr = 0.69$

$$h_{re} = \sigma\varepsilon\left(T_p{}^2 + T_a{}^2\right)\left(T_p + T_a\right) = 5.67 \times 10^{-8} \times 0.9\,(580^2 + 303^2)\,(580 + 303)$$

$$h_{re} = 19.3 \text{ W m}^{-2}\text{K}^{-1}$$

$$Gr = \frac{\beta\, g\, \Delta T\, \rho^3 L^3}{\mu^2} = \frac{9.81 \times (250-30) \times 0.854^3\ x\ 3^2}{383 \times (2.34\times10^{-5})^2} = 5.35.\,10^{10}$$

$$Gr\,Pr = 5.35.\,10^{10} \times 0.68 = 3.64 \times 10^{10}$$

From this we deduce: $h_{ce} = 1.52\,\Delta T^{1/3} = 1.52 \times 220^{0.33} = 9.0 \text{ W m}^{-2}\text{K}^{-1}$.

3) $\frac{1}{h} = \frac{S_e}{h_i S_i} + \frac{1}{h_e} = \frac{0.031}{12.9 \times 0.03} + \frac{1}{9.0+19.3}$ hence: $h = 8.6 \text{ W m}^{-2}\text{K}^{-1}$.

4) The average flue temperature is around: $\frac{400+330}{2} = 365°\text{C}$.

$$\rho = 0.553 \text{ kg m}^{-3}; c = 1{,}050 \text{ J kg}^{-1}\text{K}^{-1}$$

The fumes mass flow rate is: $\dot{m} = \rho V S = 0.553 \times 3 \times \pi \times 0.15^2 = 0.117 \text{ kg s}^{-1}$.

The air is at a constant temperature, the thermal balance of the mass of fumes comprised between x and $x + dx$ is therefore written as: $-\dot{m}c\frac{dT}{dx} = hS_e(T - T_a)dx$.

Hence:

$$T_{fs} = T_a + \left(T_{fe} - T_a\right) exp\left(-\frac{hS_e}{\dot{m}c}\right) = 30 + (400 - 30) exp\left(-\frac{8.6\times\pi\times0.301\times3}{0.117\times1{,}050}\right)$$

We deduce: $T_{fs} = 333.4°\text{C}$.

The average chimney duct gas temperature can be recalculated: $\overline{T_f} = \frac{T_{fe}+T_{fs}}{2} = \frac{400+333.4}{2} = 366.7°\text{C}$. This value is close to the value $\overline{T_f} = 365°\text{C}$ chosen initially to calculate the various parameters. If this had not been the case, we would have resumed the calculations with $\overline{T_f} = 367°\text{C}$.

4.4.6. *Modeling natural convection in double glazing*

We consider the permanent natural convection flow established between two infinitely long vertical parallel plates separated by a distance $2b$ at respective temperatures T_1 and T_2.

The device is arranged in such a way that the average fluid volume flow through a section ($x = constant$) is zero.

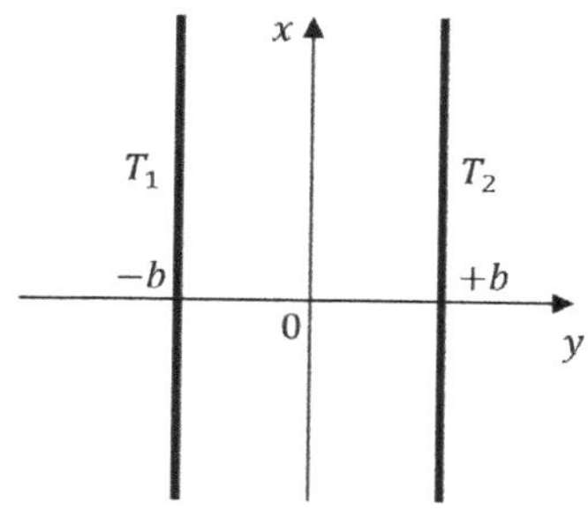

The fluid is Newtonian with dynamic viscosity μ, thermal conductivity λ and constant specific heat c_p. We admit that its density depends only on the temperature T and is written in the vicinity of $T_m = \frac{T_1+T_2}{2}$ as:

$$\rho = \rho_m[1 - \beta(T - T_m)] \text{ with: } \beta = -\frac{1}{\rho}\left(\frac{\partial \rho}{\partial T}\right)_P$$

1) Preliminaries.

1.1) Give a physical description the flow.

1.2) What is the value β in the case of an ideal gas?

2) Conservation equations.

2.1) Write the mass conservation equation assuming a priori the flow in isovolume evolution.

2.2) Write the equations of motion in general.

2.3) Write the energy conservation equation neglecting the viscous dissipation term and the term related to spatial pressure variations.

2.4) How are the previous equations simplified here, in particular because the flow and the temperature field are permanent and established in x?

3) Thermal field resolution.

3.1) Solve the energy equation and give the expression for the temperature field between the plates.

3.2) What is the Nusselt number Nu characterizing the heat exchange between the two plates using the half-space b between the plates as the characteristic length?

3.3) Is the hypothesis of 2.1 verified?

4) Dynamic field resolution.

4.1) Applying the results of the previous paragraph and using the equation of state $\rho(T)$, rewrite the equations of motion.

4.2) Prove that the function $u(y)$ is odd. What can be said about the pressure field p?

4.3) Put the following equation of motion along Ox in dimensionless form by introducing the following dimensionless quantities: $u^* = \frac{\rho_m u b}{\mu}$; $y^* = \frac{y}{b}$; $Gr = \frac{\rho_m{}^2 g b^3 \beta \Delta T}{\mu^2}$.

Give the physical meaning of the Grashof number Gr.

4.4) Give the expression of the reduced velocity profile u^*. Qualitatively represent its shape.

5) Optimization of double glazing.

We wish to use the previous results to study the optimum spacing to be given to double glazing so as to ensure maximum thermal insulation.

5.1) What conclusion do we draw from 3? Does it seem consistent with your intuition of phenomena?

5.2)

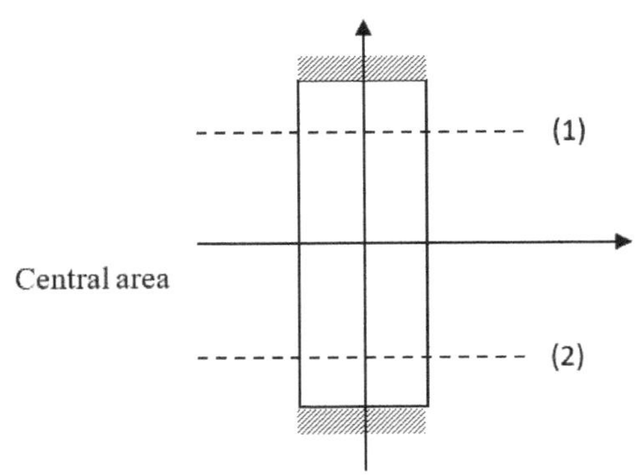

In fact, we will take into consideration the heat transfer at the ends of the double glazing, the horizontal faces of which are thermally insulated. The double glazing has a sufficiently large height ℓ ($\ell \gg b$) so that in the central part it can be considered that the flow is dynamically and thermally established. We now assume that $\rho = \rho_m$.

What is the enthalpy flow Q carried by the fluid in each of sections (1) and (2)?

5.3) Indicate on a diagram how the heat should be exchanged at the ends of the double glazing. In particular, what is the minimum value of the heat rate that must be exchanged at the ends? We will retain this value for the following.

5.4) By simply adding the heat rates exchanged over the height ℓ due to the transfer calculated in 3 and at the ends at the previous question, give the expression for the overall heat exchange coefficient h by relating all the exchanges to the side surface of the glazing. Express the global Nusselt number relative to the half-spacing b as a function of the dimensionless number $GrPr\frac{b}{\ell}$.

5.5) The height ℓ being given, calculate the value of the ratio $GrPr\frac{b}{\ell}$ leading to a minimal global heat exchange coefficient h as a function of the Rayleigh number $Ra = \frac{g\ell^3\beta\Delta T}{\nu_m a_m}$ with the height ℓ of the double glazing as the characteristic quantity and where ν_m and a_m represent, respectively, the kinematic viscosity and the thermal diffusivity of the fluid measured at the temperature T_m. What is the Nusselt number Nu?

5.6) Numerical application: $T_1 = 0°C$; $T_2 = 20°C$.

For air at $T_m = 10°C$: $\nu_m = 1.4 \times 10^{-5} m^2 s^{-1}$ and $a = 1.9 \times 10^{-5} m^2 s^{-1}$.

What is the optimal spacing of double glazing of height $\ell = 1$ m?

Corrected

1.1) The density of the hot fluid is lower than that of the cold fluid. In order to obtain a zero average flow rate, there is an upward movement of the fluid on the side of the hot plate and downward on the side of the cold plate due to buoyancy forces.

1.2) $\rho = \frac{MP}{RT}$ and $\beta = -\frac{1}{\rho}\left(\frac{\partial\rho}{\partial T}\right)_P = \frac{1}{T}$.

2.1) $\frac{D\rho}{Dt} + \rho\vec{\nabla}.\vec{v} = 0$, i.e. $\frac{\partial u}{\partial x} + \frac{\partial v}{\partial y} = 0$.

2.2) $\rho\left(\frac{\partial u}{\partial t} + u\frac{\partial u}{\partial x} + v\frac{\partial u}{\partial y}\right) = -\frac{\partial P}{\partial x} - \rho g + \mu\left(\frac{\partial^2 u}{\partial x^2} + \frac{\partial^2 u}{\partial y^2}\right)$.

$$\rho\left(\frac{\partial v}{\partial t} + u\frac{\partial v}{\partial x} + v\frac{\partial v}{\partial y}\right) = -\frac{\partial P}{\partial y} + \mu\left(\frac{\partial^2 v}{\partial x^2} + \frac{\partial^2 v}{\partial y^2}\right).$$

2.3) $\rho c\left(\frac{\partial T}{\partial t} + u\frac{\partial T}{\partial x} + v\frac{\partial T}{\partial y}\right) = \lambda\left(\frac{\partial^2 T}{\partial x^2} + \frac{\partial^2 T}{\partial y^2}\right)$.

2.4) $\frac{\partial}{\partial t} = 0$, steady state.

$\frac{\partial u}{\partial x} = \frac{\partial v}{\partial x} = \frac{\partial T}{\partial x} = 0$, established regime.

According to 2.1., we get $\frac{\partial v}{\partial y} = 0$ from where: $v = cste\ = 0$ (at the wall).

$$0 = -\frac{\partial P}{\partial x} - \rho g + \mu \frac{\partial^2 u}{\partial y^2}$$

$$0 = -\frac{\partial P}{\partial y}$$

$$0 = \frac{d^2 T}{dy^2}$$

3.1) $T(y) = T_m + \frac{\Delta T}{2}\left(\frac{y}{b}\right)$ with: $T_m = \frac{T_1+T_2}{2}$ and: $\Delta T = T_2 - T_1$.

3.2) $q = \lambda \frac{\Delta T}{2b} = h\Delta T$; $Nu = \frac{hb}{\lambda} = \frac{1}{2}$.

3.3) $\frac{D\rho}{Dt} = \frac{\partial \rho}{\partial t} + u\frac{\partial \rho}{\partial x} + v\frac{\partial \rho}{\partial y} = 0$.

The flow is isovolume.

4.1) $0 = -\frac{\partial P}{\partial x} - \rho_m[1 - \beta(T - T_m)]g + \mu \frac{\partial^2 u}{\partial y^2}$.

4.2) $0 = -\frac{\partial P}{\partial y}$.

The system is symmetric with respect to $y = 0$, the velocity is ascending on one side and descending on the other, so the velocity function $u(y)$ is odd.

Hence, in $y = 0$: $u = 0$ and $\frac{\partial^2 u}{\partial y^2} = 0$.

$$\left(\frac{\partial P}{\partial x}\right)_{y=0} = -\rho_m g \text{ and } \frac{\partial P}{\partial y} = 0, \text{ hence: } \frac{\partial P}{\partial x} = -\rho_m g$$

The pressure field is “hydrostatic”.

4.3) We then have along *Ox:* $\rho_m g\beta(T - T_m) + \mu \frac{\partial^2 u}{\partial y^2} = 0$.

Hence: $\frac{Gr}{2} y^* + \frac{d^2 u^*}{dy^{*2}} = 0$.

Gr represents the ratio between the forces due to buoyancy and viscosity forces.

4.4) $u^*(0) = 0$ and $u^*(1) = 0$.

We get by integration: $u^* = \frac{Gr}{12}\left(y^* - y^{*3}\right)$.

The indicative shape of the velocity profile is as follows:

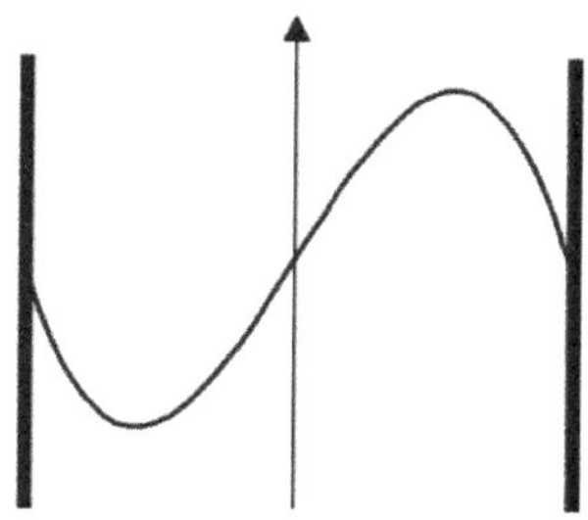

5.1) From 3, we deduce that h will be all the more low as b will be high. Intuitively speaking, it seems logical to imagine that if the thickness of the double glazing is small, the transfer by convection will be more blocked.

5.2) Let $\phi = \int_{-b}^{+b} \rho c(T - T_m)\, u dy$.

$$\phi = c\mu Gr \frac{\Delta T}{2} \int_{-1}^{+1} \left(y^* - y^{*3}\right) y^* dy^* = c\mu Gr \frac{\Delta T}{24} \times 2 \times \left(\frac{1}{3} - \frac{1}{5}\right) = \frac{c\mu Gr \Delta T}{90}.$$

5.3)

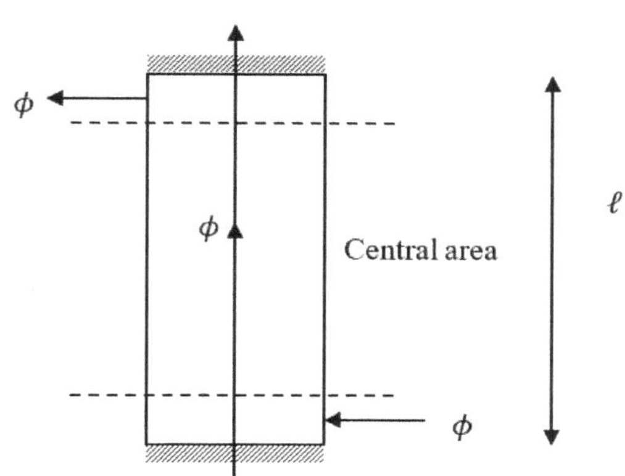

At the ends, the exchange can only take place at the bottom from the hot wall towards the colder fluid and vice versa at the top. Hence, the configuration of the exchanges shown in the figure above corresponds to the minimum flow exchanged.

5.4) $hS\Delta T = \lambda S\frac{\Delta T}{2b} + c\mu\frac{Gr\Delta T}{90}$ with $S = 1 \times \ell$ (unit depth).

Hence: $h = \frac{\lambda}{2b} + \frac{c\mu}{\ell}\frac{Gr}{90}$.

And: $Nu = \frac{hb}{\lambda} = \frac{1}{2} + \frac{b}{\ell}\frac{GrPr}{90}$.

5.5) We set: $Ra = \frac{g\ell^3\beta\Delta T}{\nu_m a_m} = \frac{\rho_m{}^2 c g\ell^3\beta\Delta T}{\lambda\mu}$.

Hence: $h = \lambda\left(\frac{1}{2b} + \frac{c\mu}{\lambda\ell}\frac{\rho_m{}^2 g b^3\beta\Delta T}{90\mu^2}\right) = \lambda\left(\frac{1}{2b} + \frac{Ra}{90}\frac{b^3}{\ell^4}\right)$.

$$\frac{dh}{db} = \lambda\left(-\frac{1}{2b^2} + \frac{Ra}{90}\frac{b^2}{\ell^4}\right)$$

$$\frac{dh}{db} = 0 \text{ for: } \left(\frac{b}{\ell}\right)^4 = \frac{15}{Ra}$$

We calculate Nu: $Nu = \frac{1}{2} + \frac{b^4}{\ell^4}\frac{Ra}{90} = \frac{1}{2} + \frac{15}{Ra}\frac{Ra}{90} = \frac{1}{2} + \frac{1}{6} = \frac{2}{3}$ and $h = \frac{2}{3}\frac{\lambda}{b}$.

We then calculate $\lambda_{eq} = 2b\,h$: $\lambda_{eq} = 2b\lambda\left(\frac{1}{2b} + \frac{Ra}{90}\frac{b^3}{\ell^4}\right) = \lambda\left(1 + \frac{Ra}{45}\frac{b^4}{\ell^4}\right) = \lambda\left(1 + \frac{Ra}{45}\frac{15}{Ra}\right) = 1.33\,\lambda$.

Convection increases the heat transfer between the two plates by at least 33%, this is the minimum increase obtained for a thickness: $b = 2\ell\left(\frac{15}{Ra}\right)^{0.25}$.

5.6) Numerical application:

$$Ra = \frac{g\ell^3\beta\Delta T}{\nu_m a_m} = \frac{9.81\times1^3\times20}{283\times1.4.10^{-5}\times1.9.10^{-5}} = 2.606\times10^9$$

$\frac{b}{\ell} = 8.71\times10^{-3}$ hence: $b = 8.71$ mm and the distance between the two plates: $2b = 17.4$ mm.

It will be noted that the optimum spacing $2b$ depends on the temperature difference.

4.4.7. Calculation of exchanges by convection in double glazing

We propose to study the heat exchanges taking place by conduction and convection between two glass plates constituting a double glazing by using a correlation established from experimental values.

We consider a vertical double glazing of height $L = 1$ m with a spacing of the plates $d = 16$ mm. We will first consider an interior temperature $T_i = 20°\text{C}$ and an exterior temperature $T_e = 0°\text{C}$.

The plates will be assumed to be isothermal, gray and opaque with respect to IR radiation, and with an emissivity. $\varepsilon_v = 0.9$. The values of the internal and external convective transfer coefficients will be taken, respectively, $h_{c,i} = 5\ \text{W m}^{-1}\text{K}^{-1}$ and $h_{c,e} = 10\ \text{W m}^{-1}\text{K}^{-1}$.

Correlation of (ElSherbiny et al. 1982) (see section A.18 of the Appendix):

$$Nu = max[Nu_1, Nu_2, Nu_3]$$

with:

$$Nu_1 = 0.0605\ Ra_d{}^{0.33};\ Nu_2 = \left\{1 + \left[\frac{0.104\ Ra_d{}^{0.293}}{1+\left(\frac{6{,}310}{Ra_d}\right)^{1.36}}\right]^3\right\}^{\frac{1}{3}};\ Nu_3 = 0.242\ \left(\frac{Ra_d d}{L}\right)^{0.272}.$$

1) Check that the difference ΔT between the temperatures of the two plates can be considered to be close to 10°C.

2) Convection occurs as soon as $Ra_d > 500\ \frac{L}{d}$. Calculate the distance d from which convection appears.

3) Calculate the equivalent thermal conductivity λ_{eq} of the air layer under these conditions. We define λ_{eq} by: $\lambda_{eq} = \frac{\phi d}{\Delta T}$, where ϕ is the heat flux transmitted by conduction and convection.

4) We propose to carry out a parametric study of the evolution of the thermal resistance R_c of conduction/convection of the air layer. Draw the curves $R_c = f(d)$ and $\frac{\lambda_{eq}}{\lambda} = f(d)$ for d varying from 10 to 30 mm for the following three temperature differences: $\Delta T = 10°\text{C}$ and $\Delta T = 20°\text{C}$ and $\Delta T = 30°\text{C}$ and interpret the results.

Corrected

1) The different flows in the system are shown in the following figure:

– $\phi_{r,i1} = \sigma\varepsilon_v(T_i^4 - T_1^4) = h_{r,i}(T_i - T_1)$, heat flux exchanged by radiation, with: $h_{r,i} = 4\sigma\varepsilon_v T_i^3 = 4 \times 5.67 \times 10^{-8} \times 0.9 \times 273^3 = 4.15\ \mathrm{W\,m^{-2}K^{-1}}$;

– $\phi_{c,i1} = h_{c,i}(T_i - T_1)$, heat flux exchanged by convection;

– $\phi_{12} = \frac{\lambda_v}{e}(T_1 - T_2)$, heat flux exchanged by conduction;

– $\phi_{r,23} = \sigma \frac{T_2^4 - T_3^4}{\frac{2}{\varepsilon_v} - 1} = h_{r,23}(T_2 - T_3)$, heat flux exchanged by radiation.

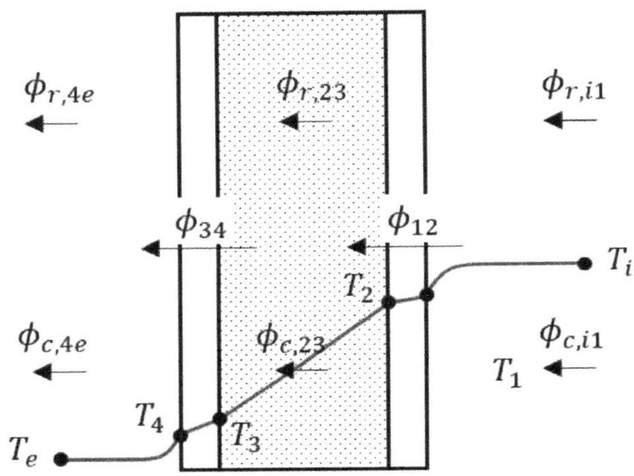

with:

– $h_{r,23} = \frac{\sigma}{\frac{2}{\varepsilon_v} - 1}\left(\frac{T_i + T_e}{2}\right)^3 = \frac{4 \times 5.67 \times 10^{-8}}{\frac{2}{0.9} - 1} \times 278^3 = 3.99\ \mathrm{W\,m^{-2}K^{-1}}$;

– $\phi_{c,23} = h_{c,23}(T_2 - T_3)$, the heat flux exchanged by conduction and convection;

– $\phi_{34} = \frac{\lambda_v}{e}(T_3 - T_4)$, the heat flux exchanged by conduction;

– $\phi_{r,4e} = \sigma\varepsilon_v(T_4^4 - T_e^4) = h_{r,e}(T_4 - T_e)$, the heat flux exchanged by radiation;

– $h_{r,e} = 4\sigma\varepsilon_v T_e^3 = 4 \times 5.67 \times 10^{-8} \times 0.9 \times 283^3 = 4.62\ \mathrm{W\,m^{-2}K^{-1}}$;

– $\phi_{c,4e} = h_{c,e}(T_4 - T_e)$, the heat flux exchanged by convection.

The global thermal resistance is written as:

$$R = \frac{1}{h_{r,i}+h_{c,i}} + \frac{\lambda_v}{e} + \frac{1}{h_{r,23}+h_{c,23}} + \frac{\lambda_v}{e} + \frac{1}{h_{r,e}+h_{c,e}}$$

Let us calculate the different temperatures assuming that the convection between the two plates increases the transfer by 30%: $h_{c,23} = 1.3\,\frac{\lambda_{air}}{e} = 1.3\,\frac{0.0248}{0.016} = 2.0\ \mathrm{W\,m^{-2}K^{-1}}$.

The total thermal resistance has the value:

$$R = \frac{1}{4.64+5} + \frac{0.004}{1} + \frac{1}{3.99+2.0} + \frac{0.004}{1} + \frac{1}{4.15+10} = 0.349\ \mathrm{K\,m^2\,W^{-1}}$$

We have: $\frac{T_i-T_e}{R} = \left(h_{r,23} + h_{c,23}\right)(T_2 - T_3)$.

Hence: $\Delta T = T_2 - T_3 = \frac{T_i-T_e}{R\left(h_{r,23}+h_{c,23}\right)} = \frac{20}{0.349\times(3.99+2)} = 9.6°\mathrm{C}$.

This value is very close to 10°C.

2) At an average temperature of 10°C, air has the following properties:

$Pr = 0.715$; $\rho = 1.248\ \mathrm{kg\,m^{-3}}$; $c = 1\,006\ \mathrm{J\,kg^{-1}K^{-1}}$; $\lambda = 0.0248\ \mathrm{W\,m^{-1}K^{-1}}$;

$\mu = 1.76 \times 10^{-5}\,\mathrm{Pa\,s^{-1}}$

Let us calculate the Rayleigh number:

$$Ra_d = Gr_d Pr = \frac{\beta g \Delta T \rho^2 d^3}{\mu^2}\frac{c\mu}{\lambda} = \frac{\beta g c \Delta T \rho^2 d^3}{\lambda\mu} = \frac{1}{283}\,\frac{9.81\times1{,}006\times10\times1.248^2\times d^3}{0.0249\times1.76\times10^{-5}} = 1.24\times10^9\times d^3$$

The condition $Ra_d > 500\,\frac{L}{d}$ is written as: $1.24\times10^9\times d^3 > 500\,\frac{L}{d}$.

Hence: $d^4 > 500\,\frac{L}{1.24\times10^9}$.

We deduce that there is no convection if $d < 25.2$ mm.

3) We will calculate $h_{c,23}$ using the proposed correlation.

$$Ra = 1.24 \times 10^9 \times d^3 = 1.24 \times 10^9 \times 0.016^3 = 5{,}032$$

$$Nu_1 = 0.0605\, Ra_d{}^{0.33} = 0.0605 \times 5{,}032^{0.33} = 1.008$$

$$Nu_2 = \left\{1 + \left[\frac{0.104\, Ra_d{}^{0.293}}{1+\left(\frac{6{,}310}{Ra}\right)^{1.36}}\right]^3\right\}^{\frac{1}{3}} = \left\{1 + \left[\frac{0.104 \times 5{,}032^{0.293}}{1+\left(\frac{6{,}310}{5{,}032}\right)^{1.36}}\right]^3\right\}^{\frac{1}{3}} = 1.024$$

$$Nu_3 = 0.242\left(\frac{Ra_d d}{L}\right)^{0.272} = 0.242 \times \left(\frac{5{,}032}{62.5}\right)^{0.272} = 0.798$$

$$Nu = max[Nu_1, Nu_2, Nu_3] = 1.024$$

$$Nu = \frac{h_{c,23} d}{\lambda} \text{ therefore: } h_{c,23} = \frac{\lambda Nu}{d} = \frac{0.0248 \times 1.024}{0.016} = 1.59 \text{ W m}^{-2}\text{K}^{-1}$$

$$h_{c,23} = \frac{\lambda_{eq}}{d}$$

Hence: $\lambda_{eq} = h_{c,23} d = 1.59 \times 0.016 = 0.0254 \text{ W m}^{-1}\text{K}^{-1}$.

4) We have: $R_c = \frac{d}{\lambda_{eq}} = \frac{1}{h_{c,23}}$ and we obtain the curves represented in the following graph:

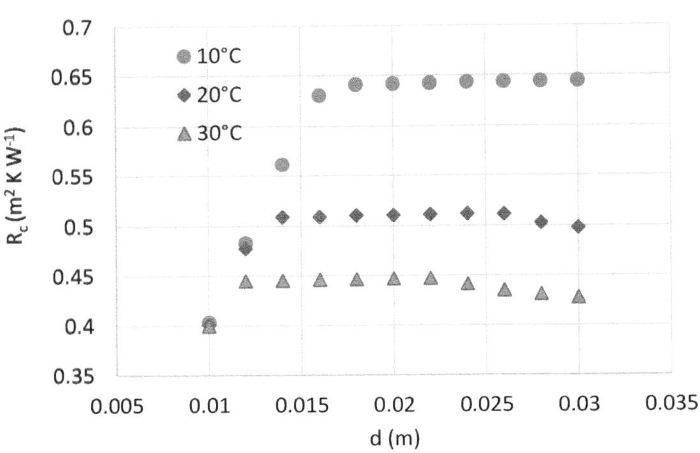

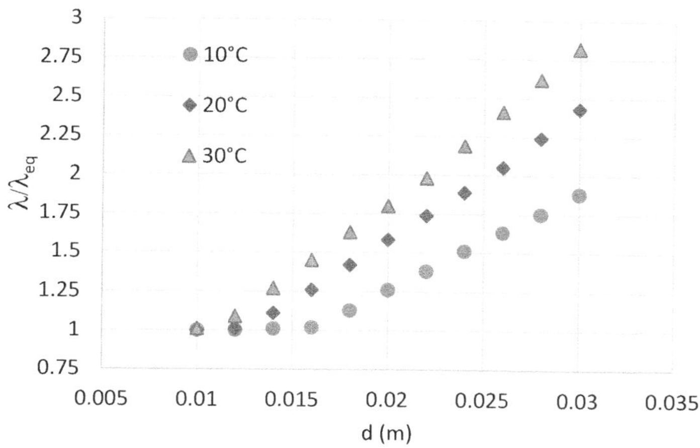

For low values of d, the thermal resistance R_c evolves linearly with d because there is no convection yet. The change in slope of the curves $\frac{\lambda_{eq}}{\lambda} = f(d)$ indicates the point where the contribution of convection becomes significant.

The curve $R_c = f(d)$ presents a very flat maximum, it is an area where when d increases the intensification of the convection is compensated by a reduction of the conduction. For large values of d, convection increases faster than conduction decreases and the overall thermal resistance decreases.

At d being constant, Ra_d increases when ΔT increases, thus convection intensifies and the overall conduction/convection thermal resistance decreases. It is observed however in all cases that this thermal resistance hardly increases any more beyond $d = 16$ mm, justifying the choice of this value in double-glazed windows.

4.4.8. *Study of an electric kettle*

An electric kettle with a volume of 1.5 ℓ is fitted with an electric resistance[1] capable of delivering a power φ_0 =2,200 W. It is similar to a 15 cm high cylinder with plastic walls of thickness $e = 2$ mm and thermal conductivity $\lambda = 0.2$ W m^{-1}K^{-1}. The electrical resistance is a stainless steel tube with a length $L = 20$ cm, with 9 mm inside diameter and 10 mm outside diameter. It is used to heat 1.5ℓ of water from 20°C to 100°C.

1) Calculate the time taken to heat water from 20°C to 100°C. Bottom losses will be neglected.

2) What is the boiling regime at 100°C?

3) Calculate the temperature difference between the tube and the water at 100°C.

4) Deduce the value of the heat transfer coefficient by convection during boiling.

5) After how long would the mass of water in the kettle be equal to only 1 ℓ if the thermostat was no longer working?

Data:

– latent heat of evaporation of water between 0 and 100°C: $\Delta H = 2{,}495.3 - 2.3457\ T$ (kJ kg^{-1}); with T in °C.

– emissivity of the outer wall of the kettle: $\varepsilon = 0.8$.

Corrected

1) It is first necessary to calculate the coefficients of transfer by convection between the external walls and the ambient air. We check that $GrPr$ is less than 10^9. The following two correlations therefore apply:

- vertical wall: $h = 1.54\left(\frac{\Delta\theta}{L}\right)^{1/4}$;

- horizontal wall hot face above: $h = 1.41\left(\frac{\Delta\theta}{D}\right)^{1/4}$.

The diameter is calculated from the volume and the height: $D = 11.28$ cm.

The lateral surface is therefore equal to: $S_v = \pi DL = \pi \times 0.2 \times 0.1128 = 0.07087\ \text{m}^2$.

1 Figure source: https://www.jeremplace.com/5670-resistance-pour-quette-2200-wattsms-5999457.html.

And the horizontal area is equal to: $S_h = \frac{\pi D^2}{4} = \frac{\pi \times 0.1128^2}{4} = 0.0100 \text{ m}^2$.

We calculate as first approximation the average coefficients for an average ΔT of 40°C:

$$h_{cv} = 1.54 \left(\frac{\Delta\theta}{L}\right)^{1/4} = 1.54 \left(\frac{40}{0.2}\right)^{1/4} = 5.34 \text{ W m}^{-2}\text{K}^{-1}$$

$$h_{ch} = 1.41 \left(\frac{\Delta\theta}{D}\right)^{1/4} = 1.41 \left(\frac{40}{0.1128}\right)^{1/4} = 5.73 \text{ W m}^{-2}\text{K}^{-1}$$

It is necessary to add to these values the radiation coefficient also calculated at an average value of 40°C assuming the emissivity of the wall equal to 0.8:

$$h_r = 5.67 \times 10^{-8}\varepsilon(T^2 + T_a{}^2)(T + T_a)$$
$$= 5.67 \times 10^{-8} \times 0.8 \times (333^2 + 293^2)(333 + 293)$$

$$h_r = 5.59 \text{ W m}^{-2}\text{K}^{-1}$$

hence finally:

$$h_v = h_{cv} + h_{rv} = 5.34 + 5.59 = 10.93 \text{ W m}^{-2}\text{K}^{-1}$$

$$h_h = h_{ch} + h_{rh} = 5.73 + 5.59 = 11.32 \text{ W m}^{-2}\text{K}^{-1}$$

The convection resistances are in series with the conduction resistance of the wall. The vertical and horizontal resistances are, respectively:

$$R_v = \frac{1}{S_v}\left(\frac{e}{\lambda} + \frac{1}{h_v}\right) = \frac{1}{0.07087}\left(\frac{0.002}{0.2} + \frac{1}{10.93}\right) = 1.29$$

$$R_h = \frac{1}{S_v}\left(\frac{e}{\lambda} + \frac{1}{h_h}\right) = \frac{1}{0.0100}\left(\frac{0.002}{0.2} + \frac{1}{11.32}\right) = 9.93$$

The thermal balance of the volume of water is written as:

$$\varphi_0 = mc\frac{dT}{dt} + (T - T_a)\left(\frac{1}{R_v} + \frac{1}{R_h}\right)$$

$$\frac{dT}{dt} = \frac{\varphi_0}{mc} - \frac{1}{mc}\left(\frac{1}{R_v} + \frac{1}{R_h}\right)(T - T_a) = A - B(T - T_a)$$

$$t = -\frac{1}{B} ln\left[\frac{A - B(T_f - T_a)}{A}\right] = 255 \text{ s}$$

2) To be in the nucleate boiling regime, the heat flux transmitted to the tube surface must be lower than the burn-out heat flux which can be calculated by:

$$\phi_{bo} = \frac{\pi \rho_v}{24} \Delta H \left[\frac{\sigma g(\rho_l - \rho_v)}{\rho_v{}^2}\right]^{1/4} \left(1 + \frac{\rho_v}{\rho_l}\right)^{1/2}$$

For water at 100°C:

$$\Delta H = 2.261 \times 10^6 \text{J kg}^{-1}$$

$$Pr_l = 1.74$$

$$\sigma = 58.8 \times 10^{-3} \text{N m}^{-1} \text{ (see values in Table 4.3)}$$

$$\rho_l = 960 \text{ kg m}^{-3}$$

$$\rho_v \approx \frac{MP}{RT} \approx \frac{0.018 \times 1.013 \times 10^5}{8.32 \times 373} = 0.588 \text{ kg m}^{-3}$$

$$C = 0.008 \text{ (see values in Table 4.4)}$$

$$s = 1 \text{ for water}$$

$$\phi_{bo} = \frac{\pi \rho_v}{24} \Delta H \left[\frac{\sigma g(\rho_l - \rho_v)}{\rho_v{}^2}\right]^{1/4} \left(1 + \frac{\rho_v}{\rho_l}\right)^{1/2}$$

$$\phi_{bo} = \frac{\pi \times 0.588}{24} \times 2.261 \times 10^6 \times \left[\frac{58.8.10^{-3} \times 9.81 \times (960 - 0.588)}{0.588^2}\right]^{1/4} \left(1 + \frac{0.588}{967}\right)^{1/2}$$

$$\phi_{bo} = 1.101 \times 10^6 \text{ W m}^{-2}$$

The flow density dissipated by a tube of 20 cm is: $\phi = \frac{\varphi_0}{\pi DL} = \frac{2{,}200}{\pi \times 0.2 \times 0.01} = 3.501 \times 10^5 \text{W m}^{-2}$.

$\phi < \phi_{bo}$ hence we are in a nucleate boiling regime.

3) The temperature difference in the nucleate boiling regime is given by the equation:

$$\frac{c_l \, \Delta T}{\Delta H (Pr_l)^s} = C \left[\frac{\phi}{\mu_l \, \Delta H} \sqrt{\frac{\sigma}{g(\rho_l - \rho_v)}}\right]^{0.33}$$

We deduce:

$$\Delta T = \frac{0.008\times 2.261\times10^6\times1.74}{4{,}216}\left[\frac{3.50\times10^5}{2.82\times10^{-4}\times2.261\times10^6}\sqrt{\frac{58.8\times10^{-3}}{9.81\,(967-0.588)}}\right]^{0.33}$$

$$\Delta T = 8.28°\text{C}$$

4) $\phi = h\Delta T$ hence: $h = \frac{\phi}{\Delta T} = \frac{3.501\times10^5}{8.28} = 4.23\times10^4 \text{W m}^{-2}\text{K}^{-1}$.

5) The quantity of heat to be supplied is: $Q = \Delta mLv + (h_v S_v + h_h S_h)\times 80$.

Thus: $Q = P\times t$ hence: $t = \frac{\Delta mLv+(h_v S_v+h_h S_h)\times80}{P} = 1{,}027$ s.

Appendices

A.1. Physical properties of some materials

	ρ kg m^{-3}	c_p J kg^{-1} K^{-1}	λ W m^{-1} K^{-1}		ρ kg m^{-3}	c_p J kg^{-1} K^{-1}	λ W m^{-1} K^{-1}
Metals and alloys				**Construction materials**			
Carbon steel	7,833	465	54	**Slate**	2,400	879	2.2
Stainless steel 15% Cr, 10% Ni	7,864	460	20	**Basalt**	2,850	881	1.6
Stainless steel 18% Cr, 8% Ni	7,816	460	16.3	**Cavernous concrete**	1,900	879	1.4
Stainless steel 25% Cr, 20% Ni	7,864	460	13	**Solid concrete**	2,300	878	1.75
Alumina			29	**Bitumen (cardboard)**	1,050	1,305	0.23
Aluminum	2,707	896	204	**Light hardwoods (dry)**	525	1,250	0.15
Silver	10,525	234	407	**Medium–heavy hardwoods (dry)**	675	1,250	0.23
Bronze 75% Cu, 25% Sn	8,800	377	188	**Very light hardwoods (dry)**	375	1,250	0.12
Bronze 92% Cu, 8% Al	7,900	377	71	**Light softwoods (dry)**	375	1,250	0.12
Graphite carbon	2,250	707	147	**Medium–heavy softwoods (dry)**	500	1,250	0.15
Silicon carbide			13	**Very light softwoods (dry)**	375	1,250	0.10
Chromium	2,118	7,160	449	**Cotta brick**	1,800	878	1.15
Constantan 60% Cu, 40% Ni	8,922	410	22.7	**Hard limestone**	2,450	882	2.4

	ρ kg m^{-3}	c_p J kg^{-1} K^{-1}	λ W m^{-1} K^{-1}
Metals and alloys			
Copper	8,954	383	386
Cupronickel 70% Cu, 30% Ni	8,900	377	29.3
Duralumin	2,787	883	164
Tin	7,304	226	64
Iron	7,870	452	73
Melting	7,849	460	59
Brass 70% Cu, 30% Zn	8,522	385	111
Magnesium	1,740	1,004	151
Gold	19,300	128	312
Platinum	21,400	140	69
Lead	11,373	130	35
Liquid sodium	930	1,381	84.5
Titanium	4,500	523	20.9
Tungsten	19,350	134	163
Zinc	7,144	384	112
Miscellaneous materials			
Asbestos	575	1,046	0.15
Asphalt	2,115	920	0.062
Rubber (natural)	1,150		0.28
Rubber (vulcanized)	1,100	2,010	0.13
Cardboard	86	2,030	0.048
Leather	998		0.159
Ice	920	2,040	1.88
Plexiglass	1,190	1,465	0.19
Porcelain	2,400	1,088	1.035
Polyethylene	929	1,830	0.46
PVC	1,459	930	0.19
Sand	1,515	800	0.2-1.0
Teflon	2,170	1,004	0.25
Wet soil	1,900	2,000	2
Dry soil	1,500	1,900	1
Glass	2,300	837	1.05
Pyrex glass	2,220	728	1.13

	ρ kg m^{-3}	c_p J kg^{-1} K^{-1}	λ W m^{-1} K^{-1}
Construction materials			
Soft limestone	1,650	879	1
Floor tile	2,400	875	2.4
Okoumé plywood	400	3,000	0.12
Pine plywood	500	3,000	0.15
Granite	2,600	881	3
Gravel (bulk)	1,800	889	0.7
Sandstone	2,500	880	2.6
Lava stone	2,350	881	1.1
Marble	2,700	881	2.5
Plaster	1,440	840	0.48
Shale	2,400	879	2.2
Insulating materials			
Balsa	140		0.054
Cotton	80	1,300	0.06
Kapok			0.035
Rockwool	20	880	0.047
	55	880	0.038
	135	880	0.041
Glass wool	8	875	0.051
	10	880	0.045
	15	880	0.041
	40	880	0.035
Expanded cork	120	2,100	0.044
Carpet	200	1,300	0.06
Polyurethane (foam)	32	1,300	0.03
	50	1,360	0.035
	85	1,300	0.045
PVC (rigid foam)	30	1,300	0.031
	40	1,300	0.041
Expanded polystyrene	12	1,300	0.047
	14	1,300	0.043
	18	1,300	0.041
Styrofoam	30		0.032

A.2. Physical properties of air and water

Properties of water at saturation							Properties of air at 1 atm						
T	ρ	c_p	λ	$10^4.\mu$	$10^7.\alpha$	Pr	T	ρ	c_p	λ	$10^4.\mu$	$10^7.\alpha$	Pr
°C	kg m^{-3}	J kg^{-1} K^{-1}	W m^{-1} K^{-1}	Pa s^{-1}	m^2 s^{-1}		°C	kg m^{-3}	J kg^{-1} K^{-1}	W m^{-1} K^{-1}	Pa s^{-1}	m^2 s^{-1}	
0	1,002	4,218	0.552	17.90	1.31	13.06	0	1,292	1,006	0.024 2	1.72	1.86	0.72
20	1,001	4,182	0.597	10.10	1.43	7.02	20	1,204	1,006	0.025 7	1.81	2.12	0.71
40	995	4,178	0.628	6.55	1.51	4.34	40	1,127	1,007	0.027 2	1.90	2.40	0.70
60	985	4,184	0.651	4.71	1.55	3.02	60	1,059	1,008	0.028 7	1.99	2.69	0.70
80	974	4,196	0.668	3.55	1.64	2.22	80	0.999	1,010	0.030 2	2.09	3.00	0.70
100	960	4,216	0.680	2.82	1.68	1.74	100	0.946	1,012	0.031 8	2.18	3.32	0.69
120	945	4,250	0.685	2.33	1.71	1.45	120	0.898	1,014	0.033 3	2.27	3.66	0.69
140	928	4,283	0.684	1.99	1.72	1.24	140	0.854	1,016	0.034 5	2.34	3.98	0.69
160	910	4,342	0.680	1.73	1.73	1.10	160	0.815	1,019	0.035 9	2.42	4.32	0.69
180	889	4,417	0.675	1.54	1.72	1.00	180	0.779	1,022	0.037 2	2.50	4.67	0.69
200	867	4,505	0.665	1.39	1.71	0.94	200	0.746	1,025	0.038 6	2.57	5.05	0.68
220	842	4,610	0.652	1.26	1.68	0.89	220	0.700	1,028	0.039 9	2.64	5.43	0.68
240	816	4,756	0.635	1.17	1.64	0.88	240	0.688	1,032	0.041 2	2.72	5.80	0.68
260	786	4,949	0.611	1.08	1.58	0.87	260	0.662	1,036	0.042 5	2.79	6.20	0.68
280	753	5,208	0.580	1.02	1.48	0.91	280	0.638	1,040	0.043 7	2.86	6.59	0.68
300	714	5,728	0.540	0.96	1.32	1.02	300	0.616	1,045	0.045 0	2.93	6.99	0.68

Physical properties between 0 and 100°C (T, temperature in °C).

For air:

$$\rho = \frac{353}{T+273} \text{ kg m}^{-3}$$

$$c_p = 1{,}008 \text{ J kg}^{-1}\text{ K}^{-1}$$

$\lambda = 7.57 \times 10^{-5}\, T + 0.0242$ W m^{-1} K^{-1}

$\mu = 10^{-5}(0.0046\, T + 1.7176)$ kg m^{-1} s^{-1}

$a = 10^{-5}(0.0146\, T + 1.8343)$ m^2 s^{-1}

$Pr = -2.54 \times 10^{-4}\, T + 0.7147$

$\beta \approx \frac{1}{T+273}$ K^{-1}

For water:

$\rho = -0.00380\, T^2 - 0.0505\, T + 1\,002.6$ kg m^{-3}

$c_p = 4{,}180$ J kg^{-1} K^{-1}

$\lambda = -9.87 \times 10^{-6} T^2 + 2.238 \times 10^{-3} T + 0.5536$ W m^{-1} K^{-1}

$\mu = 10^{-4} \times \frac{0.0003354\, T^2 - 0.07377\, T + 17.9}{8.765.10^{-5} T^2 + 0.03032\, T + 1}$ kg m^{-1} s^{-1}

$a = 10^{-7} \times (-0.00360\, T + 1.340)$ m^2 s^{-1}

$Pr = \frac{-0.0037\, T^2 + 1.387\, T + 13.06}{0.005297\, T^2 + 0.1241\, T + 1}$

$\frac{g\beta\rho^2 c_p}{\lambda\mu} = (0.0105\, T^2 + 0.477\, T - 0.0363)10^9$ K^{-1}m^{-3}

$\log_{10}\left[p_{sat\,(T+273)}\right] = 20.3182 - \frac{2795}{T+273} - 3.868 \log_{10}(T + 273)$

p_{sat} in mmHg and -50°C <T< 200°C

$L_v = 2{,}495 - 2.346\, T$ kJ kg^{-1}

0°C < T< 100°C

A.3. Conduction shape coefficient value

$\boldsymbol{\varphi = \lambda F \Delta T}$(W)

System	Diagram	Shape coefficient F	Scope
Isothermal cylinder of radius r buried in a semi-infinite medium with an isothermal surface	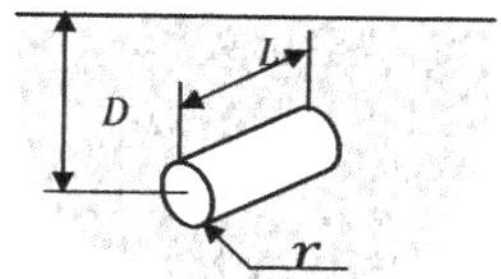	$\frac{2\pi L}{cosh^{-1}\left(\frac{D}{r}\right)}$	$L \gg r$
Isothermal sphere of radius r buried in an infinite medium	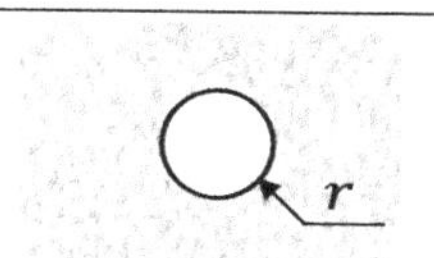	$4\ \pi r$	
Isothermal sphere buried in a semi-infinite medium with an isothermal surface	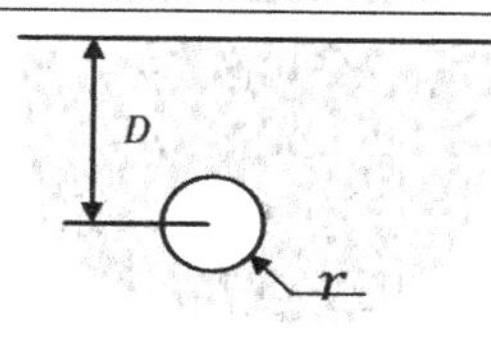	$\frac{4\pi r}{1-\frac{r}{2D}}$	
Conduction between two isothermal cylinders buried in an infinite medium	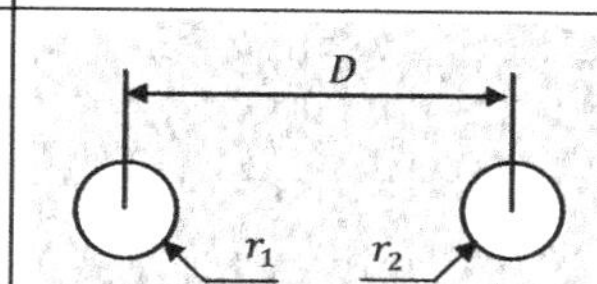	$\frac{2\pi L}{cosh^{-1}\left(\frac{D^2-r_1^2-r_2^2}{2-r_1 r_2}\right)}$	$L \gg r$ $L \gg D$
Horizontal cylinder of length L at the center of an infinite plate	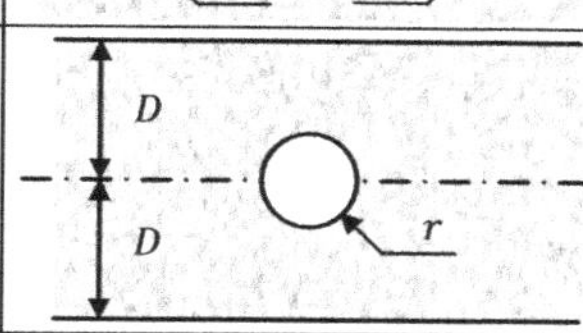	$\frac{2\pi L}{ln\left(\frac{4D}{r}\right)}$	
Isothermal cylinder of radius r placed in a semi-infinite medium	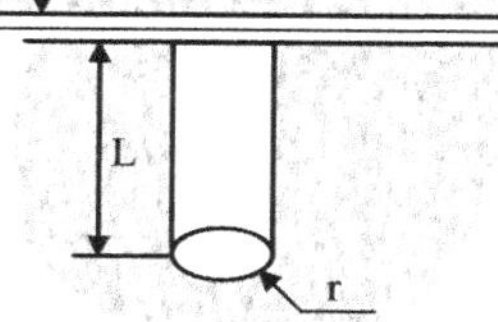	$\frac{2\pi L}{ln\left(\frac{2L}{r}\right)}$	$L \gg 2r$
Isothermal rectangular parallelepiped buried in a semi-infinite medium with an isothermal surface	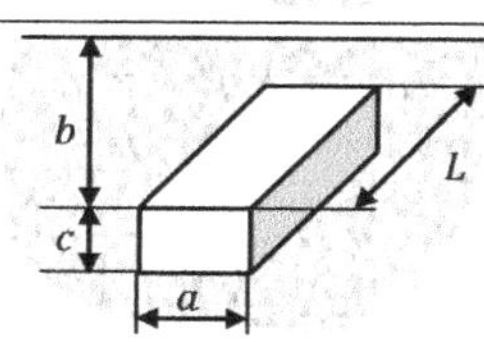	$1.685L\left[ln\left(1+\frac{b}{a}\right)\right]^{-0.59}\left(\frac{b}{c}\right)^{-0.078}$	
Cylinder in the center of a parallelepiped of square cross-section	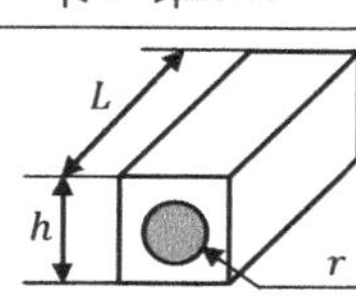	$\frac{2\pi L}{ln\left(0.54\frac{h}{r}\right)}$	$L \gg W$

System	Diagram	Shape coefficient *F*	Scope
Hollow cylinder of length *L*	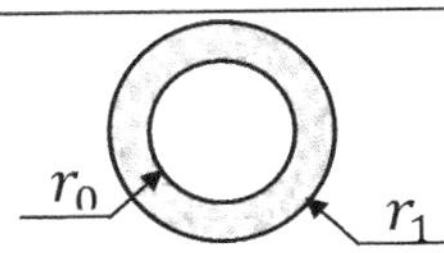	$\frac{2\pi L}{ln\left(\frac{r_1}{r_0}\right)}$	
Hollow sphere	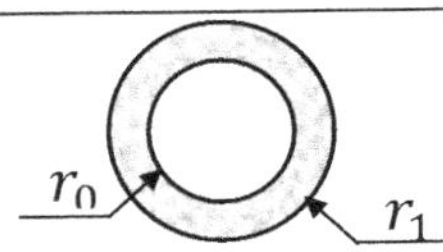	$\frac{4\pi r_0 r_1}{r_1 - r_0}$	

A.4. Fin efficiency

Hypothesis: zero flow at the tip of the fin, verified whether $\sqrt{\frac{he}{\lambda}} \ll 1$ (according to Whitaker (1983)).

Right fin	**Rectangular** $y = e$	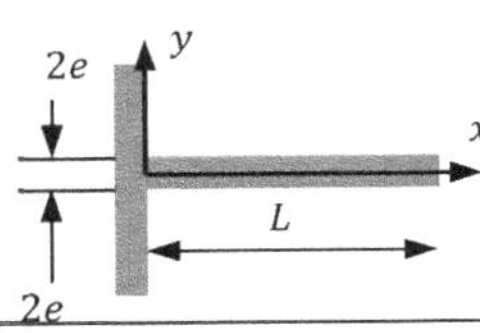	$\eta = \frac{1}{\omega L} tanh(\omega L)$ with $\omega = \sqrt{\frac{h}{\lambda e}}$
	Parabolic $y = e\left(1 - \frac{x}{L}\right)^{\frac{1}{2}}$	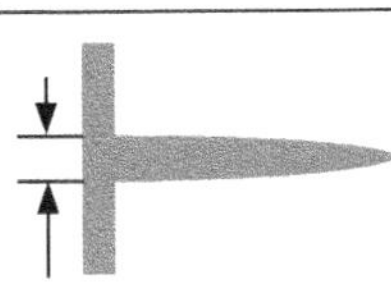	$\eta = \frac{1}{\omega L} \frac{I_{2/3}\left(\frac{4}{3}\omega L\right)}{I_{-1/3}\left(\frac{4}{3}\omega L\right)}$
	Triangular $y = e\left(1 - \frac{x}{L}\right)$	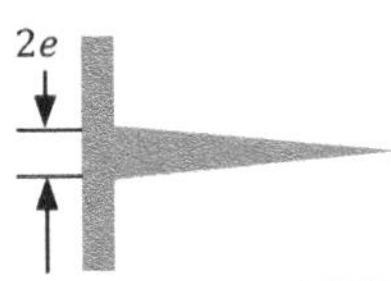	$\eta = \frac{1}{\omega L} \frac{I_1(2\omega L)}{I_0(2\omega L)}$
	Parabolic $y = e\left(1 - \frac{x}{L}\right)^2$	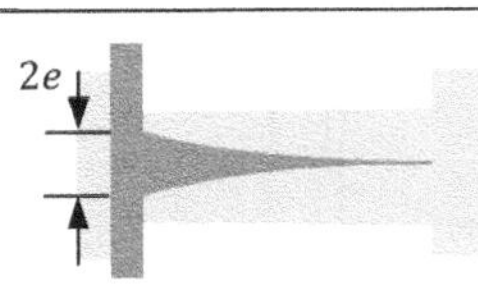	$\eta = \frac{2}{\sqrt{4(\omega L)^2 + 1} - 1}$

Circular fin	**Rectangular** $y = e$	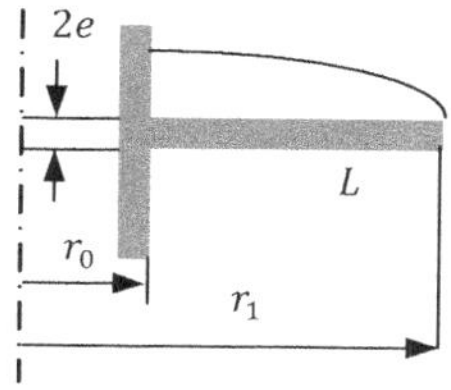	$\eta = \frac{2}{\omega L\left(\frac{r_1}{r_0}+1\right)}\frac{I_1(\omega r_0)K_1(\omega r_1) - I_1(\omega r_1)K_1(\omega r_0)}{I_0(\omega r_0)K_1(\omega r_1) - I_1(\omega r_1)K_0(\omega r_0)}$
Needles (circular cross-section)	**Rectangular** $y = e$	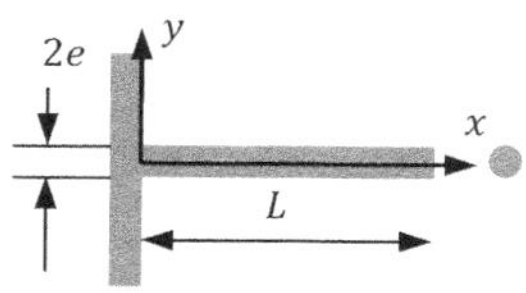	$\eta = \frac{1}{\sqrt{2}\omega L} tanh\left(\sqrt{2}\omega L\right)$
	Parabolic $y = e\left(1 - \frac{x}{L}\right)^{\frac{1}{2}}$	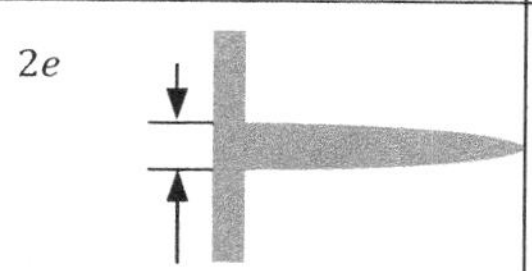	$\eta = \frac{3}{2\sqrt{2}\omega L}\frac{I_1\left(\frac{4}{3}\sqrt{2}\omega L\right)}{I_0\left(\frac{4}{3}\sqrt{2}\omega L\right)}$
	Triangular $y = e\left(1 - \frac{x}{L}\right)$	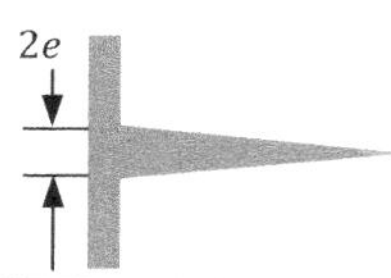	$\eta = \frac{\sqrt{2}}{\omega L}\frac{I_2\left(2\sqrt{2}\omega L\right)}{I_1\left(2\sqrt{2}\omega L\right)}$
	Parabolic $y = e\left(1 - \frac{x}{L}\right)^2$	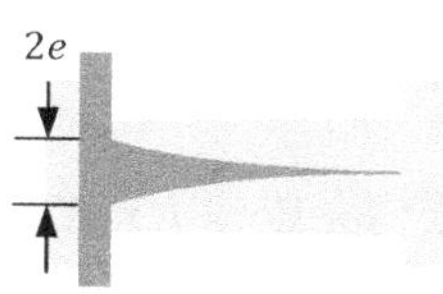	$\eta = \frac{2}{\sqrt{\frac{8}{9}(\omega L)^2 + 1} - 1}$

A.5. Bessel equations and functions

A.5.1. *Particular Bessel equations and their solutions*

$$y'' + \frac{y'}{x} + m^2 y = 0 \Rightarrow y = k_1 J_0(mx) + k_2 Y_0(mx)$$

$$x^2 y'' + xy' + (x^2 - n^2)y = 0 \Rightarrow y = k_1 J_n(x) + k_2 Y_n(x)\ (n\ integer)$$

$$y'' + \frac{y'}{x} - m^2 y = 0 \Rightarrow y = k_1 I_0(mx) + k_2 K_0(mx)$$

$$x^2 y'' + xy' - (x^2 - n^2)y = 0 \Rightarrow y = k_1 I_n(x) + k_2 K_n(x)\ (n\ integer)$$

with:

– J_n, the unmodified Bessel function of the first kind of order n.

– I_n, the modified Bessel function of the first kind of order n.

– Y_n, the unmodified Bessel function of the second kind of order n.

– K_n, the modified Bessel function of the second kind of order n.

A.5.2. *Main properties of Bessel functions*

– Recurrence:

$$J_{n+1}(u) = -J_{n-1}(u) + \frac{2n}{u} J_n(u) \qquad Y_{n+1}(u) = -Y_{n-1}(u) + \frac{2n}{u} Y_n(u)$$

$$I_{n+1}(u) = I_{n-1}(u) - \frac{2n}{u} I_n(u) \qquad K_{n+1}(u) = K_{n-1}(u) - \frac{2n}{u} K_n(u)$$

– Derivative:

$$\frac{dJ_0(u)}{du} = -J_1(u); \qquad \frac{dI_0(u)}{du} = I_1(u); \qquad \frac{dK_0(u)}{du} = -K_1(u);$$

$$\frac{dY_0(u)}{du} = -Y_1(u); \frac{d[uJ_1(u)]}{du} = uJ_0(u)$$

– Integrals:

$$\int x^{v} W_{v-1}(\alpha x)\, dx = \frac{1}{\alpha} x^{v} W_v(\alpha x) \text{ for } W = J, Y, I$$

$$\int x^{-v} W_{v+1}(\alpha x)\, dx = -\frac{1}{\alpha} x^{-v} W_v(\alpha x) \text{ for } W = J, Y, K$$

$$\int_0^R r[J_0(\alpha r)]^2 dr = \frac{R^2}{2}\{[J_0(\alpha R)]^2 + [J_1(\alpha R)]^2\}$$

A.5.3. *Asymptotic behavior of Bessel functions of order 0 and 1*

If $u \to 0$: $J_0(u) \to 1$; $J_1(u) \to \frac{u}{2}$; $Y_0(u) \to \left(\frac{2}{\pi}\right)\left[ln\left(\frac{u}{2}\right) + \gamma\right]$; $Y_1(u) \to \frac{2}{\pi u}$;

$I_0(u) \to 1$; $I_1(u) \to \frac{u}{2}$; $K_0(u) \to -ln\left(\frac{u}{2}\right) - \gamma$; $K_1(u) \to \frac{1}{u}$

If $u \to \infty$: $J_0(u) \to \sqrt{\frac{2}{\pi u}} cos\left(u - \frac{\pi}{4}\right)$; $J_1(u) \to \sqrt{\frac{2}{\pi u}} cos\left(\frac{u}{2} - \frac{\pi}{4}\right)$;

$Y_0(u) \to \sqrt{\frac{2}{\pi u}} sin\left(u - \frac{\pi}{4}\right)$; $Y_1(u) \to \sqrt{\frac{2}{\pi u}} sin\left(\frac{u}{2} - \frac{\pi}{4}\right)$;

$I_0(u), I_1(u) \to \sqrt{\frac{2}{\pi u}} exp(u)$;$K_0(u), K_1(u) \to \sqrt{\frac{\pi}{2u}} exp(-u)$

A.5.4. *Useful equations*

$$I_v(x)K_{v+1}(x) + I_{v+1}(x)K_v(x) = \frac{1}{x}$$

$$J_{v+1}(x)Y_v(x) - J_v(x)Y_{v+1}(x) = \frac{2}{\pi x}$$

A.6. Main integral transforms: Laplace, Fourier, Hankel

A.6.1. *Laplace transform*

– Definition:

$L[T(t)] = \theta(p) = \int_0^\infty exp(-pt)\, T(t)dt$ and $L^{-1}[\theta(p)] = T(t)$ (inverse transform)

– Properties:

- linearity $L[a_1T_1(t) + a_2T_2(t)] = a_1L[T_1(t)] + a_2L[T_2(t)]$, same for L^{-1};
- translation $L[exp(at)T(t)] = \theta(p - a)$; $L^{-1}[\theta(p - a)] = exp(at)T(t)$.

$L^{-1}[exp(-ap)\theta(p)] = T(t - a)$ if $t > a$

$L^{-1}[exp(-ap)\theta(p)] = 0$ if $t \leq a$

– Change of scale:

$L[T(at)] = \frac{1}{a}\theta\left(\frac{p}{a}\right)$ $L^{-1}[\theta(ap)] = \frac{1}{a}T\left(\frac{t}{a}\right)$

– Derivation:

$$L[T'(t)] = p\theta(p) - T(0) \qquad L^{-1}\left[\theta^{(n)}(p)\right] = (-1)^n t^n T(t)$$

$$L[T''(t)] = p^2\theta(p) - pT(0) - T'(0)$$

– Integration:

$$L\left[\int_0^t T(u)du\right] = \frac{\theta(p)}{p} \qquad L^{-1}\left[\int_p^\infty \theta(u)du\right] = \frac{T(t)}{t}$$

– Multiplication by t^n:

$$L[t^n T(t)] = (-1)^n \theta^{(n)}(p) \qquad L^{-1}[p\theta(p)] = T'(t) - T(0)\delta(t)$$

– Division by t:

$$L\left[\frac{T(t)}{t}\right] = \int_p^\infty \theta(u)du \qquad L^{-1}\left[\frac{\theta(p)}{p}\right] = \int_0^t T(u)du$$

– Periodic functions: $L[T(t)] = \frac{\int_0^P exp(-pt)T(t)dt}{1-exp(pP)}$.

(Period P).

A.6.2. *Complex Fourier transform*

– Definition:

$$F[T(x)] = \theta(\omega) = \frac{1}{\sqrt{2\pi}}\int_{-\infty}^{+\infty} e^{i\omega x}\, T(x)dx$$

$$T(x) = F^{-1}[\theta(\omega)] = \frac{1}{\sqrt{2\pi}}\int_{-\infty}^{+\infty} e^{i\omega x}\, \theta(\omega)d\omega$$

– Properties:

$$F\left[\frac{\partial T}{\partial x}\right] = -i\omega\theta(\omega); F\left[\frac{\partial^2 T}{\partial x^2}\right] = -\omega^2\theta(\omega)$$

A.6.3. *Fourier transform into sine and cosine*

– Definitions:

- sine:

$$F_s[T(x)] = \theta_s(\omega) = \sqrt{\frac{2}{\pi}}\int_0^{\infty} T(x)sin(\omega x)\,d$$

$$T(x) = F_s^{-1}[\theta_s(\omega)] = \sqrt{\frac{2}{\pi}}\int_0^{\infty} \theta_s(\omega)sin(\omega x)\,dx$$

- cosine:

$$F_c[T(x)] = \theta_c(\omega) = \sqrt{\frac{2}{\pi}}\int_0^{\infty} T(x)cos(\omega x)\,dx$$

$$T(x) = F_c^{-1}[\theta_c(\omega)] = \sqrt{\frac{2}{\pi}}\int_0^{\infty} \theta_c(\omega)cos(\omega x)\,dx$$

– Properties:

$$F_s\left[\frac{\partial T}{\partial x}\right] = -\omega\theta_c(\omega);\ F_c\left[\frac{\partial T}{\partial x}\right] = -\omega\theta_s(\omega) - \sqrt{\frac{2}{\pi}}T(0)$$

$$F_s\left[\frac{\partial^2 T}{\partial x^2}\right] = -\omega^2\theta_s(\omega) + \omega\sqrt{\frac{2}{\pi}}T(0);\quad F_c\left[\frac{\partial^2 T}{\partial x^2}\right] = -\omega^2\theta_c(\omega) - \sqrt{\frac{2}{\pi}}\left(\frac{\partial T}{\partial x}\right)_{x=0}$$

A.6.4. *Finite Fourier transform in sine and cosine*

– Definitions: if the temperature $T(x)$ is only defined on the interval $[0, L]$, we can use a finite Fourier transformation in sine or cosine:

$$F_s[T(x)] = \theta_s(n) = \int_0^{L} T(x)sin\left(\frac{n\pi x}{L}\right)dx$$

where $F_c[T(x)] = \theta_c(n) = \int_0^L T(x)cos\left(\frac{n\pi x}{L}\right)dx$.

$$T(x) = F_s^{-1}[\theta_s(n)] = \frac{2}{L}\sum_{n=1}^{\infty}\theta_s(n)sin\left(\frac{n\pi x}{L}\right)$$

where $T(x) = F_c^{-1}[\theta_c(n)] = \frac{1}{L}\theta_c(0) + \frac{2}{L}\sum_{n=1}^{\infty}\theta_c(n)cos\left(\frac{n\pi x}{L}\right)$.

– Properties:

$$F_s\left[\frac{\partial^2 T}{\partial x^2}\right] = \frac{n\pi}{L}[T(0) - (-1)^n T(L)] - \frac{n^2\pi^2}{L^2}\theta_s(n)$$

$$F_c\left[\frac{\partial^2 T}{\partial x^2}\right] = (-1)^n\left(\frac{\partial T}{\partial x}\right)_{x=L} - \left(\frac{\partial T}{\partial x}\right)_{x=0} - \frac{n^2\pi^2}{L^2}\theta_c(n)$$

A.6.5. *Hankel transform of order v*

– Definition: for $v > -\frac{1}{2}$:

$$H_v[T(r)] = \theta_v(\sigma) = \int_0^{\infty} rJ_v(\sigma r)T(r)dr$$

$$T(r) = H_v^{-1}[\theta_v(\sigma)] = \int_0^{\infty} \sigma J_v(\sigma r)\theta_v(\sigma)d\sigma$$

- Property:

$$H_v\left[\frac{1}{r}\frac{\partial}{\partial r}\left(r\frac{\partial T}{\partial r}\right) - \frac{v^2 T}{r^2}\right] = -\sigma^2\theta_v(\sigma)$$

at order 0:

$$H_0[T(r) = T_i] = T_i\int_0^{\infty} rJ_0(\sigma r)dr = T_i\frac{r}{\sigma}J_1(\sigma r)$$

A.7. Inverse Laplace transform

A.7.1. *Analytical method*

The Laplace transform $\theta(p)$ of the function $T(t)$ is given by: $L[T(t)] = \theta(p) = \int_0^{\infty} exp(-pt)\,T(t)dt$.

There is no general analytic formula for calculating $T(t)$ knowing $\theta(p)$. However, we know the exact expression of $T(t)$ for some particular functions $\theta(p)$, examples of which can be found in section A.7.4 (see Spiegel for more complete tables). The use of these tables, associated with the particular properties of the inverse Laplace transform recalled in section A.6, allows us to solve a certain number of cases. We will always try to decompose a complex function into a sum, product, series, etc., of simple functions that are more easily invertible.

A.7.2. *Numerical methods*

When an analytical solution cannot be found, we can use one of the following two numerical methods:

– Stehfest method: the inverse transform of the function $\theta(p)$ can be calculated by (Stehfest 1970):

$$T(t) = \frac{ln(2)}{t}\sum_{j=1}^{N} V_j \theta\left[\frac{jln(2)}{t}\right]$$

$N = 20$ (double precision):

$V_1 = -5.511463844797178 \times 10^{-6}$ $V_2 = 1.523864638447972 \times 10^{-1}$

$V_3 = -1.174654761904762 \times 10^{2}$ $V_4 = 1.734244933862434 \times 10^{4}$

$V_5 = -9.228069289021164 \times 10^{5}$ $V_6 = 2.377408778710318 \times 10^{7}$

$V_7 = -3.494211661953704 \times 10^{8}$ $V_8 = 3.241369852231879 \times 10^{9}$

$V_9 = -2.027694830723779 \times 10^{10}$ $V_{10} = 8.946482982379724 \times 10^{10}$

$V_{11} = -2.870209211471027 \times 10^{11}$ $V_{12} = 6.829920102815115 \times 10^{11}$

$V_{13} = -1.219082330054374 \times 10^{12}$ $V_{14} = 1.637573800842013 \times 10^{12}$

$V_{15} = -1.647177486836117 \times 10^{12}$ $V_{16} = 1.221924554444226 \times 10^{12}$

$V_{17} = -6.488065588175326 \times 10^{11}$ $V_{18} = 2.333166532137059 \times 10^{11}$

$V_{19} = -5.091380070546738 \times 10^{10}$ $V_{20} = 5.091380070546738 \times 10^{9}$

$N = 10$ (single precision):

$$V_1 = \frac{1}{12}; V_2 = -\frac{385}{12}; V_3 = 1279; V_4 = -\frac{46871}{3}; V_5 = \frac{505465}{6};$$

$$V_6 = -\frac{473915}{2};\ V_7 = \frac{1127735}{3};\ V_8 = -\frac{1020215}{3};\ V_9 = \frac{328125}{2};\ V_{10} = -\frac{65625}{2}$$

– Fourier method:

$$T(t) = \frac{exp(ct)}{t_{max}}\langle\frac{\theta(c)}{2} + \sum_{k=1}^{\infty}\{Re[\theta(c+j\omega_k)]cos(\omega_k t) - Im[\theta(c+j\omega_k)]sin(\omega_k t)\}\rangle$$

with: $\omega_k = \frac{k\pi}{t_{max}}$.

The infinite sum is in practice calculated for a finite number N of terms; we will generally take $N > 100$. This method requires choosing two parameters: c and t_{max}. We must ensure a posteriori that $exp(-2ct_{max})(2t_{max}) \approx 0$.

A.7.3. *Choosing a method and checking the results*

The Stehfest method is simpler to implement, because it does not require choosing certain parameters. The Fourier method can lead to a better result in the case of inversion of certain functions such as periodic functions, for example (Maillet et al. 2000; Den Iseger 2006; Toutain et al. 2011). We can also directly use the Matlab subroutine "Invlap" based on De Hoog's algorithm.

The study of the behavior of the function $\theta(p)$ at long times ($t\rightarrow\infty$ thus $p\rightarrow 0$) and at short times ($t\rightarrow 0$ thus $p\rightarrow\infty$) can lead to approximate formulas of $\theta(p)$ for which we can then find the inverse Laplace transform analytically. The comparison of these analytical solutions with the results of the numerical inversion gives an indication of the accuracy of the numerical inversion.

A.7.4. *Table of Laplace transforms*

From Spiegel (1990), $q = \sqrt{\frac{p}{a}}$.

$\theta(p) = [T(t)]$	$T(t)$
$\frac{1}{p}$	1
1	$\delta(t)$ Dirac

$\frac{1}{p^2}$	t
$\frac{1}{p^n} n = 1,2,3 \ldots$	$\frac{t^{n-1}}{(n-1)!}$
$\frac{ln(p)}{p}$	$-ln(t) - \gamma;\ \ \gamma = 0.57721$
$\frac{\omega}{p^2 - \omega^2}$	$sinh(\omega t)$
$\frac{p}{p^2 - \omega^2}$	$cosh(\omega t)$
$\frac{\omega^2}{p^2+\omega^2}$	$sin(\omega t)$
$\frac{p}{p^2 + \omega^2}$	$cos(\omega t)$
$\frac{1}{\sqrt{p}}$	$\frac{1}{\sqrt{\pi t}}$
$\frac{1}{p\sqrt{p}}$	$\frac{2}{\sqrt{\pi}}\sqrt{t}$
$\frac{b}{p(b+\sqrt{p})}$	$1 - exp(b^2 t)erfc(b\sqrt{t})$
$\frac{b}{\sqrt{p}(b+\sqrt{p})}$	$exp(b^2 t)erf(b\sqrt{t})$
$\frac{1}{p+\alpha}$	$e^{-\alpha t}$
$\frac{1}{(p+\alpha)(p+\beta)}$	$\frac{e^{-\beta t} - e^{-\alpha t}}{\alpha - \beta}$
$\frac{1}{(p+\alpha)^2}$	$te^{-\alpha t}$
$\frac{p}{(p+\alpha)^2}$	$(1-\alpha t)e^{-\alpha t}$
$\frac{p}{(p+\alpha)(p+\beta)}$	$\frac{\alpha e^{-\beta t} - \beta e^{-\alpha t}}{\alpha - \beta}$
$\frac{1}{(p+\alpha)(p+\beta)(p+\gamma)}$	$\frac{(\gamma-\beta)exp(-\alpha t) + (\alpha-\gamma)exp(-\beta t) + (\beta-\alpha)exp(-\gamma t)}{(\alpha-\beta)(\beta-\gamma)(\gamma-\alpha)}$
$\frac{p}{(p+\alpha)(p+\beta)(p+\gamma)}$	$\frac{\alpha(\beta-\gamma)exp(-\alpha t) + \beta(\gamma-\alpha)exp(-\beta t) + \gamma(\alpha-\beta)exp(-\gamma t)}{(\alpha-\beta)(\beta-\gamma)(\gamma-\alpha)}$
e^{-qx}	$\frac{x}{2\sqrt{\pi a t^3}} exp\left(-\frac{x^2}{4at}\right)$
$\frac{e^{-qx}}{q}$	$\sqrt{\frac{a}{\pi t}} exp\left(-\frac{x^2}{4at}\right)$

$\frac{e^{-qx}}{p}$	$erfc\left(\frac{x}{2\sqrt{at}}\right)$
$\frac{e^{-qx}}{pq}$	$2\sqrt{\frac{a}{\pi t}}exp\left(-\frac{x^2}{4at}\right) - xerfc\left(\frac{x}{2\sqrt{at}}\right)$
$\frac{e^{-qx}}{p^2}$	$\left(t + \frac{x^2}{2a}\right)erfc\left(\frac{x}{2\sqrt{at}}\right) - x\sqrt{\frac{t}{\pi a}}exp\left(-\frac{x^2}{4at}\right)$
$\frac{e^{-qx}}{q+h}$	$\sqrt{\frac{a}{\pi t}}exp\left(-\frac{x^2}{4at}\right) - haexp(hx + ath^2)erfc\left(\frac{x}{2\sqrt{at}} + h\sqrt{at}\right)$
$\frac{e^{-qx}}{q(q+h)}$	$aexp(hx + ath^2)erfc\left(\frac{x}{2\sqrt{at}} + h\sqrt{at}\right)$
$\frac{e^{-qx}}{p(q+h)}$	$\frac{1}{h}erfc\left(\frac{x}{2\sqrt{at}}\right) - \frac{1}{h}exp(hx + ath^2)erfc\left(\frac{x}{2\sqrt{at}} + h\sqrt{at}\right)$
$\frac{e^{-qx}}{pq(q+h)}$	$\frac{2}{h}\sqrt{\frac{a}{\pi t}}exp\left(-\frac{x^2}{4at}\right) - \frac{1+hx}{h^2}erfc\left(\frac{x}{2\sqrt{at}}\right)$ $+ \frac{1}{h^2}exp(hx + ath^2)erfc\left(\frac{x}{2\sqrt{at}} + h\sqrt{at}\right)$
$\frac{e^{-qx}}{p-\alpha}$	$\frac{1}{2}exp(\alpha t)\left\{exp\left(-\frac{x\sqrt{\alpha}}{\sqrt{a}}\right)erfc\left(\frac{x}{2\sqrt{at}} - \sqrt{\alpha t}\right)\right.$ $\left. + exp\left(\frac{x\sqrt{\alpha}}{\sqrt{a}}\right)erfc\left(\frac{x}{2\sqrt{at}} + \sqrt{\alpha t}\right)\right\}$
$\frac{e^{-qx}}{q(p-\alpha)}$	$\frac{1}{2}exp(\alpha t)\sqrt{\frac{a}{\alpha}}\left\{exp\left(-\frac{x\sqrt{\alpha}}{\sqrt{a}}\right)erfc\left(\frac{x}{2\sqrt{at}} - \sqrt{\alpha t}\right)\right.$ $\left. - exp\left(\frac{x\sqrt{\alpha}}{\sqrt{a}}\right)erfc\left(\frac{x}{2\sqrt{at}} + \sqrt{\alpha t}\right)\right\}$
$\frac{e^{-qx}}{(q+h)^2}$	-2h$-2h\left(\frac{a^3t}{\pi}\right)exp\left(-\frac{x^2}{4at}\right) + a(1 - hx - 2ah^2t)$
$\frac{e^{-qx}}{p(q+h)^2}$	$\frac{1}{h^2}erfc\left(\frac{x}{2\sqrt{at}}\right) - \frac{2}{h}\sqrt{\frac{at}{\pi}}exp\left(-\frac{x^2}{4at}\right)$ $-\frac{1}{h^2}(1 + hx + 2ah^2t)exp(hx + ah^2t)$ $\times erfc\left(\frac{x}{2\sqrt{at}} + h\sqrt{at}\right)$
$K_0(qx)$	$\frac{1}{2t}exp\left(-\frac{x^2}{4at}\right)$
$\frac{1}{p^{1/2}}K_{2n}(qx)$	$\frac{1}{2\sqrt{\pi t}}exp\left(-\frac{x^2}{8at}\right)K_n\left(\frac{x^2}{8at}\right)$
$p^{n/2}K_n(qx)$	$\frac{x^n}{a^{n/2}(2t)^{n+1}}exp\left(-\frac{x^2}{4at}\right)$

A.8. Choice of integral transforms for different configurations

According to Maillet et al. (2000).

Plate of thickness e					
No.	**Boundary condition** $x = 0$	**Boundary condition** $x = e$	**Eigenvalues** α_n	**Eigenfunctions**	**Norm**
1	$T = 0$	$T = 0$	$\frac{n\pi}{e}$	$sin(\alpha_n x)$	$\frac{e}{2}$
2	$T = 0$	$\frac{dT}{dx} = 0$	$(n+1)\frac{\pi}{e}$	$sin(\alpha_n x)$	$\frac{e}{2}$
3	$T = 0$	$-\lambda\frac{dT}{dx} + hT = 0$	Positive roots of $tan(\alpha_n e) = -\frac{\lambda}{h}\alpha_n$	$sin(\alpha_n x)$	$\frac{e}{2}\left[1+\frac{\frac{he}{\lambda}}{\left(\frac{he}{\lambda}\right)^2+(\alpha_n e)^2}\right]$
4	$\frac{dT}{dx} = 0$	$T = 0$	$(n+1)\frac{\pi}{e}$	$cos(\alpha_n x)$	$\frac{e}{2}$
5	$\frac{dT}{dx} = 0$	$\frac{dT}{dx} = 0$	$\frac{n\pi}{e}$	$cos(\alpha_n x)$	e for $n = 0$ $\frac{e}{2}$ for $n = 1, 2, \ldots$
6	$\frac{dT}{dx} = 0$	$-\lambda\frac{dT}{dx} + hT = 0$	Positive roots of $cotan(\alpha_n e) = -\frac{\lambda}{h}\alpha_n$	$cos(\alpha_n x)$	$\frac{e}{2}\left[1+\frac{\frac{he}{\lambda}}{\left(\frac{he}{\lambda}\right)^2+(\alpha_n e)^2}\right]$
7	$-\lambda\frac{dT}{dx} + hT = 0$	$T = 0$	Positive roots of $tan(\alpha_n e) = -\frac{\lambda}{h}\alpha_n$	$sin[\alpha_n](e-x)$	$\frac{e}{2}\left[1+\frac{\frac{he}{\lambda}}{\left(\frac{he}{\lambda}\right)^2+(\alpha_n e)^2}\right]$
8	$-\lambda\frac{dT}{dx} + hT = 0$	$\frac{dT}{dx} = 0$	Positive roots of $cotan(\alpha_n e) = -\frac{\lambda}{h}\alpha_n$	$cos[\alpha_n](e-x)$	$\frac{e}{2}\left[1+\frac{\frac{he}{\lambda}}{\left(\frac{he}{\lambda}\right)^2+(\alpha_n e)^2}\right]$
9	$-\lambda\frac{dT}{dx} + h_1T = 0$	$-\lambda\frac{dT}{dx} + h_2T = 0$	Positive roots of $tan(\alpha_n e) = -\frac{\lambda\alpha_n(h_1+h_2)}{(\lambda\alpha_n)^2 - h_1h_2}$	$cos(\alpha_n x) + \frac{h_1}{\lambda\alpha_n}sin(\alpha_n x)$	$\frac{1}{2e\alpha_n^2}\times\left\{\left[\left(\frac{h_1e}{\lambda}\right)^2+(\alpha_n e)^2\right]\times\left[1+\frac{\frac{h_2e}{\lambda}}{\left(\frac{h_2e}{\lambda}\right)^2+(\alpha_n e)^2}+\frac{h_2e}{\lambda}\right]\right\}$

No.	Boundary condition $x = 0$	Boundary condition $x = e$	Equation verified by the eigenvalues α_n	Eigenfunctions	Norm
Hollow cylinder of radii R_1 and R_2					
1	$T = 0$	$T = 0$	$J_0(\alpha_n R_1)Y_0(\alpha_n R_2) - Y_0(\alpha_n R_1)J_0(\alpha_n R_2) = 0$	$J_0(\alpha_n r)Y_0(\alpha_n R_2) - Y_0(\alpha_n r)J_0(\alpha_n R_2)$	$\frac{2}{\pi^2}\frac{J_0{}^2(\alpha_n R_1) - J_0{}^2(\alpha_n R_2)}{\alpha_n{}^2 J_0{}^2(\alpha_n R_1)}$
2	$T = 0$	$\frac{dT}{dx} = 0$	$J_1(\alpha_n R_2)Y_0(\alpha_n R_1) - Y_1(\alpha_n R_2)J_0(\alpha_n R_1) = 0$	$J_1(\alpha_n R_2)Y_0(\alpha_n r) - Y_1(\alpha_n R_2)J_0(\alpha_n r)$	$\frac{2}{\pi^2}\frac{J_0{}^2(\alpha_n R_1) - J_1{}^2(\alpha_n R_2)}{\alpha_n{}^2 J_0{}^2(\alpha_n R_1)}$
3	$\frac{dT}{dx} = 0$	$T = 0$	$J_0(\alpha_n R_2)Y_1(\alpha_n R_1) - Y_0(\alpha_n R_2)J_1(\alpha_n R_1) = 0$	$J_0(\alpha_n r)Y_1(\alpha_n R_2) - Y_0(\alpha_n r)J_0(\alpha_n R_2)$	$\frac{2}{\pi^2}\frac{J_1{}^2(\alpha_n R_1) - J_0{}^2(\alpha_n R_2)}{\alpha_n{}^2 J_1{}^2(\alpha_n R_1)}$
4	$\frac{dT}{dx} = 0$	$T = 0$	$J_1(\alpha_n R_1)Y_1(\alpha_n R_2) - Y_1(\alpha_n R_1)J_1(\alpha_n R_2) = 0$	$J_1(\alpha_n R_2)Y_0(\alpha_n r) - Y_1(\alpha_n R_2)J_0(\alpha_n r)$	$\frac{2}{\pi^2}\frac{J_1{}^2(\alpha_n R_1) - J_1{}^2(\alpha_n R_2)}{\alpha_n{}^2 J_1{}^2(\alpha_n R_1)}$
Solid cylinder with outer radius R					
5		$T = 0$	$J_0(\alpha_n R) = 0$	$J_0(\alpha_n R)$	$\frac{R^2}{2} J_1{}^2(\alpha_n R)$
6		$\frac{dT}{dx} = 0$	$J_1(\alpha_n R) = 0$	$J_0(\alpha_n R)$	$\frac{R^2}{2} J_0{}^2(\alpha_n R)$ if $\alpha_n \neq 0$ $\frac{R^2}{2}$ if $\alpha_n = \alpha_0 = 0$

A.9. erf function value

Properties:

$$erf(u) = \frac{2}{\sqrt{\pi}}\int_0^u exp(-t^2)dt; erfc(u) = 1 - erf(u)$$

$$ierfc(u) = \frac{1}{\sqrt{\pi}} exp(-u^2) - u[1 - erf(u)]$$

$$\frac{d}{dx}[erf(x)] = \frac{2}{\sqrt{\pi}} exp(-x^2)$$

For $x \to \infty$: $erfc(x) = \frac{exp(-x^2)}{\sqrt{\pi}}\left[\frac{1}{x} - \frac{1}{2\,x^3} + \frac{1.3}{2^2 x^5} - \frac{1.3.5}{2^3 x^7} + \cdots\right]$.

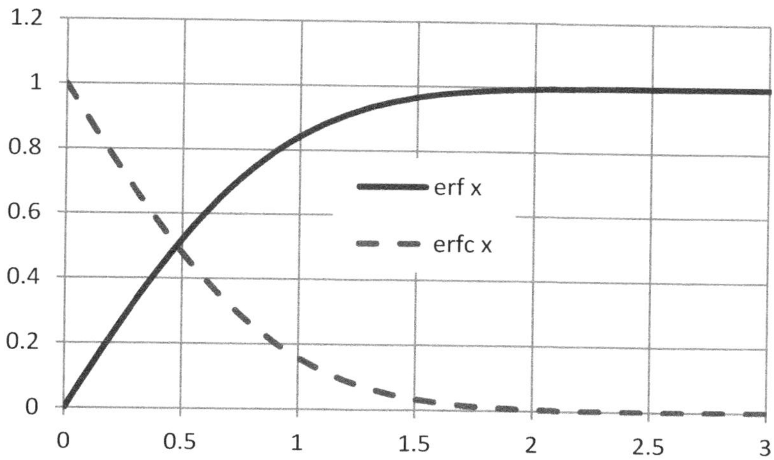

x	$erf(x)$	$erfc(x)$	$ierfc(x)$	x	$erf(x)$	$erfc(x)$	$ierfc(x)$
0	0.000000	1.000000	0.5641896	**1.1**	0.880205	0.11980	0.036465
0.05	0.056372	0.943628	0.5155995	**1.2**	0.910314	0.08969	0.026049
0.1	0.112463	0.887537	0.4698221	**1.3**	0.934008	0.06599	0.018314
0.15	0.167996	0.832004	0.4268365	**1.4**	0.952285	0.04772	0.012670
0.2	0.222703	0.777297	0.3866080	**1.5**	0.966105	0.03390	0.008623
0.25	0.276326	0.723674	0.3490886	**1.6**	0.976378	0.02362	0.005819
0.3	0.328627	0.671373	0.3142186	**1.7**	0.983790	0.01621	0.003799
0.35	0.379382	0.620618	0.2819256	**1.8**	0.989091	0.01091	0.002460
0.4	0.428392	0.571608	0.2521274	**1.9**	0.992790	0.00721	0.001563
0.45	0.475482	0.524518	0.2247329	**2**	0.995322	0.00468	0.000977
0.5	0.520500	0.479500	0.1996413	**2.1**	0.997021	0.00298	0.000602
0.55	0.563323	0.436677	0.1767460	**2.2**	0.998137	0.00186	0.000362
0.6	0.603856	0.396144	0.1559353	**2.3**	0.998857	0.00114	0.000216
0.65	0.642029	0.357971	0.1370922	**2.4**	0.999311	0.00069	0.000124
0.7	0.677801	0.322199	0.1200981	**2.5**	0.999593	0.00041	0.000072
0.75	0.711156	0.288844	0.1048325	**2.6**	0.999764	0.00024	0.000040
0.8	0.742101	0.257899	0.0911737	**2.7**	0.999866	0.00013	0.000023
0.85	0.770668	0.229332	0.0790027	**2.8**	0.999925	0.00008	0.000012
0.9	0.796908	0.203092	0.0682015	**2.9**	0.999959	0.00004	0.000007
0.95	0.820891	0.179109	0.0586561	**3**	0.999978	0.00002	0.000004
1	0.842701	0.157299	0.0502547				

A.10. Semi-infinite medium with imposed transfer coefficient

$$Tr(x,t) = \frac{T(x,t)-T_\infty}{T_i-T_\infty};\ Bi = \frac{hx}{\lambda};\ Fo = \frac{at}{x^2}.$$

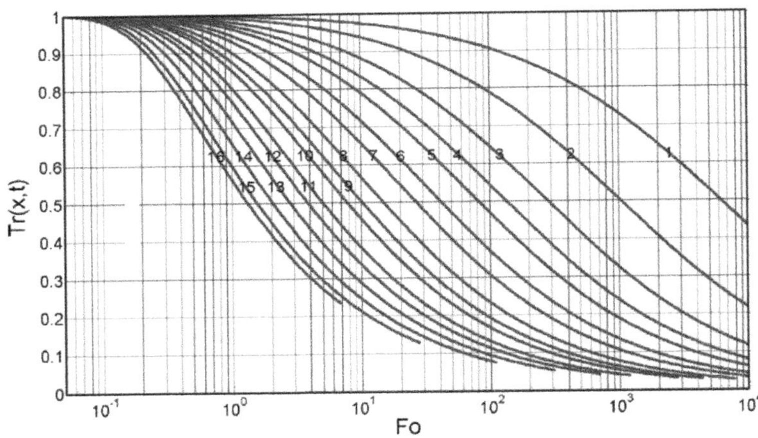

N°	Bi
1	0.01
2	0.025
3	0.05
4	0.075
5	0.1
6	0.15
7	0.2
8	0.3
9	0.4
10	0.5
11	0.75
12	1
13	1.5
14	2.5
15	5
16	10

A.11. Infinite plate of thickness 2L with imposed surface temperature

$$Tr(x,t) = \frac{T(x,t)-T_0}{T_i-T_0};\ Fo = \frac{at}{x^2};\ x = 0 \text{ on the surface.}$$

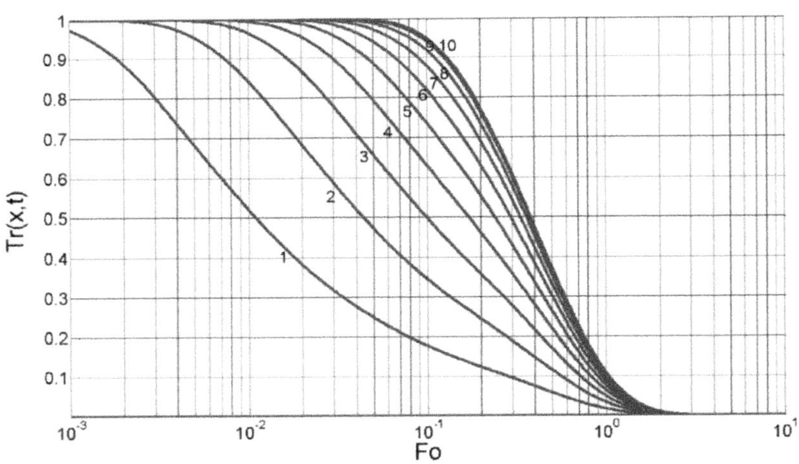

N°	x/L
1	0.1
2	0.2
3	0.3
4	0.4
5	0.5
6	0.6
7	0.9
8	0.8
9	0.9
10	1.0

A.12. Infinite plate of thickness 2L with imposed transfer coefficient

$$Tr(x,t) = \frac{T(x,t)-T_\infty}{T_i-T_\infty};\ Bi = \frac{hL}{\lambda};\ Fo = \frac{at}{L^2}.$$

Reduced temperature at the center of the plate ($x/L = 0$)

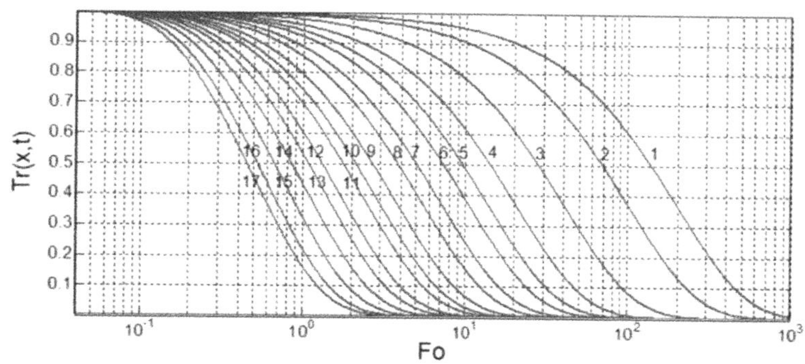

N°	Bi
1	0.005
2	0.01
3	0.025
4	0.05
5	0.075
6	0.1
7	0.15
8	0.2
9	0.3
10	0.4
11	0.5
12	0.75
13	1
14	1.5
15	2.5
16	5
17	10

Reduced temperature at the surface of the plate (x/L= 1)

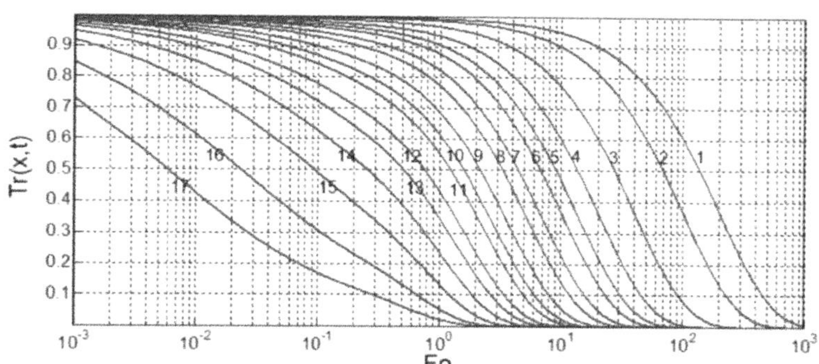

Ratio of the reduced temperature $T_r(x,t)$ to the center reduced temperature $T_0(0,t)$

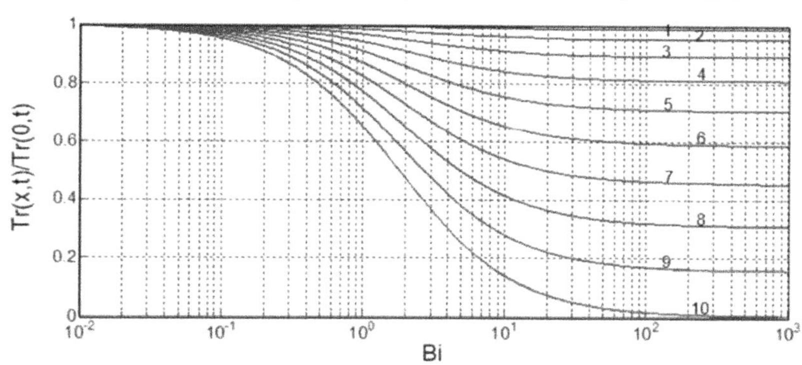

N°	x/L
1	0.1
2	0.2
3	0.3
4	0.4
5	0.5
6	0.6
7	0.9
8	0.8
9	0.9
10	1.0

A.13. Quadrupole matrices for different configurations

$q = \sqrt{\frac{p}{a}}$; I_0, I_1, K_0, K_1: Bessel functions (see section A.5).

Quadrupole associated with unidirectional transfer in a medium without energy generation (Maillet et al. 2000).

– Medium of finite thickness:

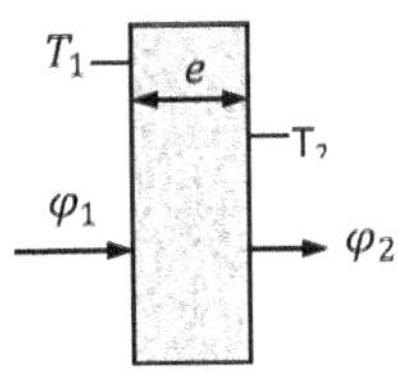

$$\begin{bmatrix}\theta_1\\ \Phi_1\end{bmatrix} = \begin{bmatrix}A & B\\ C & D\end{bmatrix}\begin{bmatrix}\theta_2\\ \Phi_2\end{bmatrix}$$

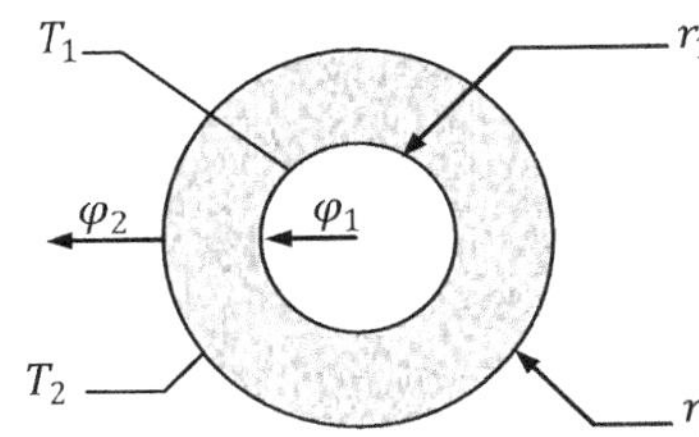

	Plane wall of thickness e	Hollow cylinder of radii r_1 and r_2	Hollow sphere of radii r_1 and r_2
A	$\cosh(qe)$	$qr_2[I_0(qr_1)K_1(qr_2) + I_1(qr_2)K_0(qr_1)]$	$\frac{r_2}{r_1}\cosh(s) - \frac{\sinh(s)}{qr_1}$
B	$\frac{\sinh(qe)}{\lambda qS}$	$\frac{1}{2\pi\lambda L}[I_0(qr_2)K_0(qr_1) - I_0(qr_1)K_0(qr_2)]$	$\frac{\sinh(s)}{\pi\lambda qr_1r_2}$
C	$\lambda qS\sinh(qe)$	$2\pi\lambda Lqr_1qr_2\begin{bmatrix}I_1(qr_2)K_1(qr_1) -\\ I_1(qr_1)K_1(qr_2)\end{bmatrix}$	$4\pi\lambda r_2\begin{bmatrix}\left(1-\frac{r_1}{r_2}\right)\cosh(s) +\\ \left(qr_1 - \frac{1}{qr_2}\right)\sinh(s)\end{bmatrix}$
D	$\cosh(qe)$	$qr_1[I_0(qr_2)K_1(qr_1) + I_1(qr_1)K_0(qr_2)]$	$\frac{r_1}{r_2}\cosh(s) - \frac{\sinh(s)}{qr_2}$

with $s = q(r_2 - r_1)$.

– Semi-infinite medium: the Laplace transform of Φ and the heat flow rate φ can be written as: $\Phi = \frac{\theta}{Z}$ with:

	Infinite wall	Semi-infinite cylinder of inner radius r_1	Semi-infinite sphere of inner radius r_1
Z	$\dfrac{1}{ES\sqrt{p}}$	$\dfrac{1}{2\pi\lambda L}\dfrac{K_0(qr_1)}{qr_1K_1(qr_1)}$	$\dfrac{1}{4\pi\lambda r_1(1+qr_1)}$

where $E = \sqrt{\lambda\rho c}$ is the thermal effusivity.

A.13.1. *Quadrupole associated with constriction resistance (Maillet et al. 2000)*

Sudden variation of the cross-section of the heat flow.

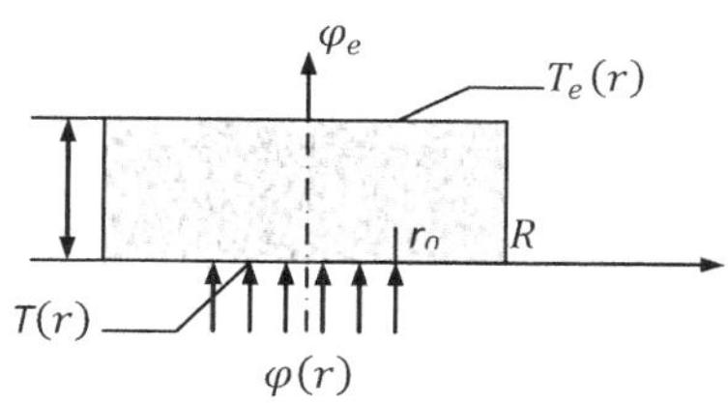

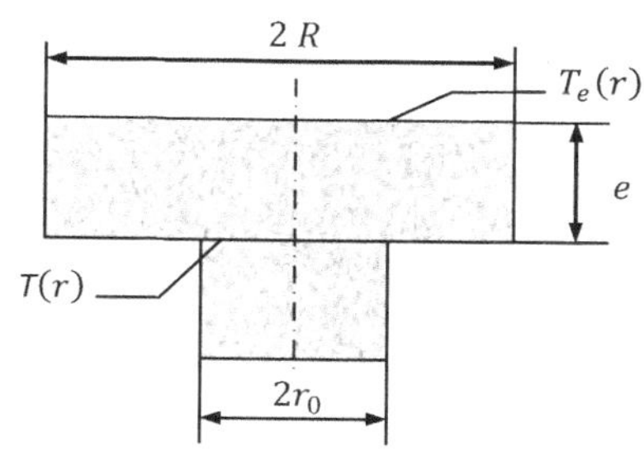

Transfer of a heat flow rate $\varphi(r)$ to the surface of a cylinder of radius R and thickness e, with $\varphi(r) = 0$ if $r > r_0$.

	Infinite$R$ and e $\boldsymbol{\Phi} = \dfrac{\overline{\boldsymbol{\theta}}_{0-r_0}}{\boldsymbol{Z}_c}$		Finite R and e $\begin{bmatrix}\boldsymbol{\theta}\\ \boldsymbol{\Phi}\end{bmatrix} = \begin{bmatrix}\mathbf{1} & Z_c\\ \mathbf{0} & \mathbf{1}\end{bmatrix}\begin{bmatrix}A & B\\ C & D\end{bmatrix}\begin{bmatrix}\boldsymbol{\theta}_e\\ \boldsymbol{\Phi}_e\end{bmatrix}$	
	$T(r) = T_0$ for $r < r_0$	$\varphi(r) = \varphi_0$ for $r < r_0$	$T_e(r) = T_{e0}$	$\varphi_e(r) = \varphi_{e0}$
$\boldsymbol{Z_c}$	$\dfrac{1}{4\lambda r_0\left(1+\frac{\pi}{4}qr_0\right)}$	$\dfrac{8}{3\pi^2\lambda r_0\left(1+\frac{8}{3\pi}qr_0\right)}$	$\displaystyle\sum_{n=1}^{\infty} F_n tanh(\gamma_n e)$	$\displaystyle\sum_{n=1}^{\infty}\frac{F_n}{tanh(\gamma_n e)}$

with $q = \sqrt{\frac{p}{a}}$ and $A = D = cosh(qe)$; $B = \frac{sinh(qe)}{\lambda q\pi R^2}$; $C = \lambda q\pi R^2 sinh(qe)$.

$$F_n = \frac{4J_1{}^2(\alpha_n r_0)}{\lambda\pi R^2 r_0{}^2 \alpha_n{}^2 \gamma_n J_0{}^2(\alpha_n r_0)} \text{ with } \alpha_n \text{solution of } J_1(\alpha_n R) = 0 \text{ and } \gamma_n = \alpha_n{}^2 + \frac{p}{a}.$$

If $\frac{e}{R} > 1$ then $tanh(\gamma_n e) = 1$.

A.13.2. *Quadrupole associated with unidirectional transfer in a medium with energy generation*

Temperature considered = average temperature of the heating element.

	Plate of thickness e	Solid cylinder of radius r and height L	Solid sphere of radius r
A	1	1	1
B	$\frac{1}{\lambda qS tanh(qe)} - \frac{1}{\rho cSep}$	$\frac{1}{2\pi\lambda L}\frac{I_0(qr)}{I_1(qr)} - \frac{1}{\rho c\pi r^2 Lp}$	$\frac{1}{4\pi\lambda r[qrcosh(qr) - 1]} - \frac{3}{4\rho c\pi r^3 p}$
C	$\rho cSep$	$\rho c\pi r^2 Lp$	$\frac{4}{3}\rho c\pi r^3 p$
D	$\frac{qe}{tanh(qe)}$	$\frac{qr}{2}\frac{I_0(qr)}{I_1(qr)}$	$\frac{(qr)^2}{3[qrcosh(qr) - 1]}$

A.14. Orthogonality of eigenfunctions

Consider the example where the eigenfunctions ψ_m and ψ_n of eigenvalues ω_m and ω_n are solutions of the following differential equations over the domain $0 < x < L$:

$$\frac{d^2\psi_m}{dx^2} + \omega_m{}^2\psi_m = 0 \quad \text{(a)}$$

$$\frac{d^2\psi_n}{dx^2} + \omega_n{}^2\psi_n = 0 \quad \text{(b)}$$

By multiplying equation (a) by ψ_n and equation (b) by ψ_m and taking the difference of the two equations which is then integrated over the domain $0 < x < L$, we get:

$$(\omega_m{}^2 - \omega_n{}^2)\int_0^L \psi_m\psi_n dx + \int_0^L \left(\psi_n \frac{d^2\psi_m}{dx^2} - \psi_m \frac{d^2\psi_n}{dx^2}\right) dx = 0 \quad \text{(c)}$$

We immediately integrate the second term of equation (c) to obtain:

$$({\omega_m}^2 - {\omega_n}^2) \int_0^L \psi_m \psi_n dx + \left[\psi_n \frac{d\psi_m}{dx} - \psi_m \frac{d\psi_n}{dx}\right]_{x=0}^{x=L} = 0 \qquad \text{(d)}$$

If the functions ψ_m and ψ_n satisfy on the boundaries of the domain, namely in $x = 0$ and $x = L$, homogeneous boundary conditions (a condition is said to be homogeneous if it is verified by $K\psi$, where K is any constant):

– homogeneous Dirichlet condition: $\psi = 0$;

– homogeneous Neumann condition: $\frac{\partial \psi}{\partial x} = 0$;

– homogeneous Fourier condition $\frac{\partial \psi}{\partial x} = h\,\psi$.

It is easy to see that the second term of (d) on the boundaries of the domain is zero. We then have:

$$({\omega_m}^2 - {\omega_n}^2) \int_0^L \psi_m \psi_n dx = 0$$

Hence: $\int_0^L \psi_m \psi_n dx = 0$ if $\omega_m \neq \omega_n$ if $m \neq n$.

A.15. Program for calculating the temperature of a plate[1]

```
% Program for calculating the temperature at the center of a plate
% Data
Ti=20; % Initial temperature
T0=40; % Temperature imposed on the surface
fi=100; % Imposed heat flux
h=1000; % Imposed transfer coefficient
lam0=1; % Thermal conductivity at 0°C
rhoc0=1e6; % Volumetric heat capacity at 0°C
a=lam0/rhoc0;
k1=1e-2; % Temperature coefficient of thermal conductivity
k2=5e-3; % Temperature coefficient of volumetric heat capacity
tf=5000; % Final time
L=0.05; % Half-thickness of the plate

% Calculation parameters;
J=21; % Number of space steps
k=4;
dx=L/(J-1); % Space step
dt=dx^2/a/k; % Time step
I=round(tf/dt ); % Number of time steps
for i=1:I
```

1 For a color version of this Appendix, see http://www.iste.co.uk/jannot/heattransfer1.zip.

```
t(i)=(i-1)*dt;
end

% Initialization of tables
T1(1:J)=Ti;
T2(1:J)=Ti;
Tar(1)=0;

for i=1:I
% Calculation for the first mesh
T2(1)=T0; % Temperature T0 imposed
% T2( 1)=
(2*dt/rhoc0/(1+k2*T1(1))/dx)*(fi+(lam0*(1+k1*T2(1))/dx)*(T1(2)-
T1(1)))+T1(1);%
Imposedheat flow density fi
% T2(1)=(2*dt/rhoc0/(1+k2*T1(1))/dx)*(h*(T0-
T1(1))+(lam0*(1+k1*T2(1))/dx)*(T1(2)-
T1(1)))+ T1(1);% Heat flow density fi imposed

% Calculation of the 2nd to the (D- 1)th mesh
for j=2:J-1
T2(j)=T1(j)+(dt/rhoc0/(1+k2*T1(j))/dx)*((lam0*(1+k1*T1(j))/dx)*(T1(j+1
)+T1(j-1)-
        2*T1(j)));
End

% Calculation for the last mesh
T2(J)=T1(J)-(2/rhoc0/(1+k2*T1(J ))/ dx)*(lam0*(1+k1*T1(J))*(T1(J)-T1
(D-1))*dt/dx);

T1=T2 ;
Tar(i+1)=(T2(11)-Ti)/(T0-Ti); % Temperature at the center
End

plot(t,Tar )
```

A.16. Conservation equations

– Nomenclature:

- c_p, heat capacity at constant pressure (J kg^{-1} K^{-1});
- c_v, heat capacity at constant volume (J kg^{-1} K^{-1});
- f, resultant force of the external field per unit of fluid mass (N kg^{-1});
- P, pressure (Pa);
- T, temperature (°C);
- u, component of the velocity along Ox (ms^{-1});
- U, internal energy per unit mass (J kg^{-1});

- V, average velocity (m s^{-1});
- v, component of the velocity along Oy (m s^{-1});
- w, component of the velocity along Oz (m s^{-1});
- ϕ, heat flux (W m^{-2});
- μ, dynamic viscosity (kg m^{-1} s^{-1});
- ν, kinematic viscosity (m^2 s^{-1});
- ρ, density (kg m^{-3}).

– Indices:

- e, entering;
- g, generated;
- s, leaving.

A.16.1. *Notations*

Consider a scalar field $P(x, y, z)$ and a vector field $\vec{V} = u\vec{\imath} + v\vec{\jmath} + w\vec{k}$. We will use the following notations:

– operator ∇: $\nabla = \begin{pmatrix} \frac{\partial}{\partial x} \\ \frac{\partial}{\partial y} \\ \frac{\partial}{\partial z} \end{pmatrix}$;

– particulate derivative of P: $\frac{DP}{Dt} = \frac{\partial P}{\partial t} + u\frac{\partial P}{\partial x} + v\frac{\partial P}{\partial y} + w\frac{\partial P}{\partial z}$;

– particulate derivative of $\vec{V}$: $\frac{DV}{Dt} = \begin{pmatrix} \frac{\partial u}{\partial t} + u\frac{\partial u}{\partial x} + v\frac{\partial u}{\partial y} + w\frac{\partial u}{\partial z} \\ \frac{\partial v}{\partial t} + u\frac{\partial v}{\partial x} + v\frac{\partial v}{\partial y} + w\frac{\partial v}{\partial z} \\ \frac{\partial w}{\partial t} + u\frac{\partial w}{\partial x} + v\frac{\partial w}{\partial y} + w\frac{\partial w}{\partial z} \end{pmatrix}$;

– divergence of $\vec{V}$: $div\vec{V} = \nabla . \mathrm{V} = \frac{\partial u}{\partial x} + \frac{\partial v}{\partial y} + \frac{\partial w}{\partial z}$;

– rotational of $\vec{V}$: $rot\vec{V} = \nabla \wedge \mathrm{V} = \begin{pmatrix} \frac{\partial w}{\partial y} - \frac{\partial v}{\partial z} \\ \frac{\partial u}{\partial z} - \frac{\partial w}{\partial x} \\ \frac{\partial v}{\partial x} - \frac{\partial u}{\partial y} \end{pmatrix}$;

– gradient of P: $\overrightarrow{gradP} = \nabla P = \frac{\partial P}{\partial x}\vec{i} + \frac{\partial P}{\partial y}\vec{j} + \frac{\partial P}{\partial z}\vec{k}$;

– Laplacian of P: $\Delta P = \nabla^2 P = \frac{\partial^2 P}{\partial x^2} + \frac{\partial^2 P}{\partial y^2} + \frac{\partial^2 P}{\partial z^2}$.

A.16.2. *Conservation of mass equation*

Let us consider the flow of a fluid and carry out a mass balance on the system constituted by the parallelepipedic element of fluid of sides Δx, Δy and Δz and of mass dm:

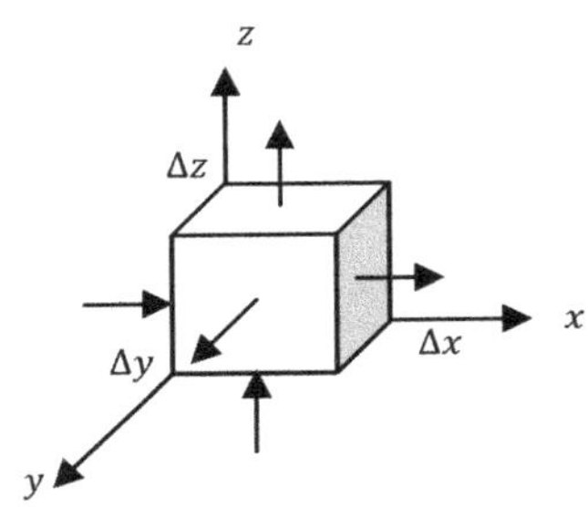

The mass conservation in this volume between times t and $t + dt$ can be written for each of the components of the mixture:

Mass + initial mass + generated mass = final mass + output mass

This balance allows us to establish the continuity equation (see fluid mechanics course):

$$\frac{\partial \rho}{\partial t} + \nabla.(\rho V) = \frac{\partial \rho}{\partial t} + div(\rho \vec{V}) = 0 \qquad [A.1]$$

where ρ is the density and $\vec{V}$ is the velocity.

A.16.3. *Equation of motion: Navier–Stokes*

We propose here to establish the equation governing the movement of a supposed monocomponent fluid.

Newton's law can be applied to an elementary volume $\Delta x\Delta y\Delta z$ of the fluid:

$$F = m\gamma$$

where γ is the fluid acceleration (m s^{-2}).

The component F_x of the force F along the direction Ox can be written as: $F_x = m\gamma_x = m\frac{du}{dt}$.

We can write u using a Taylor expansion of order 1:

$$du = \frac{\partial u}{\partial t}dt + \frac{\partial u}{\partial x}dx + \frac{\partial u}{\partial y}dy + \frac{\partial u}{\partial z}dz$$

Equation $x = vt$ allows us to write for a very short time dt during which u remains constant: $dx = udt$, $dy = vdt$, $dz = wdt$, which allows us to write du in the form:

$$du = \frac{\partial u}{\partial t}dt + \frac{\partial u}{\partial x}udt + \frac{\partial u}{\partial y}vdt + \frac{\partial u}{\partial z}wdt$$

and $\gamma_x = \frac{du}{dt} \equiv \frac{Du}{Dt} = \frac{\partial u}{\partial t} + u\frac{\partial u}{\partial x} + v\frac{\partial u}{\partial y} + w\frac{\partial u}{\partial z}$.

A.16.3.1. *Flow without viscous forces*

The forces acting on the fluid are of two types:

– forces due to an external field (gravity, for example), which apply to the entire volume; their resultant per unit mass will be denoted by f;

– pressure forces acting on the surfaces of the volume under consideration.

Newton's law Ox is written as:

$$\rho\Delta x\Delta y\Delta z\frac{Du}{Dt} = -\frac{\partial P}{\partial x}\Delta x\Delta y\Delta z + \rho f_x\Delta x\Delta y\Delta z$$

Moreover, if we write this expression for the three directions, we obtain Euler's (local) equation:

$$\rho\frac{DV}{Dt} = -\nabla P + \rho f \qquad \text{[A.2]}$$

This equation allows us to deal with the cases where the viscous forces are either absent (hydrostatic (fluid at rest)) or negligible (high velocity flows far from the wall (very low velocity gradient)).

It can also be written in integral form over a volume Λ bounded by a surface Σ:

$$\int_{\Lambda} \frac{\partial \rho}{\partial t}\, dv + \int_{\Sigma} \rho \vec{V}.\vec{n} dS = 0 \qquad [A.3]$$

where $\vec{n}$ is the normal to the surface Σ directed outwards from the volume Λ.

A.16.3.2. *Flow with viscous forces*

Compared to the previous case, it is necessary to add to the surface forces due to the pressure gradient shear forces parallel to the surfaces of the system considered. We will adopt the following notations for the shear forces (viscous friction) being applied to the surfaces of an elementary volume $dxdydz$:

– τ_{xx}, τ_{yy}, τ_{zz}: the normal component of the surface force along the directions Ox, Oy and Oz;

– τ_{xy}, τ_{yx}, τ_{xz}, τ_{zx}, τ_{yz}, τ_{zy}: the tangential component of the surface force. The first index indicates the direction normal to the surface considered; the second index indicates the direction in which the component acts.

Newton's law then makes it possible to write according to each of the three directions:

$$\rho\frac{Du}{Dt} = \rho f_x + \frac{\partial \tau_{xx}}{\partial x} + \frac{\partial \tau_{yx}}{\partial y} + \frac{\partial \tau_{zx}}{\partial z} - \frac{\partial P}{\partial x}$$
$$\rho\frac{Dv}{Dt} = \rho f_y + \frac{\partial \tau_{xy}}{\partial x} + \frac{\partial \tau_{yy}}{\partial y} + \frac{\partial \tau_{zy}}{\partial z} - \frac{\partial P}{\partial y}$$
$$\rho\frac{Dw}{Dt} = \rho f_z + \frac{\partial \tau_{xz}}{\partial x} + \frac{\partial \tau_{yz}}{\partial y} + \frac{\partial \tau_{zz}}{\partial z} - \frac{\partial P}{\partial z}$$

This formulation is valid for all fluids.

A.16.3.3. *Case of a Newtonian fluid*

A fluid is said to be Newtonian if the shear stresses are proportional to the velocity gradients; in this case, we have:

$$\tau_{yx} = \mu\left(\frac{\partial u}{\partial y} + \frac{\partial v}{\partial x}\right);\ \tau_{xz} = \tau_{zx} = \mu\left(\frac{\partial u}{\partial z} + \frac{\partial w}{\partial x}\right);\tau_{yz} = \tau_{zy} = \mu\left(\frac{\partial v}{\partial z} + \frac{\partial w}{\partial y}\right)$$

and $\tau_{xx} = \mu\left(\frac{\partial u}{\partial x} - \frac{2}{3}\nabla.V\right)$; $\tau_{yy} = \mu\left(\frac{\partial v}{\partial y} - \frac{2}{3}\nabla.V\right)$; $\tau_{zz} = \mu\left(\frac{\partial w}{\partial z} - \frac{2}{3}\nabla.V\right)$.

where μ is called the *dynamic viscosity* of the fluid, which is expressed in kg m^{-1} s^{-1}.

By replacing the stress tensor with this expression in the general equation of motion deduced from Newton's law, we finally obtain the *Navier–Stokes equation*: $\frac{DV}{Dt} = -\frac{1}{\rho}\nabla P + \nu\nabla^2 V + \frac{1}{3}\nu\nabla(\nabla.V) + f$.

$$\frac{DV}{Dt} = -\frac{1}{\rho}\nabla P + \nu\nabla^2 V + \frac{1}{3}\nu\nabla(\nabla.V) + f \qquad \text{[A.4]}$$

where $\nu = \frac{\mu}{\rho}$ is the kinematic viscosity expressed in m^2 s^{-1} and $\nabla^2 = \frac{\partial^2}{\partial x^2} + \frac{\partial^2}{\partial y^2} + \frac{\partial^2}{\partial z^2}$.

For an incompressible fluid, $\nabla.V = 0$, and we obtain a simplified form of the Navier–Stokes equation:

$$\frac{DV}{Dt} = -\frac{1}{\rho}\nabla P + \nu\nabla^2 V + f \qquad \text{[A.5]}$$

where f is the volume force due to an external field (gravity for example).

This last equation takes the following form in Cartesian coordinates:

$$\frac{\partial u}{\partial t} + u\frac{\partial u}{\partial x} + v\frac{\partial u}{\partial y} + w\frac{\partial u}{\partial z} = -\frac{1}{\rho}\frac{\partial P}{\partial x} + \nu\left(\frac{\partial^2 u}{\partial x^2} + \frac{\partial^2 u}{\partial y^2} + \frac{\partial^2 u}{\partial z^2}\right) + f_x$$
$$\frac{\partial v}{\partial t} + u\frac{\partial v}{\partial x} + v\frac{\partial v}{\partial y} + w\frac{\partial v}{\partial z} = -\frac{1}{\rho}\frac{\partial P}{\partial y} + \nu\left(\frac{\partial^2 v}{\partial x^2} + \frac{\partial^2 v}{\partial y^2} + \frac{\partial^2 v}{\partial z^2}\right) + f_y$$
$$\frac{\partial w}{\partial t} + u\frac{\partial w}{\partial x} + v\frac{\partial w}{\partial y} + w\frac{\partial w}{\partial z} = -\frac{1}{\rho}\frac{\partial P}{\partial z} + \nu\left(\frac{\partial^2 w}{\partial x^2} + \frac{\partial^2 w}{\partial y^2} + \frac{\partial^2 w}{\partial z^2}\right) + f_z$$

In the steady state, this equation can be written in the following integral form (Euler's theorem):

$$\int_\Sigma \rho\vec{V}(\vec{V}.\vec{n})\,dS = \int_\Lambda \rho\vec{f}\,dv + \int_\Sigma \vec{T}\,dS \qquad \text{[A.6]}$$

where $\vec{T}$ is the external forces (per unit area) exerted by contact on the faces of the surface Σ bounding the volume Λ.

A.16.4. *Energy conservation equation*

We consider as a system the volume of fluid with sides dx, dy, dz centered on the point of coordinates (x, y, z). We will call $(\vec{\imath}, \vec{\jmath}, \vec{k})$ the unit vectors of the axes Ox, Oy and Oz.

We apply to this system the first principle of thermodynamics between the times t and $t + dt$ during which the volume makes a displacement $\overrightarrow{dl} = \vec{V}dt$:

$$\Delta U = \Delta E_{ci} + W_v + W_p + W_c + \Delta E_{cond}$$

with:

– ΔU, the accumulation of internal and kinetic energy within the fluid contained in the volume;

– ΔE_{ci}, the internal and kinetic energy entering the system – the internal and kinetic energy leaving the system;

– W_v, the work of the viscous forces acting on the surfaces of the system;

– W_p, the work of the pressure forces acting on the faces of the system;

– W_c, the work of the forces of the external field in the volume $\Delta x\Delta y\Delta z$;

– ΔE_{cond}, the conduction energy entering the system – the conduction energy leaving the system.

We have:

$$\Delta U = \frac{\partial}{\partial t}\left(\rho U + \frac{1}{2}\rho V^2\right)\Delta x\Delta y\Delta z\, dt.$$

To calculate ΔE_{ci}, let us calculate, for example, the kinetic and internal energy entering at x through the surface perpendicular to Ox: the mass crossing this surface for a time dt is written as: $\rho dydzudt$, and the kinetic and internal energy entering has the value: $\Delta E_{ci} = \rho_x dydzu_x dt\left(U + \frac{1}{2}V^2\right)_x$.

The complete expression of ΔE_{ci} is therefore written as:

$$\Delta E_{ci} = \left\{\begin{aligned} & dydz\left[u_x\left(\rho U + \frac{1}{2}\rho V^2\right)_x - u_{x+dx}\left(\rho U + \frac{1}{2}\rho V^2\right)_{x+dx}\right] \\ & +dxdz\left[u_y\left(\rho U + \frac{1}{2}\rho V^2\right)_y - u_{y+dy}\left(\rho U + \frac{1}{2}\rho V^2\right)_{y+dy}\right] \\ & +dxdy\left[u_z\left(\rho U + \frac{1}{2}\rho V^2\right)_z - u_{z+dz}\left(\rho U + \frac{1}{2}\rho V^2\right)_{z+dz}\right] \end{aligned}\right\} dt$$

Hence:

$$u_x\left(\rho U+\tfrac{1}{2}\rho V^2\right)_x-u_{x+dx}\left(\rho U+\tfrac{1}{2}\rho V^2\right)_{x+dx}=-\frac{\partial\left[u\left(\rho U+\frac{1}{2}\rho V^2\right)\right]}{\partial x}dx.$$

Thus:

$$\begin{aligned}\Delta E_{ci}=-dxdydz&\left\{\frac{\partial\left[u\left(\rho U+\frac{1}{2}\rho V^2\right)\right]}{\partial x}+\frac{\partial\left[v\left(\rho U+\frac{1}{2}\rho V^2\right)\right]}{\partial y}\right.\\&\left.+\frac{\partial\left[w\left(\rho U+\frac{1}{2}\rho V^2\right)\right]}{\partial z}\right\}dt\\&=-dxdydz\nabla.\left[V\left(\rho U+\frac{1}{2}\rho V^2\right)\right]dt\qquad W_v=\sum\vec{F}_v.\overrightarrow{dl}\end{aligned}$$

$$W_v=\left\{\begin{aligned}&dydz\left[\left(\tau_{xx}\vec{i}+\tau_{xy}\vec{j}+\tau_{xz}\vec{k}\right).\left(u_x\vec{i}+v_x\vec{j}+w_x\vec{k}\right)-\left(\tau_{xx}\vec{i}+\tau_{xy}\vec{j}+\tau_{xz}\vec{k}\right).\left(u_{x+dx}\vec{i}+v_{x+dx}\vec{j}+wu_{x+dx}\vec{k}\right)\right]\\&+dxdz\left[\left(\tau_{yx}\vec{i}+\tau_{yy}\vec{j}+\tau_{yz}\vec{k}\right).\left(u_y\vec{i}+v_y\vec{j}+w_y\vec{k}\right)-\left(\tau_{yx}\vec{i}+\tau_{yy}\vec{j}+\tau_{yz}\vec{k}\right).\left(u_{y+dy}\vec{i}+v_{y+dy}\vec{j}+wu_{y+dy}\vec{k}\right)\right]\\&+dydx\left[\left(\tau_{zx}\vec{i}+\tau_{zy}\vec{j}+\tau_{zz}\vec{k}\right).\left(u_z\vec{i}+v_z\vec{j}+w_z\vec{k}\right)-\left(\tau_{zx}\vec{i}+\tau_{zy}\vec{j}+\tau_{zz}\vec{k}\right).\left(u_{z+dz}\vec{i}+v_{z+dz}\vec{j}+wu_{z+dz}\vec{k}\right)\right]\end{aligned}\right\}dt$$

Hence:

$$\begin{aligned}&\left(\tau_{xx}\vec{i}+\tau_{xy}\vec{j}+\tau_{xz}\vec{k}\right).\left(u_x\vec{i}+v_x\vec{j}+w_x\vec{k}\right)\\&-\left(\tau_{xx}\vec{i}+\tau_{xy}\vec{j}+\tau_{xz}\vec{k}\right).\left(u_{x+dx}\vec{i}+v_{x+dx}\vec{j}+wu_{x+dx}\vec{k}\right)\\&=\frac{\partial}{\partial x}\left(\tau_{xx}u+\tau_{xy}v+\tau_{xz}w\right)dx\end{aligned}$$

Thus:

$$W_v=dxdydz\nabla.\left(\vec{F}_v.\vec{V}\right)$$

$$\begin{aligned}W_p=\sum\vec{F}_p.\overrightarrow{dl}&\\&=P_xdydz\vec{i}.\vec{V}_xdt-P_{x+dx}dydz\vec{i}.\vec{V}_{x+dx}dt+P_ydxdz\vec{j}.\vec{V}_ydt\\&-P_{y+dy}dxdz\vec{j}.\vec{V}_{y+dy}dt+P_zdxdy\vec{k}.\vec{V}_zdt\\&-P_{z+dz}dxdy\vec{k}.\vec{V}_{z+dz}dt\end{aligned}$$

$$W_p=dxdydz\left[\frac{\partial(uP)}{\partial x}+\frac{\partial(vP)}{\partial y}+\frac{\partial(wP)}{\partial z}\right]dt=dxdydz\nabla(VP)dt$$

$$W_c=\vec{F}.\overrightarrow{dl}=\rho dxdydz\,\vec{f}.\vec{V}\,dt$$

$$\Delta E_{\text{cond}} = \left[dydz(\phi_{c_x} - \phi_{c_{x+dx}}) + dxdz\left(\phi_{c_y} - \phi_{c_{y+dy}}\right) + dydx(\phi_{c_z} - \phi_{c_{z+dz}})\right] dt$$

Hence:

$\phi_{c_x} - \phi_{c_{x+dx}} = -\frac{\partial \phi_c}{\partial x} dx$ thus: $\Delta E_{\text{cond}} = -dxdydz\left(\frac{\partial \phi_c}{\partial x} + \frac{\partial \phi_c}{\partial y} + \frac{\partial \phi_c}{\partial z}\right) dt = -dxdydz\, \nabla \phi_c\, dt.$

We arrive at *the energy equation*:

$$\frac{\partial}{\partial t}\rho\left(U + \frac{1}{2}V^2\right) + \nabla.\left[\rho V\left(U + \frac{1}{2}V^2\right)\right] = -\nabla\phi_c + \rho\vec{V}.\vec{f} - P\nabla V - \nabla.(\tau V)$$

By combining this equation with the continuity equation and the Navier–Stokes equation, we establish another form of the energy equation:

$$\rho\frac{DU}{Dt} = -\nabla\phi_c - P\nabla V + \tau{:}\nabla V \qquad [A.7]$$

where:

$\tau{:}\nabla V = \nabla.(\tau V) = \tau_{xx}\left(\frac{\partial u}{\partial x}\right) + \tau_{yy}\left(\frac{\partial v}{\partial y}\right) + \tau_{zz}\left(\frac{\partial w}{\partial z}\right) + \tau_{xy}\left(\frac{\partial u}{\partial y} + \frac{\partial v}{\partial x}\right) + \tau_{xz}\left(\frac{\partial u}{\partial z} + \frac{\partial w}{\partial x}\right) + \tau_{yz}\left(\frac{\partial v}{\partial z} + \frac{\partial w}{\partial y}\right).$

The term ϕ_c represents the increase in internal energy due to heat transfer by conduction, which is written by applying Fourier's law: $\phi_c = -\lambda\nabla T$.

Thus, $-\nabla\phi_c = \nabla.\lambda\nabla T = \lambda\nabla^2 T$, where $\nabla^2 = \frac{\partial^2}{\partial x^2} + \frac{\partial^2}{\partial y^2} + \frac{\partial^2}{\partial z^2}$ is the Laplacian.

The term $-P\nabla V$ is the increase due to pressure forces of the internal energy of the element per unit volume. This term is zero for an incompressible fluid.

The viscous dissipation term $\tau{:}V$ is a positive quantity that represents the irreversible transformation of mechanical energy into thermal energy.

The classical equations of thermodynamics allow us to write the internal energy in the form:

$$dU = c_v dT + \left[-P + T\left(\frac{\partial P}{\partial T}\right)_v\right] d\left(\frac{1}{\rho}\right)$$

allowing us to write:

$$\rho\frac{DU}{Dt}=\rho c_v\frac{DT}{Dt}+\left[-P+T\left(\frac{\partial P}{\partial T}\right)_v\right]\rho\frac{D\left(\frac{1}{\rho}\right)}{Dt}.$$

The continuity equation leads to $\rho\frac{D\left(\frac{1}{\rho}\right)}{Dt}=\nabla.V$, and to a new form of the energy equation:

$$\rho c_v\frac{DT}{Dt}=\lambda\nabla^2T-T\left(\frac{\partial P}{\partial T}\right)_v\nabla.V+\tau:\nabla V \qquad [A.8]$$

The term $\tau:\nabla.V$ is important in very high velocity flows (speed of sound) and in areas with strong velocity gradients (near the walls). It can be neglected in the other cases, which is done in what follows.

We show that equation [A.8] can also be written as:

$$\rho c_p\frac{DT}{Dt}=\lambda\nabla^2T-\left[\frac{\partial ln(V)}{\partial ln(T)}\right]_P\frac{DP}{Dt}+\tau:\nabla V \qquad [A.9]$$

If we also make the ideal gas hypothesis, we have equations $\left(\frac{\partial P}{\partial T}\right)_v=\frac{P}{T}$ and $\left[\frac{\partial ln(V)}{\partial ln(T)}\right]_P=1$.

For a fluid at constant pressure in which the viscous effects are neglected (or for a solid), equation [A.9] can be written as:

$$\rho c_p\left(\frac{\partial T}{\partial t}+V.\nabla T\right)=\lambda\nabla^2T \qquad [A.10]$$

In the case of a solid $V=0$, and we obtain:

$$\rho c\frac{\partial T}{\partial t}=\lambda\nabla^2T \qquad [A.11]$$

It is the Poisson equation that governs heat diffusion.

A.17. Correlations for transfer coefficients in forced convection

The characteristics of the fluid calculated at $T_f=\frac{T_p+T_\infty}{2}$.

<table>
<tr><th>Geometry</th><th>Correlation</th></tr>
<tr><td>Flow on a plate</td><td>$Nu(x)$: Nu at the distance x from the edge of the plane

$\overline{Nu_L}$: Nu average over the length L of the plane

Turbulent flow: ($Re > 5.10^5$ and $Pr > 0.5$)

$Nu(x) = 0.0288\ Re(x)^{0.8} Pr^{1/3}$
$\overline{Nu_L} = 0.035\ {Re_L}^{0.8} Pr^{1/3}$

Laminar flow: ($Re < 5.10^5$ and $10 > Pr > 0.5$)

$Nu(x) = 0.324\ Re(x)^{0.5} Pr^{1/3}$

$\overline{Nu_L} = 0.628\ {Re_L}^{0.5} Pr^{1/3}$</td></tr>
<tr><td>Flow in a tube</td><td>Turbulent flow: ($Re > 5{,}000$ and $100 > Pr > 0.6$)

$Nu = 0.023\ Re^{0.8} Pr^n$

$n = 0.3$ if $T_{fluid} > T_{wall}$
$n = 0.4$ if $T_{fluid} < T_{wall}$

Re calculated for $D_h = \frac{4S}{P}$, where S is the fluid cross-section and P is the fluid/wall contact perimeter.
Laminar flow: $Nu = 1.86(RePr)^{1/3} \left(\frac{D}{L}\right)^{1/3} \left(\frac{\mu}{\mu_p}\right)^{0.14}$
Valid for $RePr\frac{D}{L} > 10$, μ_p calculated at T_p</td></tr>
<tr><td>Flow perpendicular to a circular cylinder (Jacob 1949)</td><td>$Nu = CRe^n Pr^{1/3}$, velocity u_∞ calculated upstream of the tube
<table>
<tr><th>Re</th><th>C</th><th>n</th></tr>
<tr><td>0.4–4</td><td>0.989</td><td>0.330</td></tr>
<tr><td>4–40</td><td>0.911</td><td>0.385</td></tr>
<tr><td>40–4,000</td><td>0.683</td><td>0.466</td></tr>
<tr><td>4,000–40,000</td><td>0.193</td><td>0.618</td></tr>
<tr><td>40,000–250,000</td><td>0.0266</td><td>0.805</td></tr>
</table></td></tr>
<tr><td>Flow perpendicular to a non-circular cylinder (Jacob 1947)</td><td>
<table>
<tr><th>Geometry</th><th>Re</th><th>C</th><th>n</th></tr>
<tr><td rowspan="2">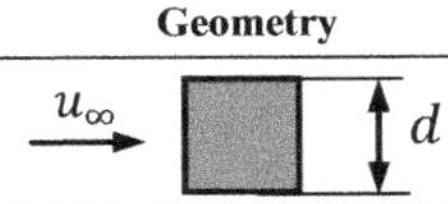</td><td>$5\ 10^3$–10^5</td><td>0.102</td><td>0.675</td></tr>
<tr><td>$4\ 10^3$–$1.5\ 10^4$</td><td>0.228</td><td>0.731</td></tr>
</table></td></tr>
</table>

Geometry	Correlation								
		$\frac{S_n}{d}$							
	$\frac{S_p}{d}$	1.25		1.5		2.0		3.0	
		C	n	C	n	C	n	C	n
		In-line arrangement							
	1.25	0.386	0.592	0.305	0.608	0.111	0.704	0.070	0.752
	1.5	0.407	0.586	0.278	0.620	0.112	0.702	0.075	0.744
	2.0	0.464	0.570	0.332	0.602	0.254	0.632	0.220	0.648
	3.0	0.322	0.601	0.396	0.584	0.415	0.581	0.317	0.608
		Staggered arrangement							
	0.16	-	-	-	-	-	-	0.236	0.636
	0.9	-	-	-	-	0.495	0.571	0.445	0.581
	1.0	-	-	0.552	0.558	-	-	-	-
Flow perpendicular to a bundle of 10 tubes (Grimson1937)	**1.125**	-	-	-	-	0.531	0.565	0.575	0.560
	1.25	0.575	0.556	0.561	0.554	0.576	0.556	0.579	0.562
	1.5	0.501	0.568	0.511	0.562	0.502	0.568	0.542	0.568
	2.0	0.448	0.572	0.462	0.568	0.535	0.556	0.498	0.570
	3.0	0.344	0.592	0.395	0.580	0.488	0.562	0.467	0.574

$Nu = CRe^n Pr^{1/3}$, velocity u_∞ calculated upstream of the tube

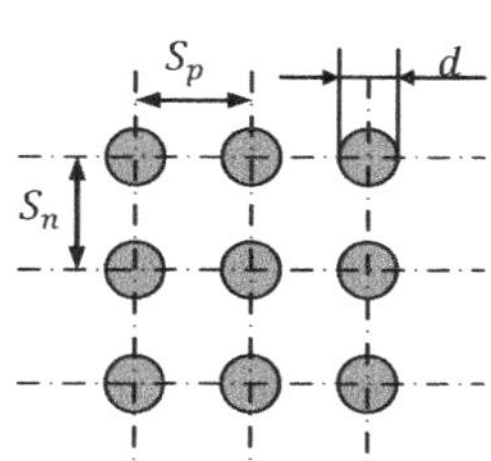

In-line arrangement

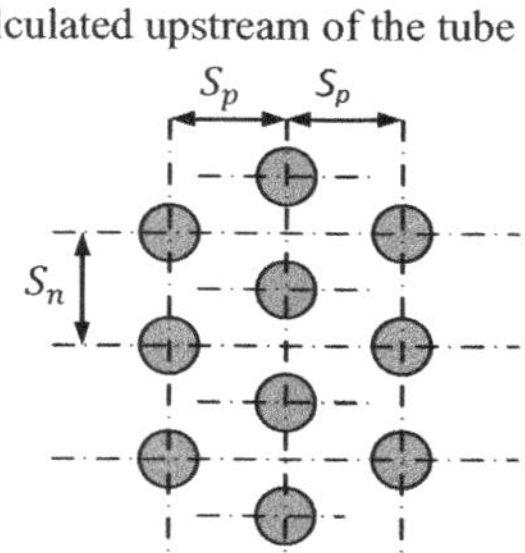

Staggered arrangement

Flow perpendicular to a bundle of n rows of tubes ($n \leq 10$) (Grimson 1937)	$N = \frac{h_n}{h_{10}}$									
Number of rows n	1	2	3	4	5	6	7	8	9	10
***N* in-line**	0.64	0.80	0.87	0.90	0.92	0.94	0.96	0.98	0.99	1.0
***N* staggered**	0.68	0.75	0.83	0.89	0.92	0.95	0.97	0.98	0.99	1.0

A.18. Correlations for transfer coefficients in natural convection

According to Holman (1990).

Correlations valid for all fluids: $Nu = C(GrPr)^m$			
Geometry	$GrPr$	C	m
Vertical plates and cylinders	10^4–10^9 10^9–10^{13}	0.59 0.021	1/4 2/5
Horizontal cylinders	10^{-10}–10^{-2} 10^{-2}–10^2 10^2–10^4 10^4–10^7 10^7–10^{12}	0.675 1.02 0.850 0.480 0.125	0.058 0.148 0.188 0.25 0.33
Upper side of a hot plate or bottom side of a cold plate	2.10^4–$8\ 10^6$ 8.10^6–10^{11}	0.54 0.15	0.25 0.33
Bottom side of a hot plate or upper side of a cold plate	10^5–10^{11}	0.27	0.25
Simplified equations for air at atmospheric pressure and at $T \approx 20°C$			
Geometry	Laminar $10^4 < GrPr < 10^9$	Turbulent $GrPr > 10^9$	
Vertical plate or cylinder	$h = 1.54\left(\frac{\Delta T}{L}\right)^{1/4}$	$h = 1.22(\Delta T)^{1/3}$	
Horizontal cylinder	$h = 1.26\left(\frac{\Delta T}{D}\right)^{1/4}$	$h = 1.52(\Delta T)^{1/3}$	
Upper side of a hot horizontal plate or bottom side of a cold plate	$h = 1.41\left(\frac{\Delta T}{L}\right)^{1/4}$	$h = 1.83(\Delta T)^{1/3}$	
Bottom side of a hot plate or upper side of a cold plate	$h = 0.71\left(\frac{\Delta T}{L}\right)^{1/4}$	$h = 0.71\left(\frac{\Delta T}{L}\right)^{1/4}$	

$$Ra = GrPr = \frac{c\beta g\Delta T\rho^2 d^3}{\lambda\mu}$$

<table>
<tr><th colspan="2">Inclined closed rectangular cell</th></tr>
<tr>
<td rowspan="2">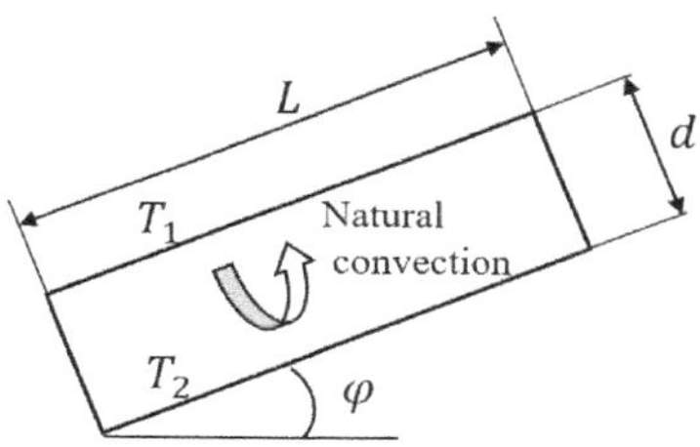
</td>
<td>Correlation of Hollands et al. (1976):

$$Nu = 1 + 1.44\left[1 - \frac{1{,}708}{Ra_d cos(\varphi)}\right]^{*}\left\{1 - \frac{1{,}708[sin(1.8\varphi)^{1.6}]}{Ra_d cos(\varphi)}\right\} + \left\{\left[\frac{Ra_d cos(\varphi)}{5{,}830}\right]^{1/3} - 1\right\}^{*}$$

If $0 < \varphi < tan^{-1}(4{,}800\ Pr), \frac{L}{d} > 12$
The quantities * are taken equal to 0 if the result of their calculation leads to a negative number.
$0 < \varphi < 90°$ if the hot plate is at the bottom.</td>
</tr>
<tr>
<td>Correlation of ElSherbiny et al. (1982):
Valid for $\frac{L}{d} < 100$ and $Ra_d < 10^6$

$$Nu = max[Nu_1, Nu_2, Nu_3]$$

with:

$$Nu_1 = 0.0605\ Ra_d^{\ 0.33}$$

$$Nu_2 = \left\{1 + \left[\frac{0.104\ Ra_d^{\ 0.293}}{1+\left(\frac{6310}{Ra_d}\right)^{1.36}}\right]^3\right\}^{\frac{1}{3}}$$

$$Nu_3 = 0.242\left(\frac{Ra_d d}{L}\right)^{0.272}$$</td>
</tr>
</table>

References

Abramovitz, M. and Stegun, I.A. (1972). *Handbook of Mathematical Functions with Formulas, Graphs and Mathematical Tables*, 9th edition. Dover Publications, Mineola.

Bernard, R., Menguy, G., Schwartz, M. (1980). *Le rayonnement solaire, conversion thermique et applications*. Technique et Documentation, Paris.

Bouvenot, A. (1980). *Transferts de chaleur*. Masson, Paris.

Bromley, L.A. (1950). Heat transfer in stable film boiling. *Chemical Engineering Progress*, 46, 221–227.

Carslaw, H.S. and Jaeger, J.C. (1959). *Conduction of Heat in Solids*, 2nd edition. Oxford University Press.

Chandrasekhar, S. (1960). *Radiative Transfer*. Dover Publications, New York.

Chassériaux, J.-M. (1984). *Conversion thermique du rayonnement solaire*. Dunod, Paris.

Chevrier, J.C. (1982). *Transferts de chaleur par conduction*. École des Mines, Nancy.

Collares-Peirera, M. and Rabl, A. (1979). The average distribution of solar radiation-correlations between diffuse and hemispherical and between daily and hourly insolation values. *Solar Energy*, 22, 155–164.

Daguenet, M. (1985). *Les séchoirs solaires : théorie et pratique*. UNESCO, Paris.

Degiovanni, A. (1977). Diffusivité et méthode flash. *Revue générale de thermique*, 185, 420–441.

Degiovanni, A. and Laurent, A. (1986). Une nouvelle technique d'identification de la diffusivité thermique pour la méthode flash. *Revue de physique appliquée*, 21, 229–237.

Den Iseger, P. (2006). Numerical transform inversion using Gaussian quadrature. *Probability in the Engineering and Informational Sciences*, 20, 1–44.

Duffie, J.A. and Beckman, W.A. (1980). *Solar Engineering of Thermal Processes*. John Wiley & Sons, New York.

El Sherbiny, S.M., Raithby, G.D., Hollands, K.G.T. (1982). Heat transfer by natural convection across vertical and inclined air layers. *Journal of Heat Transfer – Transactions of the ASME*, 104(1), 96–102. https://doi.org/10.1115/1.3245075.

Eyglunent, B. (2000). *Manuel de thermique. Théorie et pratique*. Hermes, Paris.

Gobbé, C., Iserna, S., Ladevie, B. (2004). Hot strip method: Application to thermal characterisation of orthotropic media. *International Journal of Thermal Sciences*, 23(10), 951–958.

Gosse, J. (1981). *Guide technique de thermique*. Dunod, Paris.

Grimson, E.D. (1937). Correlation and utilization of new data on flow resistance and heat transfer for cross flow of gases over tube banks. *Transactions of the ASME*, 59, 583–594.

Hay, B., Filtz, J.R., Batsale, J.C. (2002). Mesure de la diffusivité thermique par la méthode flash. *Technique de l'ingénieur*, R2, 995.

Hladik, J. (1990). *Métrologie des propriétés thermophysiques des matériaux*. Masson, Paris.

Hollands, K.G.T., Unny, T.E., Raithby, G.D. (1976). Free convective heat transfer across inclined air layers. *Journal of Heat Transfer*, 8(2), 189–193.

Holman, J.P. (1990). *Heat Transfer*, 7th edition. McGraw-Hill, New York.

Hottel, H.C. and Cohen, E.S. (1958). Radiant heat exchange in a gas-filled enclosure: Allowance for nonuniformity of gas temperature. *AIChE Journal*, 4(1), 3–14.

Howell, J.R. (1998). The Monte-Carlo method in radiative heat transfer. *Journal of Heat Transfer*, 120(3), 547–560.

Huet, O. and Celaire, R. (1986). *Bioclimatisme en zone tropicale*. La Documentation française, Paris.

IIF (1976). Les techniques du froid dans les pays chauds en développement. Guide, IIF, Paris.

Jakob, M. (1949). *Heat Transfer*, Volume 1. John Wiley & Sons, New York.

Jannot, Y. and Degiovanni, A. (2018). *Thermal Properties Measurement of Materials*. ISTE Ltd, London, and John Wiley & Sons, New York.

Jannot, Y. and Meukam, P. (2004). Simplified estimation method for the determination of thermal effusivity and thermal conductivity with a low cost hot strip. *Measurement Science and Technology*, 15, 1932–1938.

Jannot, Y., Acem, Z., Kanmogne, A. (2006). Transient hot plate method with two temperature measurements for thermal characterization of metals. *Measurement Science and Technology*, 17, 69–74.

Jeans, J.H. (1917). The equations of radiative transfer of energy. *Monthly Notices of the Royal Astronomical Society*, 78, 28–36.

Ladevie, B. (1998). Mise au point de dispositifs de caractérisation thermophysique de matériaux isolants solides ou pâteux, extension aux fluides cisaillés. PhD Thesis, ENSAM, Bordeaux.

Laraqi, N. and Monier-Vinard, E. (2013). New analytical solution for solving steady-state heat conduction problems with singularities. *Thermal Science*, 17(3), 665–672.

Maillet, D., André, A., Batsale, J.-C., Degiovanni, A., Moyne, C. (2000). *Thermal Quadrupoles*. John Wiley & Sons, New York.

Meukam, P. (2004). Caractérisation de matériaux locaux en vue de l'isolation thermique de bâtiments. PhD Thesis, Université de Cergy-Pontoise.

Meulemans, J. (2018). Caractérisation des propriétés thermiques de liquides semi-transparents à haute température. Application aux liquides silicatés. PhD Thesis, Université de Lorraine.

Nukiyama, S. (1966). The maximum and minimum values of the heat Q transmitted from metal to boiling water under atmospheric pressure. *International Journal of Heat and Mass Transfer*, 9(12), 1419–1433.

Özisik, M.N. (1973). *Radiative Transfer and Interactions with Conduction and Convection*. Wiley/Interscience Publications, New York.

Özisik, M.N. (2012). *Heat Conduction*. John Wiley & Sons, New York.

Parker, W.J., Jenkins, R.J., Butler, C.P., Abbot, G.L. (1961). Flash method of determining thermal diffusivity, heat capacity and thermal conductivity. *Journal of Applied Physics*, 32(9), 1679–1684.

Peri, G. (1977). *Les capteurs convertisseurs héliothermiques sans concentration*. Editions Européennes Thermique et Industrie, Paris.

Perrin de Brichambaut, C. and Lamboley, G. (1974). *Le rayonnement au sol et ses mesures*. Editions Européennes Thermique et Industrie, Paris.

Philipon, A. (1988). Échanges thermiques. École d'ingénieurs de l'équipement rural, Ouagadougou.

Pitts, D.R. and Sissom, L.E. (1977). *Schaum's Outline of Theory and Problems of Heat Transfer*. McGraw-Hill, New York.

Raithby, G.D. and Chui, E.H. (1990). Finite-volume method for predicting a radiant heat transfer in enclosures with participating media. *Journal of Heat Transfer*, 112(2), 415–423.

Rohsenow, W.M., Hartnett, J.P., Ganic, E.N. (1985). *Handbook of Heat Transfer Fundamentals*. McGraw-Hill, New York.

Sacadura, J.-F. (1989). *Initiation aux transferts thermiques*. Lavoisier, Paris.

Sfeir, A.A. and Guarracino, G. (1980). *Ingéniérie des systèmes solaires*. Lavoisier, Paris.

Siegel, R. and Howell, J.R. (1992). *Thermal Radiation Heat Transfer*. Hemisphere Publishing Publications, New York.

Sparrow, E.M. and Cess, R.D. (1978). *Radiation Heat Transfer*. McGraw-Hill, New York.

Spiegel, M.R. (1990). *Transformées de Laplace, cours et problèmes*. Schaum, Mequon.

Stehfest, H. (1970). Algorithm 368. Numerical inversion of Laplace transforms. *Communications of the ACM*, 13, 47–49.

Taine, J. and Petit, J.-P. (1995). *Transferts thermiques, cours et données de base*. Dunod, Paris.

Toutain, J., Battaglia, J.-L., Pradere, C., Pailhes, J., Kusiak, A., Aregba, W., Batsale, J.-C. (2011). Numerical inversion of Laplace transform for time resolved thermal characterization experiment. *Journal of Heat Transfer*, 133(4), 044504. https://doi.org/10.1115/1.4002777.

Vianou, A. and Girardey, A. (1993). Exploitation of the regular-state phenomenon in thermokinetics for the determination of the thermal properties of building materials. *High Temperatures High Pressures*, 25, 635–641.

Whitaker, S. (1983). *Fundamental Principles of Heat Transfer*. Robert E. Krieger Publishing Company Inc., Malabar

Wong, H.Y. (1977). *Heat Transfer for Engineers*. Longman, London.

Zuber, N. (1958). On the stability of boiling heat transfer. *Transactions of the ASME*, 80, 711.

Index

Summary of Volume 2

Other titles from

ISTE

in

Energy

2023

ALVAREZ-HERAULT Marie-Cécile, GOUIN Victor, CHARDIN-SEGUI Trinidad, MALOT Alain, COIGNARD Jonathan, RAISON Bertrand, COULET Jérôme
Distribution System Planning: Evolution of Methodologies and Digital Tools for Energy Transition

2022

ABOUELATTA Mohamed, SHAKER Ahmed, GONTRAND Christian
Smart Power Integration

BOUTAUD Benoit
Energy Autonomy: From the Notion to the Concepts

2021

ROBYNS Benoît, DAVIGNY Arnaud, FRANÇOIS Bruno, HENNETON Antoine, SPROOTEN Jonathan
Electricity Production from Renewable Energies (2nd edition)

2020

BOISGIBAULT Louis, AL KABBANI Fahad
Energy Transition in Metropolises, Rural Areas and Deserts

SOUALHI Abdenour, RAZIK Hubert
Electrical Systems 1: From Diagnosis to Prognosis
Electrical Systems 2: From Diagnosis to Prognosis

2019

BENALLOU Abdelhanine
Energy Transfers by Convection
(Energy Engineering Set – Volume 3)
Energy Transfers by Radiation
(Energy Engineering Set – Volume 4)
Mass Transfers and Physical Data Estimation
(Energy Engineering Set – Volume 5)

LACHAL Bernard
Energy Transition

ROBYNS Benoît, DAVIGNY Arnaud, BARRY Hervé, KAZMIERCZAK Sabine, SAUDEMONT Christophe, ABBES Dhaker, FRANÇOIS Bruno
Electrical Energy Storage for Buildings in Smart Grids

2018

BENALLOU Abdelhanine
Energy and Mass Transfers: Balance Sheet Approach and Basic Concepts
(Energy Engineering Set – Volume 1)
Energy Transfers by Conduction
(Energy Engineering Set – Volume 2)

JEMEÏ Samir
Hybridization, Diagnostic and Prognostic of Proton Exchange Membrane Fuel Cells: Durability and Reliability

RUFFINE Livio, BROSETA Daniel, DESMEDT Arnaud
Gas Hydrates 2: Geoscience Issues and Potential Industrial Applications

VIALLET Virginie, FLEUTOT Benoit
Inorganic Massive Batteries
(Energy Storage – Batteries, Supercapacitors Set – Volume 4)

2017

BROSETA Daniel, RUFFINE Livio, DESMEDT Arnaud
Gas Hydrates 1: Fundamentals, Characterization and Modeling

LA SCALA Massimo
From Smart Grids to Smart Cities: New Challenges in Optimizing Energy Grids

MOLINA Géraldine, MUSY Marjorie, LEFRANC Margot
Building Professionals Facing the Energy Efficiency Challenge

SIMON Patrice, BROUSSE Thierry, FAVIER Frédéric
Supercapacitors Based on Carbon or Pseudocapacitive Materials
(Energy Storage – Batteries, Supercapacitors Set – Volume 3)

2016

ALLARD Bruno
Power Systems-on-Chip: Practical Aspects of Design

ANDRE Michel, SAMARAS Zissis
Energy and Environment

DUFOUR Anthony
Thermochemical Conversion of Biomass for the Production of Energy and Chemicals

2015

CROGUENNEC Laurence, MONCONDUIT Laure, DEDRYVERE Rémi
Electrodes for Li-ion Batteries
(Energy Storage – Batteries, Supercapacitors Set – Volume 2)

LEPRINCE-WANG Yamin
Piezoelectric ZnO Nanostructure for Energy Harvesting
(Nanotechnologies for Energy Recovery Set – Volume 1)

ROBYNS Benoît, FRANÇOIS Bruno, DELILLE Gauthier, SAUDEMONT Christophe
Energy Storage in Electric Power Grids

ROSSI Carole
Al-based Energetic Nanomaterials
(Nanotechnologies for Energy Recovery Set – Volume 2)

TARASCON Jean-Marie, SIMON Patrice
Electrochemical Energy Storage
(Energy Storage – Batteries, Supercapacitors Set – Volume 1)

2013

LALOUI Lyesse, DI DONNA Alice
Energy Geostructures: Innovation in Underground Engineering

2012

BECKERS Benoit
Solar Energy at Urban Scale

ROBYNS Benoît, DAVIGNY Arnaud, FRANÇOIS Bruno, HENNETON Antoine, SPROOTEN Jonathan
Electricity Production from Renewable Energies

2011

GAO Fei, BLUNIER Benjamin, MIRAOUI Abdellatif
Proton Exchange Membrane Fuel Cell Modeling

MULTON Bernard
Marine Renewable Energy Handbook

2010

BRUNET Yves
Energy Storage

2009

SABONNADIÈRE Jean-Claude
Low Emission Power Generation Technologies and Energy Management

SABONNADIÈRE Jean-Claude
Renewable Energy Technologies

Printed and bound by CPI Group (UK) Ltd, Croydon, CR0 4YY

13/09/2026

14961047-0007